M000222636

Beautiful, Simple, Exact, Crazy

Beautiful, Simple, Exact, Crazy: Mathematics in the Real World

Apoorva Khare and Anna Lachowska

Yale UNIVERSITY PRESS

New Haven and London

Copyright © 2015 by Yale University.

All rights reserved.

This book may not be reproduced, in whole or in part, including illustrations, in any form (beyond that copyright permitted by Sections 107 and 108 of the U.S. Copyright Law and except by reviewers for the public press), without written permission from the publishers.

Yale University Press books may be purchased in quantity for educational, business, or promotional use. For information, please e-mail sales.press@yale.edu (U.S. office) or sales@yaleup.co.uk (U.K. office).

Printed in the United States of America.

Library of Congress Control Number: 2015935012
ISBN 978-0-300-19089-2 (pbk. : alk. paper)

A catalogue record for this book is available from the British Library.

This paper meets the requirements of ANSI/NISO Z39.48-1992 (Permanence of Paper).

10 9 8 7 6 5 4 3 2 1

Contents

Preface vii

Acknowledgments xv

1 Algebra: The art and craft of computation 1

2 Velocity: On the road 19

3 Acceleration: After the apple falls 39

4 Irrational: The golden mean and other roots 59

5 Exponents: How much would you pay for the island of
   Manhattan? 79

6 Logarithms I: Money grows on trees, but it takes time . 97

7 Logarithms II: Rescaling the world 107

8 *e*: The queen of growth and decay 129

9 Finite series: Summing up your mortgage, geometrically 149

10 Infinite series: Fractals and the myth of forever 167

11 Estimation: What is your first guess? 193

12 Modular arithmetic I: Around the clock and the calendar 213

13 Modular arithmetic II: How to keep (and break) secrets 237

14 Probability: Dice, coins, cards, and winning streaks . . . 261

15 Permutations and combinations: Counting your choices . 277

16 Bayes' law: How to win a car... or a goat 297

17 Statistics: Babe Ruth and Barry Bonds 325

18 Regression: Chasing connections in big data 341

The bigger story 353

Solutions to odd-numbered exercises 357

Practice exams 423

Index 457

Preface

The idea for this book arose out of an introductory mathematics course, "Mathematics in the Real World", that the authors co-designed and have taught at Yale and Stanford. The purpose of the course is to familiarize students whose primary interests lie outside of the sciences with the power and beauty of mathematics. In particular, we hope to show how simple mathematical ideas can be applied to answer real-world questions.

Thus we see this as a college-level course book that can be used to teach basic mathematics to students with varying skill levels. We discuss specific and relevant real-life examples: population growth models, logarithmic scales, personal finance, motion with constant speed or constant acceleration, computer security, elements of probability, and statistics. Our goal is to combine the right level of difficulty, pace of exposition, and scope of applications for a curious liberal arts college student to study and enjoy.

Additionally, the book could find use by high school students and by anyone wishing to study independently. The prerequisite is only a high school course in algebra. We hope our book can help readers without extensive mathematical training to analyze datasets and real-world phenomena, and to distinguish statements that are mathematically reasonable from those only pretending to be.

Philosophy and goals

Imagine the following dialog in a high school algebra class.

Teacher: "Find the sum of the geometric series $1 + 5 + 5^2 + \ldots + 5^{20}$."

Student (looking out the window and thinking that life is so much bigger than math, and wondering why the class has to suffer through this tedious, long, and pointless computation): "Can't we just type it into a calculator?"

Whatever the teacher's answer, the student knows that he is right at least on one count – life is indeed bigger than math. It is also bigger than physics, history, and sociology. But for some reason the irrelevance of mathematics is much easier to accept. It is not acceptable to say, "I do not care who was the first president of the United States," but it seems just fine to say, "I do not care about the purpose of a geometric series." It is not acceptable to brag, "I'm illiterate!" But many people feel justified telling friends that they cannot understand mathematics. Yet the consequences of refusing to learn even the most basic mathematical ideas are dire both for the individual and for society.[1] As a society, we need to make collective decisions based on information provided by the media and other sources of variable reliability – and the quality of such decisions depends on our understanding of logic and statistics. As individuals, we have to cope with personal finance and Internet security. We have to know how to estimate the chances an event will occur, and how to interpret lots of new information. From a less practical viewpoint, mathematics adds another dimension (or two, or twenty-three[2]) to the way we see the world, which might be a source of inspiration for a person of any occupation.

The student we just encountered likely will come to college to major in the humanities or social sciences, and is part of the target audience for this book. Thus, our main goal is to convince our reader that mathematics can be easy, its applications are real and widespread, and it can be amusing and inspiring.

To illustrate, let us return to our geometric series example. Mathematics is known to have formulas for everything, so it is not surprising that it has a formula for the sum $1 + 5 + 5^2 + \ldots + 5^n$ for any natural n. What might be more surprising is that this formula requires no derivatives, integrals, or trigonometric identities – and the computation (given in Chapter 9) takes only one line.

Practical applications of this simple mathematical concept are everywhere: for instance, it governs the payment of mortgages and car loans for millions of people. There is more! Everyone knows that $\frac{1}{3} = 0.333333\ldots$, but the fact that *any* repeating decimal, say, $0.765765765\ldots$, can be easily written as a fraction of two integers follows from the formula for the geometric series. The same formula lies in the foundation of Zeno's paradox of Achilles and a tortoise, covered in Chapter 10, and two centuries later, another Greek, Archimedes, used a geometric series to calculate the area

---

[1] We will not elaborate on this here, referring the reader, for example, to *Innumeracy* by J.A. Paulos, published by Hill and Wang, New York, 2001.

[2] According to string theory, the world might have 26 dimensions.

under a parabolic arc. Why should we care? If understanding history is a good enough reason, we can recall that by some accounts, Archimedes also took part in the defense of his town of Syracuse during the siege by the Romans (214-212 BC), and might have used his computation to design a parabolic mirror to set the enemy's fleet on fire. Now, fast-forward to the twentieth century. "How Long Is the Coast of Britain? Statistical Self-Similarity and Fractional Dimension" was the title of the article, published in *Science* in 1967 by the French-American mathematician Benoit Mandelbrot, that opened a new area of study – fractal geometry. The answer to the question in the title depends on how closely we look, and the closer we look, the greater the coast's length becomes. The same is true for many fractal curves that are easy to describe in words but impossible to draw *precisely*. One of the central tools used to understand these objects is the geometric series.

This is just a glimpse of the scope of applications of one simple mathematical idea – from personal finance, to philosophical puzzles, to fractals, objects of such breathtaking beauty that they make the boundary between art and science disappear.

This example encapsulates our philosophy for the book: we would like to show that there is a lot of simple mathematics relevant to people's everyday lives and their creative aspirations.

There is a way to teach a future artist, and there is a way to teach a layperson to appreciate art. In the latter case, the student is not required to be able to draw and paint like a master, but only to see the beauty in the works of others. Our intention for this book is more ambitious: we hope that it can serve as a guide to the world of mathematics, and also as an inspiration for readers to try their hand at developing and solving mathematical models for their own needs. From there it is only one more step to seeing the world from a mathematical point of view.

The book includes many examples and practice problems designed to gradually build students' proficiency and encourage their involvement with the material. The ultimate goal is to make even the students who are "numerically shy" at the start of a course comfortable applying their mathematical skills in a wide range of situations, from solving puzzles to analyzing statistical data.

We also hope that they become "math-friendly," admitting that mathematics can be interesting and cool. We feel that such comfort levels are indeed achieved when we teach this course at Yale and Stanford – based on our discussions with students, their exams and homework, and their feedback from the course.

Contents and structure

Here are the most important features that distinguish our approach.

First, we neither discuss nor assume any knowledge of calculus or trigonometry. There are quite a few simple concepts in mathematics that do not depend on this particular knowledge but have enormous importance in the real world, as a vast number of pertinent applications show.

Second, our preference is for a self-contained linear exposition, devoid of digressions and asides. We also refrain from presenting an overload of pictures, data analysis, or complicated examples. This book should be viewed as supporting material for a first course in college-level mathematics; there is ample opportunity to analyze more complex examples and phenomena in future studies.

The choice of topics in the book was governed by the following principles:

1. The mathematics involved should be simple, accessible to a student with no experience beyond elementary high school algebra, and explainable within a couple of pages.

2. These simple mathematical concepts should generate a wide range of practical, impressive, or amusing real-world applications.

Accordingly, each chapter of the book, starting with Chapter 2, is divided into two parts:

1. The first, shorter part contains the necessary mathematics: definitions, statements, explanations, examples. This part is supposed to be *studied*, or read slowly, with the reader occasionally doing suggested computations. It can also play the role of lecture notes, if the book is used in teaching a course. Finally, the clearly distinct math section can be used as a reference when reading about applications.

2. The second part, which constitutes most of the chapter, is an exploration of various real-world applications. It contains questions posed and answered by means of the mathematical tools presented in the first part. The second part has little or no mathematical argument and is designed to be read leisurely.

The math sections are there for studying the necessary mathematics. The applications sections are for reading about ways this math can be used. By separating these two activities, we hope to promote learning; instead of students half paying attention while reading fifty pages of a typical chapter of a basic math textbook, we require intense concentration on just a few pages, and then they can enjoy reading the remaining ten or fifteen pages of applications. We hope that having short math and long applications sections will impress on the students that even when the amount of mathematics to be learned is limited, the potential rewards for this effort can be large.

The structure allows readers to choose their own way through the book: some might skip the math sections entirely, and look only for the applications; they will not learn the mathematics but they might learn what it is good for. Others might like the math sections but choose to read only about some of the suggested applications. In some sense, the book is a hopscotch game where the reader sets the rules.

Our strategy for the choice of topics resulted in an array of subjects, presented in chapters that are only loosely interrelated. This means that even though the structure of the book is linear, a reader (or a teacher) can skip some parts of the book or change their order. The chapter dependencies are:

$$(1), \quad (2-3-4), \quad (5-6-7-8), \quad (9-10),$$

$$(11), \quad (12-13), \quad (14-15-16), \quad (17-18).$$

This makes for greater flexibility in teaching the course. For example, one author taught chapters 1-3, 5, 6, 8, 9-10, 12, 14, 16, 17-18, while the other author taught chapters 2-10, 12-16. Moreover, instructors can introduce their own topics and design additional material based on the conceptual structure presented in the book.

The book is intended, in particular, for "artists and poets," so we include several references to works of literature and art. Chapter 2 contains a mathematical analysis of a short story by an Italian writer, D. Buzzati, which is based on the repeated application of a constant speed motion model. A mathematical perspective on the development of music scales is given in Chapter 7. The same mathematical concept, a logarithm, can be used to illustrate our egocentricity: a famous *New Yorker* cover by S. Steinberg, *View of the World from 9th Avenue*, can be interpreted as a logarithmic scale of distances. In connection with the radiocarbon dating method (Chapter 8), we discuss some of the mysteries of the famous Voynich manuscript, an undeciphered early fifteenth-century text whose author, language, and subject are unknown.

The relation between nature and mathematics is another source of philosophical and poetic inspiration. Logarithmic scales allow us to grasp, at least to some extent, the soul-chilling range of magnitude of distances in the universe. Chapter 10 contains a discussion of the possibility of encountering infinity in reality. And in Chapter 4, the growth patterns in plants are shown to be related to mathematical properties of a certain irrational number, the golden ratio, whose cultural significance is also remarkable.

Perhaps the most important part of the book consists of the wide variety of examples, practice problems, and exercises of varying levels of difficulty. We tried to make sure that all of the mathematics in this book is reinforced by studying concrete examples. Here are the main highlights of the problem-solving component of the book:

- Each chapter contains a number of concrete examples. Many of them are discussed in detail and solved in the text (**Examples**).

- Others are grouped in clusters according to topics within a chapter, and provided with answers (**Practice problems**). These are usually easy questions designed to check students' basic understanding of the concepts presented in the chapter.

- Still others are confined to the end of a chapter and are supposed to be solved independently (**Exercises**). Answers and solutions to the odd-numbered exercises are given at the end of the book.

- In addition we provide sample midterms and final exams and their solutions. For those engaged in independent study, these can be used for self-evaluation and review.

Some sections of the text, examples, and exercises are marked with an asterisk. This indicates that the material discussed in this part of the book is more challenging or abstract, and can be safely skipped at first reading.

Target audience

Our primary goal is to provide a book for an introductory college-level course in mathematics and its applications. Thus, we hope to reach undergraduates looking to balance their humanities and social sciences education with a touch of math. Many colleges have an academic requirement for their students to diversify their choice of classes and include at least a few "hard-science" or "quantitative reasoning" courses. A course based on this book can be one of them. We hope that while helping students to complete their academic requirements, the book will also significantly benefit them in their future careers.

When the course "Mathematics in the Real World" was introduced at Yale University, it received twice as many applications as there was room for, and the course remains quite popular. In course evaluations and thank-you notes, students mention that the course has helped them acquire basic mathematical literacy and confidence, and instead of being scary, it was an enjoyable experience. We hope that this book will elicit similar reactions from some of our readers.

The course has also been introduced at Stanford University. Our long-term hope is that this book will contribute to the regular curriculum in many colleges and universities.

We also hope that our book will be helpful for those who would like to specialize in natural sciences or economics, but who lack some background in pre-calculus and calculus. For them, the book might serve as a first step on their way to more advanced mathematics.

The book might also be useful for advanced high-school students who are interested in real-world applications of the concepts they learn at school. We estimate that the level of exposition is suitable for most high schools, and the book can be used either by teachers, to supplement the standard program, or by students as extracurricular reading.

In addition, the universal appeal of the topics and the minimal mathematical prerequisites needed to understand this text make for a significantly wider audience. This book should be usable for independent study by busy adults who wish to improve their understanding of math (the book is short!). It might also have some appeal for busy but more experienced math fans, who could leisurely scan it for literary references and unusual applications of math concepts.

Finally, we expect the book to be attractive to international audiences. The authors are of Indian and Russian origin, and we tried to give the book an international flavor. Our examples and cultural references are drawn from all over the world.

In this book we tried to keep a balance between being instructional, being entertaining, and being practical. We hope that this approach will help us promote mathematics as an art, a skill, a language, a way of thinking, a game of puzzles, and in general a worthy activity, to a wide audience of readers.

Acknowledgments

We are grateful to Steve Orszag (1943–2011), who suggested the idea of the course "Mathematics in the Real World" to us, and to Michael Frame at Yale University, who was a source of continuous help and inspiration in developing the course and writing the book. We would like to thank Roger Howe, Andrew Casson, Frank Robinson, and Bill Segraves at Yale, as well as Bala Rajaratnam, Ravi Vakil, and Brian Conrad at Stanford, for their generous assistance in introducing the course at these universities.

We would like to thank our parents and our teachers in high school, college, and graduate school, who taught us to appreciate the depth and beauty of mathematics, and encouraged us, by their own example, to share it with others.

It is a pleasure to thank our students for whom the course was created, whose curiosity and enthusiasm helped shape this book.

We are grateful to our editor, Joseph Calamia, for his careful reading of the manuscript and for multiple suggestions and corrections, which significantly improved the exposition, and to Nancy Hulnick and Mary Pasti, whose thorough editing helped refine the text. We would also like to thank the reviewers of the manuscript for their remarks and suggestions.

Special thanks go out to our families for their continuous support, encouragement, and patience during our work on the book.

Chapter 1

Algebra: The art and craft of computation

In this book, we hope to show you that a wide variety of real-world problems and applications can be tackled systematically, comprehensively, and relatively simply by using just a few mathematical formulas and techniques. In order to introduce the mathematics and then to apply it to the real world, it is essential to be able to work with mathematical expressions and quantities in a systematic manner. Thus, we first need to be comfortable with basic operations like adding or multiplying polynomials; solving equations and systems of (linear or other) equations; and choosing an optimal way to simplify an algebraic expression. Developing these techniques is the goal of this chapter.

Sometimes these techniques produce unexpected results which some of you may have seen as "magic tricks." For instance, you can check that multiplying two consecutive odd numbers (or consecutive even numbers) yields one less than a perfect square (e.g., $5 \cdot 7 + 1 = 36 = 6^2$, $10 \cdot 12 + 1 = 121 = 11^2$). Is this *always* the case, or can we find two consecutive odd or even numbers for which this phenomenon does not occur? Note that it is impossible – even for the biggest computer – to *verify* this for all integers in finite time, because there are infinitely many numbers. But, as we will see, there is a simple way to perform just one calculation – and it will do the job for every single case.

Similarly, multiplying three successive integers and adding the middle integer to this product always yields a perfect cube! Why? Once again, we will see in the exercises in this chapter how one calculation reveals the answer for all possible cases.

Thus, the purpose of this chapter is to discuss mathematical techniques which will then be used throughout the remainder of this book. We will see real-world applications of linear and quadratic equations in Chapters 2, 3, and 4.

The distributive law

We start by discussing one of the most important principles used in simplifying and manipulating mathematical expressions. As a first example, consider the following elementary computation:

$$2 \cdot 39 + 2 \cdot 61 \; = \; ?$$

One way to solve this equation is to explicitly calculate that $2 \cdot 39 = 78$, $2 \cdot 61 = 122$, and then to add them and obtain 200. This solution involves carrying out two multiplications (neither of which seems trivial enough to do mentally) followed by one addition. However, here is a simpler approach, which involves only one addition and then one trivial multiplication: we first add: $39 + 61 = 100$, and then multiply the result by 2:

$$2 \cdot 39 + 2 \cdot 61 = 2 \cdot (39 + 61) = 2 \cdot 100 = 200.$$

Why does the first equality hold in the preceding equation? This is the principle of distributivity. It says that $2 \cdot 39 + 2 \cdot 61 = 2 \cdot (39 + 61)$. More generally, the *distributive law* says that given any three real numbers a, b, c,

$$a \cdot b + a \cdot c = a \cdot (b + c), \qquad a \cdot b - a \cdot c = a \cdot (b - c).$$

The distributive law is at the heart of many algebraic manipulations and simplifications which will be used throughout the book. Thus it is important that we understand it – as well as its implications – thoroughly. Here are further applications of the distributive law.

Example 1.1. Compute (without using a calculator):

1. $17 \cdot 3 + 3 \cdot 283$.

2. $3 \cdot 3 + 3 \cdot 17 + 20 \cdot 17$.

Solution: While the problem itself is elementary, our goal here is to see how to solve it using the distributive law, and make it easy enough to perform the computation mentally. For the first part, we compute using the distributive law:

$$17 \cdot 3 + 3 \cdot 283 = 3 \cdot 17 + 3 \cdot 283 = 3 \cdot (17 + 283) = 3 \cdot 300 = 900.$$

For the second part, we group terms two at a time, using the distributive law:

$$(3 \cdot 3 + 3 \cdot 17) + 20 \cdot 17 = 3 \cdot (3 + 17) + 20 \cdot 17 = 3 \cdot 20 + 20 \cdot 17$$
$$= 20 \cdot (3 + 17) = 20 \cdot 20 = 400.$$

$\square$

A more advanced version of this rule is when *both* factors are sums or differences. For instance, we can multiply out $101 \cdot 201$ without using a calculator, as follows:

$$101 \cdot 201 = (100 + 1) \cdot (200 + 1) = (100 + 1) \cdot 200 + (100 + 1) \cdot 1$$
$$= 200 \cdot (100 + 1) + 101 = 200 \cdot 100 + 200 \cdot 1 + 101$$
$$= 20,000 + 200 + 101 = 20,301.$$

You may have seen a variation of this method:

$$101 \cdot 201 = (100 + 1) \cdot (200 + 1) = 100 \cdot 200 + 100 \cdot 1 + 1 \cdot 200 + 1 \cdot 1 = 20,301.$$

This is sometimes called the "FOIL" method – essentially, it is simply the distributive law applied twice – once to the terms in the first factor, and once to the terms in the second.

Variables

A useful feature of the distributive law is that it applies equally well to *variables* . Thus if x, y are any unknown real numbers, then we once again have (for instance): $2 \cdot (x + y) = 2x + 2y$. The distributive law can also be used to "contract" expressions into more compact forms. Here is an example: the expression

$$(x^2 - 2)(x + 2) - 3x^2 + 6$$

looks somewhat unwieldy. However, look closely: the last two terms can be written using the distributive law as $-3(x^2 - 2)$. Now applying the distributive law again, we get

$$(x^2 - 2)(x + 2) - 3x^2 + 6 = (x^2 - 2)(x + 2) + (x^2 - 2) \cdot (-3)$$
$$= (x^2 - 2)(x + 2 - 3) = (x^2 - 2)(x - 1),$$

and so just like long expressions involving numbers (as in Example 1.1), variable expressions can also be simplified.

An advantage of using variables is that once verified, a statement involving variables automatically holds for *every* real value assumed by these variables. For instance, consider the following "magic trick."

Example 1.2. *Get your number back!* Here's a magic trick: start with any number, add 2 to it, multiply the sum by 5, subtract 10 from the product, and divide the difference by 5 – and lo and behold! You get your original number back.

Is it possible to explain what is going on without having to verify whether or not this works for every single starting number?

Solution: Indeed it is. Suppose we start with the number x – as opposed to a specific value, we use a variable because it can be set equal to any number. Then the effect of the given operations on the initial value can be explicitly computed:

$$x \to x + 2, \qquad x + 2 \to 5(x + 2) = 5x + 10,$$
$$5x + 10 \to (5x + 10) - 10 = 5x, \qquad 5x \to (5x)/5 = x,$$

and indeed, we get the original number back as the trick claims. (Note that the second operation uses the distributive law.) Thus we have explained the magic trick for every starting number x. Of course, if you want to repeat the calculations corresponding to a specific starting number x, simply set x everywhere above equal to that starting number. $\qquad \square$

The above example shows that it is indeed possible to use variables to explain general phenomena, or to solve "word problems" using simple mathematics. We now mention an example which will help us answer the question posed at the beginning of this chapter.

Example 1.3. *The sum times the difference.* One very useful identity involves taking two numbers a, b, and multiplying their sum times their difference. Let us compute the result using the distributive law:

$$(a + b) \cdot (a - b) = a^2 + ba - ab - b^2 = a^2 - b^2.$$

This means that the sum and the difference of *any* two real numbers a and b multiply to yield the difference between their squares.

Conversely, this identity can help compute the product of *any* two numbers: take the square of the number exactly in the middle, and subtract the square of the half of the difference between them. Indeed, if the numbers to multiply are $(a+b)$ and $(a-b)$, then the number in the middle is a, and half of the difference between them is b, and our statement reads

$$(a + b) \cdot (a - b) = a^2 - b^2,$$

which is the sum times the difference formula. It works especially well for two integers that are equidistant from a number whose square is easy to compute.

For instance, this identity enables us to compute $297 \cdot 303$ without too much fuss (or a calculator):

$$297 \cdot 303 = (300 - 3) \cdot (300 + 3) = 300^2 - 3^2 = 90,000 - 9 = 89,991.$$

$\square$

Example 1.4. *Multiplying consecutive odd or even numbers.* We now return to the example mentioned at the beginning of the chapter: multiplying two consecutive odd (or even) numbers always seems to yield one less than a perfect square. To check this for every pair of consecutive odd/even numbers at once, we replace these numbers by variables: denote the number between the two consecutive odd/even numbers as n. Then the two numbers that we are to multiply, are $(n + 1)$ and $(n - 1)$. Their product, by the previous example, is

$$(n + 1) \cdot (n - 1) = n^2 - 1^2 = n^2 - 1.$$

This is precisely one less than the square of the middle number n, as we claimed. $\square$

This general calculation also provides a method to compute such products: take the square of the number in between, and subtract 1. For instance, if we want to compute the product of the two successive odd numbers $101 \cdot 99$, then we simply use $n = 100$ in the above calculation, to get $100^2 - 1^2 = 9,999$.

Solving equations

Often in the real world, we see that the same quantity can be expressed in different ways using different physical units. For instance, the same distance can be given in yards, miles, or meters. Or, the same temperature can be expressed in Celsius and in Fahrenheit (or even in Kelvin) degrees. In solving questions involving such changes, a simple tool that is useful is a *linear equation*. Here is an example.

Example 1.5. What temperature has the same numerical value when measured in Fahrenheit and Celsius?

Solution: Denote the unknown temperature in Celsius, say, by T. Then this same temperature in Fahrenheit is given by: $32 + \frac{9}{5}T$. The conditions of the problem imply that we are to equate these two expressions. In formulas, we have

$$T = 32 + \frac{9}{5}T \quad \Longrightarrow \quad -32 = \frac{9}{5}T - T = \left(\frac{9}{5} - 1\right)T = \frac{4}{5}T$$
$$\Longrightarrow \quad T = (-32) \cdot \frac{5}{4} = -40.$$

We conclude that $-40°\mathrm{F} = -40°\mathrm{C}$ (and no other temperature has this property). $\qquad\square$

The above equation $T = 32 + \frac{9}{5}T$ is an example of a linear equation. It models a linear relationship, or dependency, between two varying quantities (in this case, the temperature in Celsius and in Fahrenheit). A linear equation is characterized by the way the variable appears in it: it can either stand alone, or be multiplied by a numeric coefficient. The variable can be denoted by any letter, say, x, T, w, y, A, and so on. All other terms of the equation are numbers.

Such dependencies are ubiquitous in the real world – for instance, one can use linear equations to model production costs (involving an overhead amount and manufacturing cost per unit), simple interest in banking, distance traveled by a vehicle or a jogger with constant velocity, or conversions between different units of temperature, money, weight, distance, and so on. For example, one can determine which temperature in Fahrenheit is twice (and which is half) the number that it equals in Celsius. Here is another real-world example which can be solved using linear equations.

Example 1.6. *Taxicab fares.* Suppose in a given city, cab drivers charge an initial fare of $3, followed by an additional charge of $2 per mile. What is the fare if the passenger travels for 10 miles? How many miles has the passenger traveled if the fare is $11? *Answer:* $23 and 4 miles, respectively. The linear equation here is: $fare = 3 + 2 \cdot miles$. $\qquad\square$

Another class of equations that is easy to solve involves equating a product to zero. For instance: *find all possible values of the variables a, b, c such that $a \cdot b \cdot c = 0$*. The answer is that if three numbers are nonzero, their product cannot be zero; hence at least one of the variables a, b, c must equal zero in this case. Thus, the solution is that $a = 0$ and b, c are arbitrary real numbers; or $b = 0$ and a, c are arbitrary; or $c = 0$ and a, b are arbitrary.

Example 1.7. Using the above idea, find all solutions to the following equations.

1. $(x - 1)(x + 2) = 0$.

2. $(x - 1)^2(x + 3)(y - 2) = 0$.

3. $y^2 - 1 = 99$.

Solution: The first two equations are easy to solve, using the previous reasoning. Thus for the first equation, either $x - 1 = 0$ or $x + 2 = 0$, whence $x = 1$, or $x = -2$. For the second equation, we write the equation as $(x - 1) \cdot (x - 1) \cdot (x + 3) \cdot (y - 2) = 0$. Hence at least one of the four factors is zero, which leads to: $x = 1$, and y is any number, or $x = -3$ and y is any number, or $y = 2$ and x is any number. (We allow for both $x = 1$ and $y = 2$ to occur simultaneously.)

Finally for the last equation, we add 1 to both sides to get: $y^2 = 100$. Hence $y = 10$ or -10. The other way to see this is to use Example 1.3. To do so, subtract 99 from both sides to get:

$$y^2 - 100 = 0 \quad \Longrightarrow \quad y^2 - 10^2 = 0 \quad \Longrightarrow \quad (y + 10)(y - 10) = 0.$$

Hence we obtain the complete set of solutions: $y = -10$ or $y = 10$. For convenience we write $y = \pm 10$. □

Polynomials: Solving quadratic equations

Now we move from simple dependencies to more involved ones. Using variables allows us to define a useful class of expressions called *polynomials*. A polynomial is an expression in which several (finitely many) powers of x can appear, either alone or multiplied by numbers. For instance, $f(x) = 3x^2 - 1$ is equal to $f(x) = 3 \cdot x^2 + (-1) \cdot x^0$. The number next to each power of x (or of the one variable that is used) is called its *coefficient*. For instance, the coefficients of x^0, x^1, x^2, and x^3 in $3x^2 - 1$ are $-1, 0, 3, 0$, respectively.

The highest power of x whose coefficient in a polynomial $p(x)$ is nonzero is called the *degree* of the polynomial, and is denoted by $\deg(p)$. A polynomial is said to be constant, linear, quadratic, cubic, quartic, and so on, if its degree is, respectively, $0, 1, 2, 3, 4$, and so on. A linear equation can be expressed as the condition that a linear polynomial in the variable equals zero: $ax + b = 0$ for some real numbers a, b. In general, given a polynomial $p(x)$, one is often interested in determining the set of x such that $p(x)$ equals a given value – in other words, solving a polynomial equation.

In the remainder of this chapter, we will learn how to solve some polynomial equations which are more involved than linear equations. For instance, using the distributive law (or the FOIL method), you can check that the first of the equations in Example 1.7 was an example of a second-degree polynomial equation in x – i.e., a *quadratic equation*: $x^2 + x - 2 = 0$. Other examples of quadratic equations are $x^2 - 3x + 2 = 0$, or $2x^2 - 8x + 8 = 0$, or more generally,

$$ax^2 + bx + c = 0$$

for some real numbers a, b, c with $a \neq 0$. (Note that if $a = 0$ then the equation becomes a linear equation.)

All quadratic equations have two, one, or no real roots. Here is the formula for the *solutions* to the general quadratic equation $ax^2 + bx + c = 0$:

$$x = \frac{-b \pm \sqrt{b^2 - 4ac}}{2a}.$$

This rule provides the answer in all cases whenever a is nonzero. Recall that the square root can be taken only of nonnegative values. From the formula, we see that there are three possible cases, depending on the quantity under the square root $b^2 - 4ac$, which is called the *discriminant*.

- If the discriminant $b^2 - 4ac$ is positive, the quadratic equation has two distinct real roots, mentioned in the above formula.

- If the discriminant is zero, i.e., $b^2 = 4ac$, then the quadratic equation has exactly one real (repeated) root: $x = -b/2a$.

- If the discriminant is negative (i.e., $b^2 < 4ac$), there are *no* real roots of the equation.

Let us solve some quadratic equations to illustrate the above formula.

Example 1.8. Find all (real) roots of the following equations:

1. $2T^2 - 3T + 1 = 0$.

2. $x^2 - 2x + 1 = 0$.

3. $y^2 + 1 = 0$.

Solution:

1. Even though the variable name is T and not x, the same formula for the roots applies – for, as Shakespeare wrote, *that which we call a rose by any other name would smell as sweet.* Thus, we use $a = 2, b = -3, c = 1$. Hence $b^2 - 4ac = 1$, and we compute:

$$T = \frac{-(-3) \pm \sqrt{1}}{2 \cdot 2} = \frac{3 \pm 1}{4} = \{4/4, 2/4\} = \{1, 1/2\}.$$

Indeed, one can verify that both $T = 1$ and $T = 1/2$ satisfy $2T^2 - 3T + 1 = 0$.

2. Using the formula for $a = 1, b = -2, c = 1$, we obtain:

$$x = \frac{-(-2) \pm \sqrt{(-2)^2 - 4 \cdot 1 \cdot 1}}{2 \cdot 1} = \frac{2 \pm 0}{2} = 1.$$

Thus the second equation has a unique root: $x = 1$.

3. In this case we do not even need to use the formula to verify that there are no real roots, because the equation can be rewritten as : $y^2 = -1$. Now there is no real number whose square is negative.

Indeed, if we do use the formula for $a = 1, b = 0, c = 1$, then we obtain:

$$y = \frac{0 \pm \sqrt{0^2 - 4 \cdot 1 \cdot 1}}{2 \cdot 1} = \pm\sqrt{-4}/2,$$

and hence the equation has no real roots, because there is no real number whose square is -4. This confirms what we had observed above, without using the formula.   □

Here are more examples.

Practice problem 1.9. Solve the following quadratic equations:

1. $x^2 - 9 = 0$. *Answer:* $x = 3, -3 = \pm 3$.

2. $10y^2 + 3y - 1 = 0$. *Answer:* $y = -\frac{1}{2}, \frac{1}{5}$.

There is no reason why we should always get nice-looking integer or fractional solutions to quadratic equations. For instance, the equation $x^2 = 2$, which is the same as $x^2 - 2 = 0$, has *irrational* solutions $x = \pm\sqrt{2} = \pm 1.4142139\ldots$. These are the two real numbers whose square is 2, and neither of them can be expressed as the ratio of two integers. This kind of solution of quadratic equations will be considered in detail in Chapter 4.

Practice problem 1.10. For each of the following quadratic equations, first check if the discriminant $b^2 - 4ac$ is positive, zero, or negative. Then compute the two, one, or zero real solutions, respectively.

1. $x^2 - 3x + 1 = 0$. *Answer:* Discriminant $= (-3)^2 - 4 \cdot 1 \cdot 1 = 5 > 0$, so the equation has two real solutions: $x = \frac{3 \pm \sqrt{5}}{2}$.

2. $3z^2 - \pi z + 1 = 0$. *Answer:* Discriminant $= (-\pi)^2 - 4 \cdot 3 \cdot 1 \simeq -2.13 < 0$, so the equation has no real solutions.

Verification of the quadratic formula*[1]

We end this discussion with a small algebraic calculation. Namely – let us check that the roots in the general formula *do* satisfy the quadratic equation. If $x_\pm = (-b \pm \sqrt{b^2 - 4ac})/2a$ denote the two roots, then it is relatively tedious to compute $ax_\pm^2 + bx_\pm + c$ and check that for both roots $x_\pm$ we get zero. A slightly more efficient alternative is to clear the denominators and check that the relation

$$2ax_\pm + b = \pm\sqrt{b^2 - 4ac}$$

implies the original quadratic equation.

Squaring both sides and using the distributive law, we get

$$b^2 - 4ac = (2ax_\pm + b)^2 = 4a^2x_\pm^2 + b^2 + 4abx_\pm.$$

Move the left-hand terms over to the right and simplify:

$$4a^2x_\pm^2 + 4abx_\pm + b^2 - b^2 + 4ac = 0 \quad \Longrightarrow \quad 4a^2x_\pm^2 + 4abx_\pm + 4ac = 0.$$

Finally, divide both sides by $4a$, because we know that a is nonzero. This immediately yields: $ax_\pm^2 + bx_\pm + c = 0$, and hence the claimed roots do indeed satisfy the quadratic equation.

---

[1]Here and throughout the book, we will mark with an asterisk the part of the material that is more challenging or abstract, and can be skipped at first reading.

Quadratic-type equations

In the previous section, we saw how to solve a quadratic equation in a variable x. We can apply the same technique to solve more complicated equations that look similar to quadratic equations, by reducing them to that form. For example, the equation $x^4 - 2x^2 + 1 = 0$ is not quadratic, but *quartic*, i.e., of degree 4. Or $x^{200} - 2x^{100} + 1 = 0$ is in fact an equation of degree 200. However, they both look somewhat similar – and in fact, they can both be solved using the formula for roots of quadratic equations.

Example 1.11. Solve the two equations in the previous paragraph.

Solution: Let us begin with the first equation, $x^4 - 2x^2 + 1 = 0$. Note that x^4 is the square of x^2, so if we denote x^2 by a new variable y, then the equation changes to: $y^2 - 2y + 1 = 0$. Now apply the general formula for the roots of a quadratic equation (or see Example 1.8(2)) to obtain: $y = 1$. Substituting back for x, we obtain that $x^2 = 1$, and finally $x = \pm 1$ are the roots of the equation.

For the second equation, $x^{200} - 2x^{100} + 1 = 0$, we again make a substitution: $y = x^{100}$. This leads us to the same *quadratic* equation in y, as in the preceding paragraph. Therefore $y = 1$, i.e., $x^{100} = 1$. The only real numbers whose 100th power is 1 are: $x = \pm 1$. (Thus there are only two real roots to this equation of degree 200.)   □

Here are some other equations that can be solved using the formula for the roots of quadratic equations.

Practice problem 1.12. Solve the following equations.

1. $y^5 - 4y^4 + 3y^3 = 0$. *Solution:* First take y^3 common, so either it is zero or the remaining quadratic factor is zero. Solve it to get: $y = 1, 3$, or $y = 0$.

2. $\frac{1}{y^2} - \frac{1}{y} - 2 = 0$. *Solution:* Set $x = 1/y$; then $x^2 - x - 2 = 0$. Solving, $x = -1, 2$, whence $y = 1/x = -1, 1/2$.

Number of real roots

Looking at the previous two examples, you might ask the following question: how many real nth roots does a real number have? For instance, 8 has two square roots, while -8 has no square roots. On the other hand, both 8 and -8 have exactly one cube root each: 2 and -2, respectively.

Here is the answer: if n is even (square root, fourth root, and so on), a negative number has *no* real nth roots, while a positive number has exactly two real nth roots. (And 0 has exactly one nth root: itself.) On the other hand, if n is odd, then every real number has a unique real nth root.

Solving a quadratic equation by completing the square*

We saw above that, given a quadratic equation $ax^2 + bx + c = 0$ (with $a \neq 0$), the two numbers

$$x_\pm = \frac{-b \pm \sqrt{b^2 - 4ac}}{2a}$$

indeed satisfy the equation whenever $b^2 \geq 4ac$. Now we will show that $x_\pm$ are the *only* two possible roots to the equation. Let us first demonstrate how to do this in a special case.

Example 1.13. Solve the equation $x^2 + \frac{5}{2}x - 6 = 0$ without using the general formula.

Solution: Recall that if we can rewrite the equation as a product of factors, we will be able to find the roots by setting each of the factors equal to zero (see Example 1.7). To factorize a general quadratic polynomial, there is a useful trick called *completing the square*, which can be used to write the quadratic polynomial as the sum of a real number and the square of a linear polynomial. The trick relies on the following general formula (which can be verified by using the distributive law): for any two real numbers x, y,

$$(x + y)^2 = x^2 + 2xy + y^2.$$

Now suppose we are given a quadratic polynomial like $x^2 + \frac{5}{2}x - 6$. We first consider only the linear and quadratic terms, namely $x^2 + \frac{5}{2}x$, and ask: what number should one add to this in order to obtain a square of the form $(x+y)^2$? We see that the term $\frac{5}{2}x$ should correspond to the term $2yx$, which means that $y = \frac{5}{4}$. In other words, y is half of the coefficient for x, and we add y^2 to obtain $(x+y)^2$. Thus if we add $(\frac{5}{4})^2 = \frac{25}{16}$, we will get $x^2 + \frac{5}{2}x + \frac{25}{16}$, and you should verify, using the distributive law (i.e., the FOIL method), that this equals $(x + \frac{5}{4})^2$.

Now we obtained a square of a linear polynomial, but we added $\frac{25}{16}$ to our original equation and dropped the last term -6. To recover the original equation, we must compensate for this. Thus, we obtain:

$$0 = x^2 + \frac{5}{2}x - 6 = x^2 + \frac{5}{2}x + \frac{25}{16} \quad - \frac{25}{16} - 6 = \left(x + \frac{5}{4}\right)^2 - \frac{121}{16}.$$

In the last line we used the equality $-6 - \frac{25}{16} = \frac{-96-25}{16} = -\frac{121}{16}$. We notice that $\frac{121}{16}$ is a complete square: $\frac{121}{16} = \left(\frac{11}{4}\right)^2$.

The second step is to use the formula $a^2 - b^2 = (a+b)(a-b)$ derived in Example 1.3, to obtain:

$$0 = \left(x + \frac{5}{4}\right)^2 - \frac{121}{16} = \left(x + \frac{5}{4}\right)^2 - \left(\frac{11}{4}\right)^2$$

$$= \left(x + \frac{5}{4} + \frac{11}{4}\right) \cdot \left(x + \frac{5}{4} - \frac{11}{4}\right) = (x + 4)\left(x - \frac{3}{2}\right).$$

We conclude that the only possible solutions are $x = -4$ and $x = \frac{3}{2}$. You can check that the general formula for the roots of a quadratic equation gives the same two solutions.   □

The above technique works when the coefficient of x^2 is 1. What about a general quadratic equation $ax^2 + bx + c$ with $a \neq 0$? Simply divide by a first:

Example 1.14. Solve the general quadratic equation $ax^2 + bx + c = 0$, by completing the square.

Solution: First divide both sides by $a \neq 0$ to obtain:

$$x^2 + \frac{b}{a}x + \frac{c}{a} = 0.$$

As discussed in the previous example, we have to add and subtract the square of the half of the coefficient in front of x, which equals $\left(\frac{b}{2a}\right)^2$. We obtain:

$$0 = x^2 + \frac{b}{a}x + \frac{c}{a} = x^2 + \frac{b}{a}x + \left(\frac{b}{2a}\right)^2 - \left(\frac{b}{2a}\right)^2 + \frac{c}{a} = \left(x + \frac{b}{2a}\right)^2 - \frac{b^2}{4a^2} + \frac{c}{a}.$$

(Note here that $(b/(2a))^2 = b^2/(2a)^2 = b^2/(4a^2)$.) Now we can combine the last two terms by taking the common denominator:

$$-\frac{b^2}{4a^2} + \frac{c}{a} = \frac{-b^2 + 4ac}{4a^2} = \frac{-(b^2 - 4ac)}{4a^2} = -\left(\frac{\sqrt{b^2 - 4ac}}{2a}\right)^2.$$

This yields:

$$0 = x^2 + \frac{b}{a}x + \frac{c}{a} = \left(x + \frac{b}{2a}\right)^2 - \left(\frac{\sqrt{b^2 - 4ac}}{2a}\right)^2.$$

Note that this is a difference of squares if and only if the discriminant $(b^2 - 4ac)$ is a nonnegative number, so that we can take the square root. If the discriminant is zero, we get a complete square equal to zero, and a single solution $x = -\frac{b}{2a}$. If it is negative, we have a complete square equal to a negative number, which means that no real x satisfies the equation. Finally, in case the discriminant is positive, we can apply the difference of squares formula. Then we get

$$
\begin{aligned}
0 &= \left(x + \frac{b}{2a} - \frac{\sqrt{b^2 - 4ac}}{2a} \right) \cdot \left(x + \frac{b}{2a} + \frac{\sqrt{b^2 - 4ac}}{2a} \right) \\
&= \left(x + \frac{b - \sqrt{b^2 - 4ac}}{2a} \right) \cdot \left(x + \frac{b + \sqrt{b^2 - 4ac}}{2a} \right),
\end{aligned}
$$

and we conclude that the two roots to the quadratic equation are as claimed: $x_\pm = \dfrac{-b \pm \sqrt{b^2 - 4ac}}{2a}$. □

EXERCISES

Question 1.1. Expand or contract (factorize) the following expressions, using the distributive law. All variables below denote real numbers.

1. $(a - 2)(b - 2)$.

2. $(x + 1)(2x - 3)$.

3. $(y^2 + 1)(y^2 - 1)$.

Question 1.2. Expand or contract the following expressions using the distributive law.

1. $(1 + z + z^2 + z^3)(1 - z)$.

2. $T^2 - sT + 2T - 2s$, where s is some fixed real number.

3. $AB - BC + CD - DA$.

Question 1.3. Suppose you consider any two consecutive integers. Then the difference between their squares is always an odd number. Can you prove this fact in general?

Question 1.4. Suppose you multiply any three consecutive integers, and add the middle integer to this product. Then you will always get a perfect cube. Why is this? Denote the middle integer by n and carry out this computation using n to show why this holds for *every* integer n, all at once. *Hint:* Decide for yourself in which order you would like to multiply the three integers.

Question 1.5. Solve the following equations for the unknown variables.

1. $x^4 = 25$.

2. $(x-1)^2(x-2)(x+3) = 0$.

3. $y^2 + 8y + 15 = 0$.

4. $A^2 - 3A - 4 = 0$.

Question 1.6. Solve the following equations for the unknown variables.

1. $y^3 = 27$.

2. $(z-22)^3(2z+27)(z+3)^{17} = 0$.

3. $s^2 + 10s + 20 = 0$.

4. $c^4 + 3c^2 - 4 = 0$. *Hint:* Remember that squares are never negative.

Question 1.7. The quadratic equation $y^2 = 6y$ can be solved directly using the distributive law, and also using the general formula. Solve the equation in both possible ways, and verify for yourself that the set of solutions is the same regardless of how you solve the equation.

Question 1.8. For what value(s) of b does the equation $x^2 = bx - 7$ have a unique solution?

Question 1.9. Solve the quadratic equations.

1. $2\beta^2 + 22\beta + 60 = 0$.

2. $y^2 - 3y + 3 = 0$.

Question 1.10. Solve the quadratic equations.

1. $2x^2 - 4\sqrt{5}x + 10 = 0$.

2. $2z^2 - \pi z + 1 = 0$.

Question 1.11. Evaluate the roots of each of the following equations, or show using the discriminant that the corresponding quadratic equation has no solution.

1. $x^2 - 7x + 2 = 0$.

2. $3y^2 - 6y + 12 = 0$.

3. $x^2 + 4x - 1 = -5$.

4. $x^4 - 3x^2 + 2 = 0$. *Hint:* Try to solve the equation for $y = x^2$ first.

Question 1.12. For each of the following equations, compute the roots or show that there are no roots.

1. $2\alpha^2 = 7\alpha - 1$.

2. $2t^2 - 4t + 8 = 5$.

3. $2t^2 - 4t - 3 = 0$.

4. $x^3 - 7x^2 + 6x = 0$. *Hint:* First try to factor this polynomial.

Question 1.13. The following equations may not seem like quadratic equations, but each of them can be reduced to one. Solve them.

1. $D^5 - 16D^4 + 60D^3 = 0$. *Hint:* Take out a common factor.

2. $t - 5\sqrt{t} + 6 = 0$. *Hint:* This is a quadratic polynomial – in which variable?

Question 1.14. The following equations may not seem like quadratic equations, but each of them can be reduced to one. Solve them.

1. $y^{17} - 2y^{15} + y^{13} = 0$. *Hint:* After taking out a common factor, we get a quadratic polynomial – in which variable?

2. $z^{16} - 6z^{11} + 9z^6 = 0$.

3. $\dfrac{1}{y^7} - \dfrac{3}{y^4} + \dfrac{1}{y} = 0$. *Hint:* Note here that $\frac{1}{y}$ cannot be zero for any real number y. Thus, we can take $\frac{1}{y}$ common and cancel it.

Question 1.15. Suppose the product of two successive integers is 30. Find both the numbers. *Hint:* Suppose the lesser of the two numbers is N. Can you write down what the other number should be, and then an equation that describes the given problem?

Question 1.16. One side of a rectangle is three times as long as the other side. The area of the triangle is 192 square inches (in^2). Find the lengths of the sides.

Question 1.17. The two sides of a right triangle which meet at the right angle are 5 centimeters (cm) apart in length. The area of the triangle is 75 square centimeters (cm^2). Find the lengths of the sides.

Question 1.18. Here is another magic trick: take a whole number (say, between 1 and 10), add 4 to it, then multiply again by the whole number, and finally add 4. Call the result R.

1. Is R a perfect square? If it is, can you suggest why it is always a perfect square, no matter what whole number you start with? *Hint:* If the whole number is N, what is R? Can you factor R, e.g., by completing the square?

2. Now take the square root of R, and subtract the original whole number. You will always end up with 2! Why is this?

Question 1.19. Make up your own magic trick using a similar technique. For instance, come up with a series of "confusing" operations to get from a number to its negative, or to thrice itself. (Check that the trick always works.)

Chapter 2

Velocity: On the road

The simplest kind of motion is the motion of an object along a straight line with a constant speed: a car moving along a road at sixty miles per hour, a pedestrian walking at three miles per hour. However, if multiple objects are moving with constant speeds along the same path, and they are allowed to change direction, the mathematical situation becomes more intricate. In this chapter, we consider a range of examples of systems of moving objects drawn from everyday experience, our imagination, and, in one case, literary fiction.

MATH

The mathematical model of the rectilinear constant speed motion of one object can be formulated in one line:

$$s = v \cdot t,$$

where s is the distance from the starting point covered by the object moving along a straight line with speed v during the time t. The three quantities should be measured in compatible units, meaning that if the distance is in miles (mi), and the time in hours (h), then the speed should be taken in miles per hour (mph). The same formula can be read in two other ways:

$$v = \frac{s}{t} \qquad \text{and} \qquad t = \frac{s}{v}.$$

In fact, this model describes a larger class of processes. If the speed of an object is not constant, but the *average* speed is known to be v, then the distance covered in time t is given by the same formula $s = v \cdot t$. Moreover, the path does not need to be straight. If v is the (average) speed of motion along any given path, then $t = \frac{s}{v}$ is the time needed to cover the distance s measured *along* the path.[1]

Example 2.1. How long does it take to get to a town 230 miles away along a road, driving at an average speed of 60 miles per hour? We compute:

$$t = \tfrac{s}{v} = \tfrac{230}{60} \quad \simeq 3.833 \,\text{h}$$
$$= 3\,\text{h}\,50\,\text{min}.$$

□

Now suppose there are two objects moving with the constant speeds v_1 and v_2 along the same path. Will they ever meet, and how soon?

To answer this question, it is convenient to introduce the notion of *relative speed*, defined as the rate of change of the distance between the two objects. Thus, if the two objects are moving along the same path in the same direction, then the relative speed v_{12} is the difference between their speeds:

$$v_{12} = |v_1 - v_2| = \begin{cases} v_1 - v_2, & \text{if } v_1 \geq v_2; \\ v_2 - v_1 & \text{if } v_1 < v_2. \end{cases}$$

If the faster moving object is behind, they are getting closer at the rate v_{12}, and if the slower one is behind, they are getting farther apart at the same rate.

Figure 2.1: Moving in the same direction.

Getting closer   Getting farther apart

Now, if the two objects are moving in opposite directions, then the relative speed v_{12} is the sum of their speeds:

$$v_{12} = v_2 + v_1.$$

---

[1]Motion along a curve requires an acceleration directed to the center of the curve. But we are interested in the speed of motion along a path, and we assume it to be constant.

Depending on the position of the objects, they are getting closer or farther apart at the rate v_{12}.

Figure 2.2: Moving in opposite directions.

Getting closer   Getting farther apart

The time needed for the distance between the two objects to decrease or increase by s is given by the formula

$$t = \frac{s}{v_{12}}.$$

In particular, if the objects are getting closer, and the initial distance between them is s, then $t = \frac{s}{v_{12}}$ is the time before they meet.

Example 2.2. Suppose your friend is 30 miles ahead of you on a highway, moving forward at a constant speed of 50 miles per hour. Can you catch up with him before he gets to the next town, 120 miles ahead of you, without breaking the speed limit of 65 miles per hour? Equivalently, assume you are moving at a constant speed of 65 miles per hour. When will you catch up with your friend: before or after he passes the town 120 miles ahead?

65 mph   **50 mph**

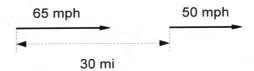

30 mi

Solution: We have two objects moving in the same direction, with $v_1 = 65$ mph and $v_2 = 50$ mph, the faster object behind. The initial distance between them is $s = 30$ mi. The time before they meet is

$$t = \frac{s}{v_{12}} = \frac{s}{v_1 - v_2} = \frac{30}{65 - 50} = 2 \text{ h}.$$

In 2 hours, you will be $2 \cdot 65 = 130$ miles ahead, already past the town situated 120 miles ahead. You cannot catch up with your friend before he passes the town. $\qquad\square$

APPLICATIONS

Constant speed motion with a change of direction

If the objects moving along the same path are allowed to change directions, the model becomes more complicated and interesting patterns can emerge. Here are a couple of examples.

Example 2.3. Two bees, Yolanda (Y) and Zoe (Z), fly non-stop between two parallel walls 30 feet (ft) apart. Yolanda starts at 9:00:00 from the western wall and flies with a constant speed of 2 feet per second (ft/sec). Zoe starts at 9:00:10 from the eastern wall and flies with a constant speed of 3 feet per second. When will Yolanda meet Zoe for the first, second, and third time?

Solution: Let us find the time of the first meeting. At 9:00:00, only Yolanda is moving:

Figure 2.3

9:00:00

30 ft

   At 9:00:10, when Yolanda is $2 \cdot 10 = 20$ feet from the western wall, Zoe starts flying west at a speed of 3 feet per second from the eastern wall.

   To meet, they have to cover the distance $s = 30 - 20 = 10$ ft, moving with the relative speed $v_{12} = v_1 + v_2 = 2 + 3 = 5$ ft/sec. This will take $t = \frac{s}{v_{12}} = \frac{10}{5} = 2$ sec. Therefore, the first meeting will happen 2 seconds after Zoe starts flying, or at 9:00:12.

   Clearly, Yolanda and Zoe will meet again if they continue flying between the walls. To figure out the time of the next meeting, we have to (1) find the moment of time when Zoe is at the western wall; (2) find the

Figure 2.4

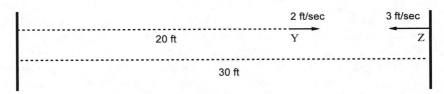

position and direction of motion of Yolanda at this moment; (3) divide the distance between the bees by their relative speed. If we want to find the time of the third meeting, we will have to do it all over again.

Luckily, there exists a more efficient approach. The picture below shows that the sum of the distances Yolanda and Zoe need to cover between any two consecutive meetings is 60 feet.

Figure 2.5

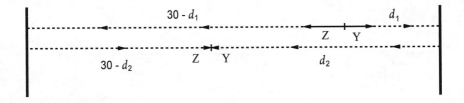

Indeed, if Yolanda has to cover the distance d_1 to the eastern wall, and the distance d_2 from the eastern wall to the meeting point, then Zoe will have to cover the distance $30 - d_1$ to the western wall, and $30 - d_2$ from the western wall to the meeting point. Together, they will have to cover

$$d_1 + d_2 + (30 - d_1) + (30 - d_2) = 60 \text{ ft.}$$

The relative speed of the two bees is $v_{12} = 2 + 3 = 5$ ft/sec. The time between the consecutive meetings is

$$t = \frac{60}{v_{12}} = \frac{60}{5} = 12 \text{ sec.}$$

Therefore, Yolanda and Zoe will meet at 9:00:12, 9:00:24, 9:00:36, and so on every 12 seconds.   □

Example 2.4. In the previous example, we can ask where exactly between the walls the meetings of Yolanda and Zoe will take place. Let us measure the distances from the eastern wall. The first meeting happens 2 seconds after Zoe starts flying, at the distance $2 \cdot 3 = 6$ ft from the eastern wall. The subsequent meetings take place every 12 seconds, so all we need to do is to find the position of, say, Zoe at these times. The distance covered by Zoe in 12 seconds is $12 \cdot 3 = 36$ ft. So the distance covered by Zoe by the time of the kth meeting with Yolanda is

$$s = 6 + 36 \cdot (k - 1) \text{ ft.}$$

When $k = 1$, the distance is 6 feet from the eastern wall.

Now we need to take into account the changes of direction. Any multiple of 60 feet brings Zoe back to the eastern wall, so it does not count as far as the position is concerned. Furthermore, if the remainder of the distance is less than 30 feet, then Zoe is situated at that distance from the eastern wall and is moving west. If the remainder is greater than 30 feet, then Zoe went all the way to the western wall and turned around. For example, if $s = 42$, then she is 12 feet from the western wall, or $30 - 12 = 18$ feet from the eastern wall, and moving east. We have the following table for Zoe's position and direction of motion at the kth meeting:

k	s	s up to a multiple of 60	distance to east wall	Zoe's direction
1	6	6	6	←
2	42	42	18	→
3	78	18	18	←
4	114	54	6	→
5	150	30	30	← then →
6	186	6	6	←

The diagram in Figure 2.6 shows the positions and directions of motion of both bees at the time of the first six meetings.

The fifth meeting takes place at the western wall, where both bees instantaneously change direction from west to east. Notice that even though the position of the fourth and the first meetings is the same, 6 feet from the eastern wall, the directions of motion of the bees are different. But the sixth meeting is an exact repetition of the first one: both bees are 6 feet from the eastern wall and Zoe is moving west, and Yolanda east. Therefore, the next meeting will be an exact repetition of the second, and

Figure 2.6: First six meetings

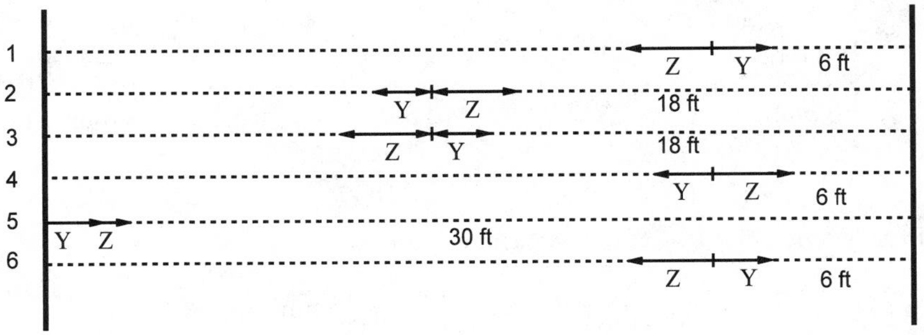

so on, as long as the bees keep flying. The meetings will take place at 6, 18, or 30 feet from the eastern wall. □

Two trains and a bee

Now imagine that in the previous example the walls were allowed to move as well. A well-known model with just one bee flying between two moving obstacles is considered below.

Example 2.5. Two trains, A from the west and B from the east, are approaching along the same straight train track, each moving at a constant speed of 5 miles per hour. When the distance between the trains is 10 miles, a bee starts flying east from the head of train A. The bee moves at a constant speed of 6 miles per hour until it hits train B, then it turns around and flies west with the same constant speed until it hits train A, and so on until the two trains meet and the bee gets smashed between them (now we are in trouble with the animal rights activists). What is the total distance covered by the bee until the crash?

Solution: We will start with a less ambitious question, namely, when and where will the bee meet train B for the first time? The diagram above shows the initial positions and speeds of the moving objects: the bee goes east at $v_1 = 6$ mph, train B goes west at $v_2 = 5$ mph, and the initial distance between them is $s_0 = 10$ mi. They will meet in

$$t_1 = \frac{10\,\text{mi}}{5 + 6\,\text{mph}} = \frac{10}{11}\,\text{h.}$$

Figure 2.7

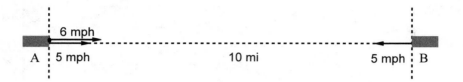

In this time, the bee will cover the distance

$$s_1 = 6\,\text{mph} \ \cdot\ \frac{10}{11}\,\text{h} = \frac{60}{11}\,\text{mi}.$$

Now we know the position of the bee at the moment when it changes direction and starts flying west. The diagram shows this moment.

Figure 2.8

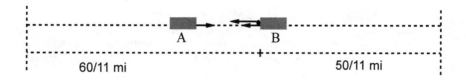

Next we can ask when and where the next meeting of the bee with train A will occur. In $\frac{10}{11}$ of an hour, train A has moved $5 \cdot \frac{10}{11} = \frac{50}{11}$ mi east, and is situated $\frac{60}{11} - \frac{50}{11} = \frac{10}{11}$ mi west from the bee. The time to their meeting is

$$t_2 = \frac{\frac{10}{11}\,\text{mi}}{5 + 6\,\text{mph}} = \frac{10}{11 \cdot 11} = \frac{10}{121}\,\text{h}.$$

In this time the bee will cover an additional distance

$$s_2 = 6\,\text{mph} \ \cdot\ \frac{10}{121}\,\text{h} = \frac{60}{121}\,\text{mi}.$$

In the first two moves, the bee covered $s_1 + s_2 = \frac{60}{11} + \frac{60}{121} = \frac{660+60}{121} = \frac{720}{121}$ mi. To find the total distance covered by the bee until is gets smashed we would have to compute and sum up infinitely many more such distances, and the whole approach is starting to look somewhat hairy. For the moment we will drop the question, and return to it in Chapter 10, where we will learn how to compute such infinite sums.

However, there is another, more efficient solution: the bee is flying non-stop at a constant speed of 6 miles per hour until the trains meet. The time before the meeting of the trains is

$$t = \frac{10}{5+5} = 1 \text{ h}.$$

Therefore, the total distance covered by the bee before the crash is

$$6 \text{ mph} \cdot 1 \text{ h} = 6 \text{ mi}.$$

$\square$

A traveler and messengers

We know of at least one excellent example of an elaborate system of objects moving at constant speed that appears in fiction. Here is a short synopsis of the novella "The Seven Messengers" by an Italian writer, Dino Buzzati.

The prince, the narrator of the story, sets out to explore his father's kingdom, hoping to find its boundaries. He and his knights start from the capital and move in one direction (due south, or so they hope) at a constant speed of 40 leagues per day. After two days of travel, the prince sends his first messenger – Alessandro – back to the capital. The next six messengers, Bartolomeo, Caio, Domenico, Ettore, Federico, and Gregorio, are sent back, respectively, after three, four, five, six, seven, and eight days of travel. All messengers move at a constant speed of 60 leagues per day. Having reached the capital, each messenger immediately starts back along the same path to catch up with the prince. Having reached the prince, the messenger immediately starts back for the capital, and so on. Thus, the seven messengers oscillate between the capital and the prince, while the prince is moving farther and farther away. The intervals between the arrivals of the messengers grow, until one day the prince realizes that the next return of a messenger will be the last one he will live to see. To simplify the diagrams, we assume in Figures 2.9 and 2.10 that the southern direction is to the right.

In the course of the story, the narrator makes multiple numerical statements, and a natural reaction for an inquisitive mind is to check whether or not they make sense. Any mathematician would definitely have an itch to do so; but even a reader with the sole interest in literature, presumably, would like to know if the author is careful about all the details of his creation, or neglectful, or intentionally misleading. On multiple occasions, the narrator mentions the number of days, months, and years that have passed since the beginning of the journey and between the consecutive returns of the messengers.

Figure 2.9: Messenger leaves for the capital.

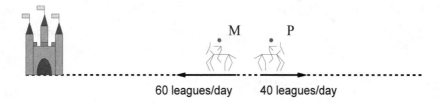

Figure 2.10: Messenger starts back to catch up with the prince.

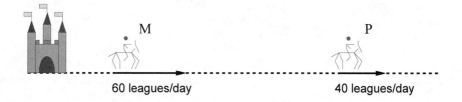

The interesting feature of this particular work of fiction is that all the numerical statements in it are verifiable. We can figure out the position of each of the messengers at any moment of time (in days since the start of the expedition) and, in particular, check the numerical statements contained in the story.

Let d be the number of days elapsed before a messenger (M) was first sent back. Let t denote the number of days M needs to catch up again with the prince (P). Then we have the equation:

$$\text{distance traveled by } M = 2 \left(\begin{array}{c} \text{distance to the capital} \\ \text{when M leaves} \end{array} \right) +$$

$$+ \left(\begin{array}{c} \text{distance traveled by P} \\ \text{since M left} \end{array} \right)$$

or, taking into account the given speeds of P and M,

$$60t = 2 \cdot 40d + 40t.$$

From this equation we find that $t = 4d$ is the time taken by M to catch up with P. By the time he catches up, $d + 4d = 5d$ days would have passed since the start of the expedition. Now we only need to plug in $d = 2$ for Alessandro, $d = 3$ for Bartolomeo, and so on, up to $d = 8$ for the last messenger, Gregorio, to obtain the times of their return to the prince. We find that they reunite with the prince after 10, 15, 20, 25, 30, 35, and 40 days, respectively.

A messenger is then immediately sent back to the capital; so, for instance, Alessandro is sent back the second time after 10 days since the start of the expedition. By the same argument as before, we see that the second return of the messengers will occur after $5 \cdot 5d = 25d$ since the start of the expedition. Next, they will return after $5 \cdot 25d = 125d$, $5 \cdot 125d = 625d$, and so on. For example, Alessandro ($d = 2$) will return to the camp after $2 \cdot 5 = 10$, $2 \cdot 25 = 50$, $2 \cdot 125 = 250$, $2 \cdot 625 = 1,250$, etc., days.

Here is the timetable (in days since the beginning of the journey) of the first five consecutive returns of each messenger to the prince:

	d	$5d$	$25d$	$125d$	$625d$
Alessandro	2	10	50	250	1,250
Bartolomeo	3	15	75	375	1,875
Caio	4	20	100	500	2,500
Domenico	5	25	125	625	3,125
Ettore	6	30	150	750	3,750
Federico	7	35	175	875	4,375
Gregorio	8	40	200	1,000	5,000

This table contains enough information to check the numerical claims of the narrator. Let us consider each one of them:

1. *"...it was sufficient to multiply by five the days elapsed so far to know when the messenger would catch up with us."* This is exactly the formula we derived above: if a messenger leaves the prince after d days since the start of the expedition, he returns to the prince after $5d$ days.

2. After fifty days, the interval between the messenger's consecutive returns increases to twenty-five days. This is exactly the contents of the second and the third columns of the table.

3. After six months of travel, this interval increases to over four months. We need to translate six months into days. There is an ambiguity here: we do not know which months they are, resulting in a number of days between 181 (January – June) and 184 (July – December). But in any case, the interval increases to at least four months (125 days) a little later – after 250 days, or over eight months of travel. Because the narrator does not state that the change occurred immediately after six months, the claim is legitimate.

4. After four years of travel, this interval increased to over twenty months. To convert four years into days, we have to take into account one leap year that happens once in four years. We have $3 \cdot 365 + 366 = 1,461$ days. Twenty months contain a full year and eight months, resulting in approximately $365 + 4 \cdot 31 + 4 \cdot 30 = 609$ days. From the last column of the table we deduce that after $1,250$ days of travel, the interval increased to 625 days, which is consistent with the statement.

5. The present return of Domenico to the narrator occurs exactly after *"eight years, six months, and fifteen days of uninterrupted travel..."* Looking at the fourth line of the table, we easily find a suitable time for this return of Domenico: it happens $3,125$ days since the start of the expedition, which we have to compare with the claimed eight years, six months, and fifteen days. Taking into account two leap years, and considering the possible range in the number of days in six months, 181 to 184, we get $6 \cdot 365 + 2 \cdot 366 + 181 + 15 = 3,118$, or $6 \cdot 365 + 2 \cdot 366 + 184 + 15 = 3,121$. In any case, the claimed interval is four to seven days shorter than we expected. We will return to this discrepancy later and suggest possible explanations.

6. *"For almost seven years I had not seen him. [Domenico]."* According to the table, the previous return of Domenico occurred 625 days after the start of the expedition; that is, $3,125 - 625 = 2,500$ days ago. We find $2,500/365 \simeq 6.849$ yr, or almost seven years.

7. *"I will again see Domenico only when thirty-four years have passed."* According to the formula, the next return of Domenico will happen after $5 \cdot 3,125 - 3,125 = 4 \cdot 3,125 = 12,500$ days, and $12,500/365 \simeq 34.246$ yr, or after thirty-four years.

8. After the present return of Domenico, *"... Ettore ... will reach me ... in a year and eight months."* After that, the narrator is not going

to send Ettore back to the capital, because he won't live long enough to see his next return. The interval between the present return of Domenico and the return of Ettore according to the last column of the table is $3,750 - 3,125 = 625$ days. We have already computed that one year and eight months contain approximately 609 days, so the first statement is true. Now let us find the interval between this and the next return of Ettore: $5 \cdot 3,750 - 3,750 = 4 \cdot 3,750 = 15,000$ days. This corresponds to $15,000/365 \simeq 41.1$ yr, the time the narrator does not hope to live long enough to see.

Therefore, we have confirmed all statements of the narrator, except (5). Clearly, the author was well aware and careful about the mathematics involved in the story. For example, claim (1) confirms our formula that determines the dates of the consecutive returns of the messengers. For the discrepancy in (5), we can suggest the following explanations.

A practical explanation

The author had found the number $3,125$ days and needed to convert it into years, months, and days. If we suppose that he counted each year to contain 366 days, and each February 29 days, as in the leap year, then the computation confirms the claim of the narrator: $366 \cdot 8 + 182 + 15 = 3,125$, where 182 days holds for the six months, January through June, counted in a leap year. It might be that the author chose this inaccurate but quick way to obtain the answer.

A poetic explanation

Alternatively, we might suppose that the few missing days were intentionally subtracted by the author. Domenico catches up with the expedition just a few days sooner than expected. In fact, the narrator knows that this might happen at the next return of Domenico, who will *"ask why I have made so little progress in the meantime."* We can conjecture that the progress of the narrator has already slowed down, without him noticing it. The prince and his knights now cover, probably, a little less than 40 leagues per day, and this has resulted in his messenger catching up with him sooner than expected. The melancholy of the story is amplified by the fact that the narrator realizes that his old age will inevitably diminish his powers, and eventually interrupt his quest, but he cannot see that his decline has already started and is already affecting his progress.

A geographical explanation

Let us have a look at the distances implied by the story. How far can one ride a horse on Earth without changing direction? In other words, what is the longest distance on land along one direction? Here are some examples:

- East Coast–West Coast distance in the US is approximately 3,000 miles, or 4,828 kilometers.

- Longest continuous distance on land along a longitude : 7,590 kilometers (Northern Russia to Southern Thailand, 99° east).

- Longest continuous distance on land along a latitude: 10,726 kilometers (Western France to Eastern China, 48° north.)

- Longest distance on land along any great circle: 13,573 kilometers (Liberia to China).

To find the distance covered so far by the prince and his knights (supposedly, they always move due south), we need to multiply 40 leagues by 3,120 (approximately) days of travel. A league is an ancient measure of distance that varies from country to country, but is approximately equal to the distance a person can walk in an hour. We suppose that Dino Buzzati, being an Italian writer, would use the Roman league, equal to 1.4 mi, or 2.2225 km. In this case, the distance traveled by the prince at the time of the narration is $2.2225 \cdot 40 \cdot 3120 = 277,368$ km. No distance on land along one direction on Earth is that long, which suggests either a fantastic setting, or that the expedition, despite the hopes of the narrator, is not moving in the same direction. Let us choose middle ground: suppose that the expedition is taking place on Earth, but the arrangement of continents allows for an indefinite movement in one direction on land. Then we can imagine the path of the expedition as a tight spiral as in Figure 2.11 starting at the equator and circling around multiple times.

   This model gives rise to another explanation for the discrepancy between the calculated time of Domenico's return (3,125 days) and the time reported by the narrator (between 3,118 and 3,121 days). The concept outlined below first came to the attention of astronomers in the Renaissance times, about the time the story of the seven messengers might be imagined to have occurred.

International date line

Recall the history of the first world circumnavigation headed by Ferdinand Magellan. He started to sail his five ships due west from the coast

Figure 2.11: Circling around the globe

of Spain on September 20, 1519. Almost three years later, on September 6, 1522, the eighteen survivors of his original crew (Magellan himself was killed in March 1521 in the Philippines) and his only surviving ship, *Victoria*, returned to Spain. But the ship's log had the arrival date marked as September 5, 1522. The log was recorded with utmost care and accuracy; in particular, the leap year 1520 was taken into account. The mystery of a missing day was discussed by the leading scientists of the time, among them the Venetian astronomer Gasparo Contarini, who suggested the correct explanation: moving westward, and sailing a whole circle around the Earth, you gain one day; moving eastward, you lose it. It was not until the nineteenth century when the International Date Line was established in the sense and position it has now: an imaginary line between the north and south poles at approximately 180° east separating Russia and Asia from the Americas, and one calendar day on Earth from the next. Now a person crossing the International Date Line traveling eastbound has to subtract a day; when traveling westbound, add a day.

If we suppose for a moment that the expedition was moving west-west-south instead of south (and was able to move continuously on land), the missing days can be explained by the same effect. A Renaissance setting of Buzzati's story suggests that the narrator (just like Magellan's crew) might not have known about the necessity to add a day for each complete circle when traveling westward on an Earth-like planet. Let us estimate the distances. The radius of Earth $R_{\oplus}$ is between $6,353$ and $6,384$ kilometers (larger at the equator). Therefore, the circumference at the equator is approximately $2\pi R_{\oplus} \simeq 40,090$ km. By the time of the narration, prince's expedition has covered $277,368$ kilometers. Dividing this distance by the circumference of the equator, we get $277,368/40,090 \simeq 6.9$. Therefore, six days (more than six if the narrator was moving along a higher latitude; less than six if he was deviating from the straight westward direction) might have to be added to the number of days in the narrator's log to obtain the actual number of days elapsed in the capital, or for a messenger who moves east and west, back and forth. The date discrepancy between a round-the-globe traveler and a stationary observer was one of the most remarkable successfully resolved mysteries of its time. Even if this argument was not intended by the author, it fits nicely with the story's Renaissance setting and its enigmatic character.

Messengers in space

With the advent of the Internet, the idea of sending a human messenger to deliver a letter might have lost some of its practical value. Nevertheless, there are real-world phenomena that essentially enact the story of the traveler and his messengers.

Can you think of a real-world situation where an object is moving away from an observer, and the messengers oscillate between them? A space probe such as *Voyager* is an example. This is an instrument, roughly a radio telescope, propelled by a rocket into deep outer space to send back to Earth information about remote planets and stars. The messengers are the photon particles that carry information between the space ship and the command center on Earth. There are multitudes of them, and their speed is the speed of light (in kilometers per second): $300,000 = 3 \cdot 10^5$ km/sec. The gravitational and relativistic effects have to be taken into account, which makes the trajectories curve and the time flow differently for different moving objects. The distances are much bigger, too: the probes *Voyager 1* and *Voyager 2*, launched in 1977, are now exploring the outermost layer of the heliosphere, the region of space dominated by the Sun. To give a rough idea of the scale of the distances, the approximate size of the Solar System is 4.5 billion kilometers ($4.5 \cdot 10^9$ kilometers). But the essence of the story of the messengers remains the same: *as the probe moves farther away from us, the messages we get from it become more and more outdated.* For example, on August 27, 2003, Mars came closest to the Earth in the previous 60,000 years. On that day a snapshot sent to Earth from the orbit of Mars took about 186 seconds (or 3.1 minutes) to arrive, while a snapshot sent from the edge of the Solar System communicates an observation made by the probe about $14,700$ seconds (about 4 hours) ago.

In fact, because the universe is expanding, any star is an object moving away from us, and its visible light is a messenger from it. Let us try to estimate how outdated our picture of the stars might be. The approximate diameter of our Galaxy is $100,000$ light-years, or $9.4 \cdot 10^{17}$ km, and we are about $27,000$ light-years away from the galactic center. This means that we see the stars at the opposite edge of our Galaxy the way they looked about $77,000$ years ago. But we still receive their messengers, their beams of light, and maybe in another hundred thousand years they might be able to see the Solar System as it is now, and us in it.

EXERCISES

Question 2.1. James Bond is in Spyburg, 140 miles from an international border, and a villain is in Villainburg, 20 miles closer to the border along the same road. At noon, the villain starts driving toward the border at a constant speed of 80 miles per hour.

1. If James Bond drives at 100 miles per hour, when is the latest he should leave Spyburg to overtake the villain before he crosses the border?

2. If James Bond leaves Spyburg at 12:30 pm, what is the minimum average speed he has to maintain to overtake the villain before he crosses the border?

Question 2.2. Liz and Pat, who live 33 miles apart, want to go biking together. At 10 am they load their bikes in their cars and start driving toward each other's houses, Liz going at 50 miles per hour and Pat at 60 miles per hour. At the meeting point, they park their cars and immediately start riding their bikes at 18 miles per hour in the direction of Liz's house.

1. When will they arrive at Liz's house?

2. After spending an hour at Liz's house, they ride their bikes back to the parked cars. If Liz immediately starts driving home at 50 miles per hour, when will she arrive?

Question 2.3. Car A moves along a road at a constant speed of 60 miles per hour. Five miles ahead of it, car B moves in the same direction at a constant speed of 45 miles per hour. Thirty miles ahead of B, car C moves *in the opposite direction* at a constant speed of 55 miles per hour.

1. Which car will be first to meet B, A or C?

2. At the moment when B first meets with one of the cars, how far from it is the remaining car?

3. How much time will elapse between the meeting of car B with the first and the second cars?

Question 2.4. At 10 am car A starts moving east along a road at a constant speed of 50 miles per hour. At the same time 55 miles to the east of car A, car B starts moving west at 60 miles per hour, and car C starts moving east at 48 miles per hour. The moment cars A and B meet, B changes direction and starts moving east at 60 miles per hour.

1. When will cars B and C meet?

2. At the moment when B and C meet, how far behind is car A?

Question 2.5. In Example 2.3, suppose that Zoe starts from the eastern wall at 9:00:00, and Yolanda starts at the western wall at 9:00:05. The distance between the walls and the speeds of the bees are the same as before. Find the times of their first, second, and third meetings.

Question 2.6. In Example 2.3, suppose that Zoe starts from the eastern wall at 9:00:00, and Yolanda starts at the western wall at 9:00:05. The distance between the walls and the speeds of the bees are the same as before.

1. Determine where exactly between the walls the bees meet for the first, second, and third times.

2. Assuming that the bees fly forever, find all possible positions of their meetings.

Question 2.7. Starting from Pigeonville, a traveler walks at a constant speed along a straight road. After 1 hour of walking, he sends a trained pigeon with a letter back to Pigeonville. The pigeon flies at a constant speed of 8/5 times the speed of the traveler. Upon reaching Pigeonville, the pigeon immediately turns back to catch up with the traveler. When does the pigeon catch up with the traveler?

Question 2.8. Starting from Pigeonville, a traveler walks at a constant speed along a straight road. After 1 hour of walking, he sends a trained pigeon with a letter back to Pigeonville. The pigeon flies at a constant speed of 6/5 times the speed of the traveler. The moment the pigeon reaches Pigeonville, the reply to the letter is given to a trained dog, who starts running immediately to catch up with the traveler. The speed of the dog is twice the speed of the traveler. How soon will the traveler get the reply?

Question 2.9. Solve Question 2.7 with 8/5 replaced by a positive parameter k. What happens if k is less than 1?

Chapter 3

Acceleration: After the apple falls

While we can model constant speed motion by linear equations, we need more complicated quadratic equations to describe motion involving a constant *acceleration* – for example an apple falling from a tree. According to legend, this was the sight that inspired Sir Isaac Newton to write down his laws of motion and gravity, leading to the equations that govern the behavior of objects in the presence of a constant force.

In this chapter we will use some of the basic algebraic methods developed in Chapter 1 to describe accelerated motion of objects. The mathematical tool we will need to describe the position of an object moving with a constant acceleration is a quadratic equation. The acceleration may be caused by a force such as gravity for falling objects, or by braking force for vehicles slowing down on a road.

MATH

In Chapter 2, we considered objects moving at a constant speed along a straight line. This applies reasonably well to the motion of a particle (or an asteroid) in deep outer space, a ball rolling along a horizontal, frictionless surface, or a car in cruise control mode on a level freeway.

Now suppose we want to study the motion of a rock falling vertically downward. In this case, the velocity changes over time under the effect of the Earth's gravity. For the same reason, a rock thrown vertically upward begins to slow down until it reaches a maximum height, at which point its velocity reverses in direction and it starts to fall down, slowly increasing in speed as it accelerates to the ground. Some of the numerical characteristics of this familiar process may seem surprising. For instance, a rock thrown into the air with an initial velocity of 32 feet per second reaches the maximum height of 16 feet. If we double the initial velocity to 64 feet per second, the maximum height is 64 and not 32 feet, as one might expect.

The crucial observation to understanding this kind of motion is Newton's postulates that (1) the vertical velocity of a falling object changes linearly with time, and (2) the coefficient of this change, called the *gravitational acceleration*, is universal for all objects near the surface of the earth.

Let us write this statement in the form of an equation. Let v_0 be the initial (upward) vertical velocity of an object (if the velocity is directed downward, we will assume it is negative), and let $v(t)$ be its velocity at time t. Then the postulates state that for an object moving near the surface of the earth, the difference $v_0 - v(t)$ is proportional to the time interval t with the coefficient given by the gravitational acceleration.

Figure 3.1

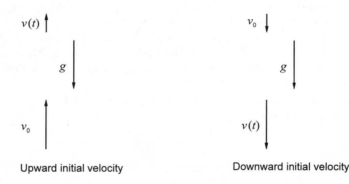

Upward initial velocity   Downward initial velocity

We have
$$v(t) - v_0 = -gt,$$
where
$$-g \simeq -9.8 \text{ m/sec}^2 \simeq -32 \text{ ft/sec}^2$$
is the (downward) gravitational acceleration on Earth.

Equivalently, we have the following formula relating the vertical velocity of an object with time:

$$v(t) = v_0 - gt.$$

We would like to know the position of the object (along a vertical line) whose velocity is given by this expression. We cannot multiply the velocity $v(t)$ by time as we did in Chapter 2, because now the velocity is not constant. However, we can find the *average* velocity over time t. Because the velocity is changing linearly, the average velocity is half the sum of the initial and the final velocities, as is shown in the diagram below. It also equals the velocity at half-time $\frac{1}{2}t$.

Figure 3.2

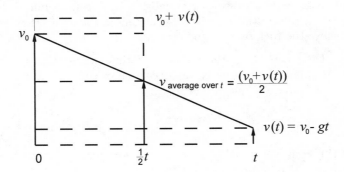

$$v_{\text{average over } t} = \frac{1}{2}\left(v_0 + v(t)\right) = \frac{1}{2}\left(v_0 + v_0 - gt\right) = v_0 - \frac{gt}{2}.$$

Now we can compute the position of the object as it depends on time. Let s_0 be the initial position (measured upward from a certain level, for example, from the ground):

$$s(t) = s_0 + (v_{\text{average over } t})t = s_0 + v_0 t - \frac{gt^2}{2}.$$

Along with the initial postulate that the acceleration $a(t)$ is constant, we obtain a familiar set of equations, often called the *equations of motion*:

$$a(t) = -g,$$
$$v(t) = v_0 - gt,$$
$$s(t) = s_0 + v_0 t - \frac{1}{2}gt^2.$$

These equations allow us to find the position at any moment of time of an object moving with constant acceleration, its initial velocity, or how long it takes for the object to reach any given elevation. In the latter case, we may need to solve a quadratic equation.

While working with these equations, we have to pay attention to the physical dimensions of the acceleration $a(t)$, velocity $v(t)$, and displacement $s(t)$. They should all be expressed either in the metric (m, m/sec, m/sec^2) or in the English (ft, ft/sec, ft/sec^2) system of measure.

Example 3.1. A rock is thrown vertically upward from an initial elevation of 1 meter with an initial velocity of 4 meters per second. How long will it take before the rock falls on the ground?

Solution: Denote by t_{final} the moment of time when the rock hits the ground. At this moment, the vertical position of the rock is $s(t_{\text{final}}) = 0$. Therefore, we have an equation for t_{final}:

$$0 = 1 + 4t_{\text{final}} - \frac{1}{2}9.8 \cdot t_{\text{final}}^2 \quad \Longrightarrow \quad -4.9t_{\text{final}}^2 + 4t_{\text{final}} + 1 = 0.$$

To solve this equation for t_{final}, we apply the formula for a general solution of a quadratic equation derived in Chapter 1:

$$ax^2 + bx + c = 0, \quad a \neq 0 \quad \Longrightarrow \quad x = \frac{-b \pm \sqrt{b^2 - 4ac}}{2a}.$$

Substituting $a = -4.9$, $b = 4$, and $c = 1$, we obtain:

$$t_{\text{final}} = \frac{-4 \pm \sqrt{4^2 - 4 \cdot (-4.9) \cdot 1}}{2 \cdot (-4.9)} \simeq \frac{-4 \pm 6}{-9.8} \simeq 1 \text{ sec.}$$

Here we have discarded the negative answer. It will take the rock about 1 second to fall to the ground.   □

The equations of motion are applicable in a wide range of problems. For instance, they describe the vertical motion of objects under the influence of any gravitational force, with the value of the gravitational acceleration

g determined according to the given situation (see Examples 3.12 and 3.13 for the vertical motion of objects on the Moon).

More generally, the same equations describe the motion of an object along a straight line with a constant acceleration caused by any external force. In this case, we have to replace g by the acceleration a determined by the given force.

APPLICATIONS

In the rest of this chapter, we will apply the equations of motion to objects moving under some form of constant force.

Vertical motion near the surface of the Earth

The problems on the motion of objects influenced by the Earth's gravity may involve solving equations for the initial velocity, or some measure of time, or the position of the object at a given moment of time. We start with the example given at the beginning of the chapter.

Example 3.2. If you throw a rock upward from the ground with the initial velocity 32 feet per second, it will reach a maximum height of 16 feet. What is the maximum height attained by a rock thrown upward with an initial velocity 64 feet per second?

Solution: Contrary to a possible first guess, it is not 32 feet. Let us express the maximum height reached by the rock in terms of its initial velocity. At the instant when the rock is at its highest, its velocity is zero because it has just finished traveling upward and is about to start falling under gravity. Thus, we first solve for the time $t_{\max}$ at which the rock reaches the maximum height given that $v(t_{\max}) = 0$:

$$v(t_{\max}) = v_0 - g t_{\max} = 0 \qquad \Longrightarrow \qquad t_{\max} = \frac{v_0}{g}.$$

The elevation of the rock at $t_{\max}$ is

$$s(t_{\max}) = 0 + v_0 t_{\max} - \frac{1}{2} g t_{\max}^2 = v_0 \frac{v_0}{g} - \frac{1}{2} g \left(\frac{v_0}{g} \right)^2 = \frac{v_0^2}{g} - \frac{1}{2} \frac{v_0^2}{g} = \frac{1}{2} \frac{v_0^2}{g}.$$

We conclude that the highest elevation $s_{\max}$ reached by an object thrown from the ground with an initial velocity v_0 is

$$s_{\max} = s(t_{\max}) = \frac{1}{2} \frac{v_0^2}{g}.$$

Let us check the first statement: plugging in $v_0 = 32$ ft/sec and $g = 32$ ft/sec^2, we have

$$s_{\max} = \frac{1}{2} \cdot \frac{32^2 (\text{ft/sec})^2}{32 \text{ ft/sec}^2} = 16 \text{ ft}.$$

Now for $v_0 = 64$ ft/sec and the same value of g, we get

$$s_{\max} = \frac{1}{2} \cdot \frac{64^2 (\text{ft/sec})^2}{32 \text{ ft/sec}^2} = 64 \text{ ft}.$$

The rock will reach the height of 64 feet. $\square$

The above example illustrates the fact that our model is non-linear: a twofold change in the initial velocity results in a change in the maximum height by a factor of four. The phenomenon is not merely theoretical: see Example 3.14 for real-world consequences of the same phenomenon in highway driving safety.

Practice problem 3.3. An arrow is shot vertically up from the ground and reaches its maximum elevation after $t_{\max} = 2.5$ sec.

1. Find the initial velocity of the arrow. *Answer:* 80 ft/sec.

2. Find the maximum height reached by the arrow. *Answer:* 100 ft.

In the next example the starting point of the object is situated above the ground.

Example 3.4. Suppose a stone is thrown upward from an initial height of 4 feet above the ground and an initial velocity of 8 feet per second. Find out (a) when the stone reaches the maximum height and what this height is; (b) when the stone reaches a height of 4.5 feet; and (c) the velocities at all of these heights.

Solution:
(a) To compute the maximum height, just like in the previous example, we find

$$0 = v(t_{\max}) = v_0 - gt_{\max} = 8 - 32t_{\max}.$$

It easily follows that $t_{\max} = \frac{1}{4}$ sec. At this time, the height is

$$s(t_{\max}) = s_0 + v_0 t_{\max} - \frac{1}{2}gt_{\max}^2 = 4 + 8\left(\frac{1}{4}\right) - 16\left(\frac{1}{4}\right)^2 = 5 \text{ ft}.$$

(b) When does the stone reach 4.5 feet? To compute the time, we use the equation for the position of the object

$$s(t) = s_0 + v_0 t - \frac{1}{2}gt^2.$$

In this equation, there are four variables used: $s(t), s_0, v_0, t$. Out of these four, we know three: $s_0 = 4$ ft, $v_0 = 8$ ft/sec, $s(t) = 4.5$ ft. The time t is therefore a solution of the quadratic equation

$$4.5 = 4 + 8t - 16t^2.$$

Let us first rewrite it in the standard form:

$$16t^2 - 8t + \frac{1}{2} = 0.$$

We will use the formula for the solutions of a quadratic equation $at^2 + bt + c = 0$, with $a = 16, b = -8, c = \frac{1}{2}$. The solutions are

$$\frac{-b \pm \sqrt{b^2 - 4ac}}{2a} = \frac{8 \pm \sqrt{64 - 4 \cdot 16 \cdot (1/2)}}{32} = \frac{4(2 \pm \sqrt{2})}{32} = \frac{2 \pm \sqrt{2}}{8}.$$

(In the second equality we used $\sqrt{32} = \sqrt{16 \cdot 2} = \sqrt{16} \cdot \sqrt{2} = 4\sqrt{2}$.) We have obtained two solutions – two moments of time at which the height of the stone is 4.5 feet: $t \simeq 0.146$ sec and $t \simeq 0.854$ sec. Both answers are meaningful: the stone has a height of 4.5 feet once when it is traveling upward, and once again when it is falling down.

(c) We have already concluded that the velocity of the stone at time t_{max} is zero.

The velocity of the stone at $(2 \pm \sqrt{2})/8$ seconds can be computed by using the equation $v(t) = v_0 - gt$. Thus,

$$v((2 + \sqrt{2})/8) = 8 - 32(2 + \sqrt{2})/8 = -4\sqrt{2} \simeq -5.657 \text{ ft/sec},$$

and similarly, $v((2 - \sqrt{2})/8) = 4\sqrt{2} = 5.657$ ft/sec. Notice the physical meaning of the opposite signs: at the first time-point $t = (2 - \sqrt{2})/8$ sec, the stone is moving upward, hence the positive velocity of 5.657 ft/sec. At the later time-point $t = (2 + \sqrt{2})/8$ sec, the stone is falling down, hence the negative velocity of -5.657 ft/sec. □

In the next example, the initial velocity of the object has to be found based on a given point in its trajectory.

Example 3.5. A cannonball is fired upward from a toy cannon on the ground, and it reaches a height of 32 feet in 2 seconds. Find (a) the initial velocity, and (b) the times at which it reaches 20 feet.

Solution: For this problem we have $s_0 = 0$ ft, while v_0 is unknown.
(a) We use the equation for the position $s(t)$, with given time $t = 2\,\mathrm{sec}$ and position $s(t) = s(2) = 32$ ft. Thus,

$$32 = s(2) = 0 + v_0 \cdot 2 - 16 \cdot 2^2,$$

which simplifies to: $2v_0 = 32 + 64 = 96\,\mathrm{ft/sec}$. Hence $v_0 = 48$ ft/sec.
(b) Having determined $v_0 = 48$ ft/sec, we now use the same equation to solve for the unknown time(s) t at which $s(t) = 20$ ft. We compute:

$$20 = 0 + 48t - 16t^2 \qquad \Longrightarrow \qquad 16t^2 - 48t + 20 = 0.$$

We can now solve for the roots of the quadratic equation, or first divide all terms by 16 in order to work with smaller numbers. If we do so, the equation reduces to: $t^2 - 3t + \frac{5}{4} = 0$. Now apply the formula for the roots, with $a = 1, b = -3, c = \frac{5}{4}$, to obtain:

$$t = \frac{-(-3) \pm \sqrt{(-3)^2 - 4 \cdot 1 \cdot (5/4)}}{2} = \frac{3 \pm \sqrt{4}}{2} = 0.5, \quad 2.5 \text{ sec.}$$

Thus, the cannonball is 20 feet high at $t = 0.5$ sec and $t = 2.5$ sec.   □

Bouncing balls

Suppose you drop a tennis ball from a certain height to the ground. How high will it bounce? Let us consider a couple of cases.

Example 3.6. Imagine a rubber ball that bounces without any energy loss, so that its velocity right after it hits the ground is the negative of its velocity right before it hits the ground. If such a ball is dropped down from height s_0, how high will it bounce back?

Solution: First we have to find the time t_{impact} it takes for the ball to fall, and its velocity at the impact v_{impact}. With the initial velocity zero, the position equation gives

$$s(t_{\text{impact}}) = s_0 - \frac{1}{2}gt^2_{\text{impact}} = 0,$$

Figure 3.3

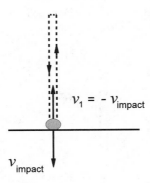

$$v_1 = -v_{\text{impact}}$$

$$v_{\text{impact}}$$

because at time t_{impact} the ball is on the ground. Then solving for v_{impact} gives

$$\frac{1}{2}gt_{\text{impact}}^2 = s_0, \quad \Longrightarrow \quad t_{\text{final}} = \sqrt{\frac{2s_0}{g}},$$

where we have discarded the negative solution. Because the initial velocity was zero, the velocity at time t_{impact} is

$$v_{\text{impact}} = v(t_{\text{impact}}) = 0 - gt_{\text{impact}} = -g\sqrt{\frac{2s_0}{g}} = -\sqrt{2gs_0}.$$

By assumption, the velocity v_1 after the ball hits the ground is the negative of v_{impact}. This is the initial velocity for the bounce:

$$v_1 = -v_{\text{impact}} = \sqrt{2gs_0}.$$

Now we can use the formula obtained in Example 3.2 for the maximum height attained by the ball with a given initial velocity:

$$s_{\text{max}} = \frac{1}{2} \cdot \frac{v_1^2}{g} = \frac{1}{2} \cdot \frac{2gs_0}{g} = s_0.$$

The ball will bounce back to the exact same height s_0. $\square$

This result may look surprising to you, and for a good reason: in reality any rubber ball loses energy, and therefore velocity, as it hits the ground. Here is a more realistic example.

Example 3.7. A tennis ball is dropped from a height of 8 feet. Each time it bounces off the ground, its vertical velocity reverses direction and loses one quarter of its magnitude. Find the maximum height attained by the ball on the first and the second bounce.

Figure 3.4

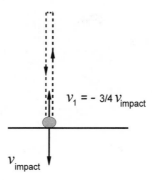

$v_1 = -\,3/4\,v_{\text{impact}}$

v_{impact}

Solution: The downward velocity v_{impact} before the ball hits the ground for the first time is determined by the same formula as in the previous example, $v_{\text{impact}} = -\sqrt{2gs_0}$. Then the upward velocity for the first bounce v_1 is three quarters of it, taken with the positive sign:

$$v_1 = \frac{3}{4}\sqrt{2gs_0}.$$

This is the initial velocity for the first bounce. The maximum height s_1 attained at the first bounce with the initial velocity v_1 is

$$s_1 = \frac{1}{2}\frac{v_1^2}{g} = \frac{1}{2}\frac{(\frac{3}{4})^2 \cdot 2 \cdot g \cdot s_0}{g} = \left(\frac{3}{4}\right)^2 \cdot s_0 = \frac{9}{16} \cdot 8 = 4.5 \text{ ft.}$$

To find the height attained on the second bounce, we have to repeat the above computation with the initial height of the drop $s_1 = 4.5$ ft. We can avoid going through the computation by noticing that the maximum height of the first bounce s_1 is related to the initial height s_0 by the formula

$$s_1 = \left(\frac{3}{4}\right)^2 \cdot s_0.$$

Because we are solving the exact same question with s_0 replaced by s_1, the maximum height of the second bounce s_2 is given by the formula

$$s_2 = \left(\frac{3}{4}\right)^2 \cdot s_1 = \left(\frac{3}{4}\right)^2 \cdot \left(\frac{3}{4}\right)^2 \cdot s_0 = \left(\frac{3}{4}\right)^4 \cdot s_0 = \frac{81}{256} \cdot 8 = \frac{81}{32} \simeq 2.53 \text{ ft.}$$

Now we could easily compute the maximum height attained in any subsequent bounce of the ball: each time it decreases by a factor of $(\frac{3}{4})^2$. $\quad\square$

Practice problem 3.8. In the conditions of Example 3.7, what is the time interval between (a) the first and the second bounce of the ball, (b) the second and the third bounce?
Answer: (a) $t_1 = \frac{3}{2}\sqrt{\frac{2s_0}{g}} \simeq 1.06 \text{ sec}$, (b) $t_2 = \frac{3}{4}t_1 \simeq 0.8 \text{ sec.}$

Examples with more than one moving object

Systems with more than one object moving under the influence of gravity lead to more complicated models. For example, you can consider two balls launched from different heights with different initial velocities. Newton's equations of motion allow us to find moments of time and positions where the balls might meet.

Example 3.9. A yellow tennis ball is dropped from a height of 5 meters, and at the same moment of time a white tennis ball is thrown upward from the ground. Find the initial upward velocity of the white ball if the balls meet exactly halfway, at a height of 2.5 meters above the ground.

Solution: Let us denote by $s_y(t)$ and $s_w(t)$ the position of the yellow and the white balls, respectively, as they change with time. Then we have

$$s_y(t) = s_0 - \frac{1}{2}gt^2, \qquad s_w(t) = v_0 t - \frac{1}{2}gt^2,$$

where $s_0 = 5$ m is the initial height of the yellow ball, and v_0 is the unknown initial velocity of the white one. At the moment t_{meet} of the meeting the balls have the same vertical position:

$$s_0 - \frac{1}{2}gt_{\text{meet}}^2 = v_0 t_{\text{meet}} - \frac{1}{2}gt_{\text{meet}}^2.$$

Then we have

$$s_0 = v_0 t_{\text{meet}} \qquad \Longrightarrow \qquad t_{\text{meet}} = \frac{s_0}{v_0}.$$

Figure 3.5

This formula is familiar from the constant speed motion model. It implies that the time before the balls meet is the same as it would be in the absence of gravity. This happens because gravity affects both balls in the same way, so that their *relative* position is independent of it. But the position of their meeting s_{meet} with respect to the ground level, of course, depends on g:

$$s_{\text{meet}} = s_{\text{y}}(t_{\text{meet}}) = s_0 - \frac{1}{2}gt^2_{\text{meet}} = s_0 - \frac{g}{2} \cdot \left(\frac{s_0}{v_0}\right)^2.$$

We are given that this height is half of s_0. Then

$$s_{\text{meet}} = \frac{1}{2}s_0 = s_0 - \frac{g}{2} \cdot \left(\frac{s_0}{v_0}\right)^2 \quad \Longrightarrow \quad \frac{g}{2} \cdot \left(\frac{s_0}{v_0}\right)^2 = \frac{1}{2}s_0 \quad \Longrightarrow \quad \frac{s_0}{v_0^2} = \frac{1}{g}.$$

This leads to

$$v_0^2 = s_0 \cdot g \quad \Longrightarrow \quad v_0 = \sqrt{s_0 \cdot g} = \sqrt{5 \cdot 9.8} = \sqrt{49} = 7\,\text{m/sec},$$

where we have discarded the negative answer for the velocity. The white ball was launched upward with a velocity of 7 m/sec.   □

Example 3.10. A black rubber ball is thrown upward from the ground with an initial velocity $v_0 = 12\,\text{ft/sec}$. At the same time, a white rubber ball is thrown upward from a height of 2 feet with an initial velocity $\frac{1}{2}v_0 = 6\,\text{ft/sec}$. When and at what height will the balls meet?

Figure 3.6

Solution: Denote by $s_b(t)$ and $s_w(t)$ the position function of the black and the white balls as they depend on time. Then for the meeting time t_{meet} we have the equation

$$s_b(t_{meet}) = s_w(t_{meet}) \quad \Longrightarrow \quad v_0 t_{meet} - \frac{1}{2}gt_{meet}^2 = s_0 + \frac{1}{2}v_0 t_{meet} - \frac{1}{2}gt_{meet}^2.$$

The part with the acceleration cancels from the equation and we obtain

$$\frac{1}{2}v_0 t_{meet} = s_0 \quad \Longrightarrow \quad t_{meet} = \frac{2s_0}{v_0} = \frac{2 \cdot 2}{12} = \frac{1}{3} \text{ sec.}$$

Now the position of the meeting can be calculated as $s_b(t_{meet})$:

$$s_b(t_{meet}) = v_0 \cdot \frac{2s_0}{v_0} - \frac{g}{2} \cdot \left(\frac{2s_0}{v_0}\right)^2 = 2s_0 - \frac{2gs_0^2}{v_0^2} = 2 \cdot 2 - \frac{2 \cdot 32 \cdot 4}{144} \simeq 2.22 \text{ ft.}$$

The balls will meet after $\frac{1}{3}$ of a second at a height of 2.22 feet. □

Fountains and waterfalls

Generalizing further, we can consider examples with more than two objects moving under the influence of gravity. In fact, the same equations allow us to model the behavior of a continuous stream of particles, like a fountain or a waterfall.

Example 3.11. By comparing the surrounding buildings, you can estimate the height of the Jet d'Eau fountain in Geneva to be approximately

140 meters. What is the initial velocity of the stream of water at the base of the fountain?

Figure 3.7: Jet d'Eau fountain in Geneva (photo by Anna Lachowska).

Solution: This setting may require contemplation. A stream of water is continuous, whereas Newton's equations of motion are designed for a single object. What exactly is the "object" that moves here? We can overcome this difficulty by thinking of the drops of water as individual objects launched at the nozzle with the same initial velocity v_0 – this is the unknown upward velocity of the stream of water at the base of the fountain. Then we can apply the formula for the maximum height attained by an object thrown upward from the ground (see Example 3.2):

$$s_{\max} = s(t_{\max}) = \frac{1}{2}\frac{v_0^2}{g}.$$

In the case of the Jet d'Eau, we have $s_{\max} = 140\,\text{m}$, $g = 9.8\,\text{m/sec}^2$. Then

$$140 = s_{\max} = \frac{v_0^2}{2g} = \frac{v_0^2}{19.6}.$$

Solving for v_0, we obtain $v_0^2 = 19.6 \cdot 140 = 2,744$. Therefore $v_0 \simeq$ 52.4 m/sec.   □

According to Wikipedia, the speed of water at the nozzle in the Jet d'Eau is 200 kilometers per hour, which is approximately 55.6 meters per second. Indeed, in reality the speed must be higher than our estimation because of the air resistance to the motion of the water (which we had ignored).

Soccer on the Moon

The strength of Newton's law of gravitational attraction, and hence the equations of motion, lies in their generality. In particular, any celestial body exerts gravitational force, and an object placed at a given distance from the source of gravity will experience a constant acceleration that is the same for all objects and depends only on the mass of the source of gravity. Here is an amusing, if not very realistic, example.

Example 3.12. In the animated film *A Grand Day Out*, Wallace and Gromit are having a picnic on the Moon. Wallace kicks a soccer ball and waits 7 seconds for it to fall. When the ball does not come back in 7 seconds, Wallace walks on. How long should he have waited for the ball to return?

Solution: We need some additional input. First, we need to know the value of the gravitational acceleration on the surface of the Moon. Online sources provide the number $g_{\mathrm{moon}} = 1.6\,\mathrm{m/sec}^2$. Second, we don't know exactly how hard Wallace kicked the ball. You can try emailing Wallace or his creators, but we found it easier first to look for estimates of how fast a soccer ball can be kicked in theory. We quickly found that Lisbon's left-back Ronny Heberson holds the record of nearly 132 miles per hour, or 59 meters per second, and that the average for professionals is about half as fast. Now, Wallace is no professional; in fact, we suspect him to be more of a spectator than a participant when it comes to sports. We won't be too far off if we assume he kicked the ball at about 15 meters per second. Now we have all the ingredients to estimate the time before the ball should return to the surface (of the Moon!).

We can take the height of the kick to be the zero level, $s_0 = 0$. We want to find the time t it takes the ball to return to the same level: $s(t) = 0$. According to the equations of motion,

$$0 = 0 + v_0 t - \frac{g_{\mathrm{moon}} t^2}{2},$$

where $v_0 = 15\,\text{m/sec}$, and $g_{\text{moon}} = 1.6\,\text{m/sec}^2$. Solving for t, we have

$$0 = t\left(v_0 - \frac{g_{\text{moon}}t}{2}\right) \qquad \Longrightarrow \qquad \left[\begin{array}{l} t = 0; \\ v_0 - \frac{gt}{2} = 0. \end{array}\right.$$

The first solution, $t = 0$, corresponds to the moment of the kick. We are interested in the second solution given by the equation

$$v_0 = \frac{g_{\text{moon}}t}{2} \qquad \Longrightarrow \qquad 2v_0 = g_{\text{moon}}t \qquad \Longrightarrow \qquad \frac{2v_0}{g_{\text{moon}}} = t.$$

Plugging in the numbers, we get

$$t = \frac{2v_0}{g_{\text{moon}}} = \frac{2 \cdot 15\,\text{m/sec}}{1.6\,\text{m/sec}^2} = 19\,\text{sec}.$$

Wallace should have waited for 19 seconds.

For comparison, let us see how long it would take for the ball kicked with the same initial velocity to fall to Earth. The equation is the same; the only difference is in the value of g – on Earth it is 9.8 meters per second squared. We have

$$t = \frac{2v_0}{g} = \frac{2 \cdot 15\,\text{m/sec}}{9.8\,\text{m/sec}^2} \simeq 3\,\text{sec}.$$

No wonder Wallace got impatient – he waited for more than twice the time the ball would take to fall down to Earth! (Whereas he should have waited for about *six* times the time.)   □

We can also compute the maximum height the soccer ball will reach on the Moon.

Example 3.13. Wallace kicked a soccer ball from the surface of the Moon upward with an initial velocity of 15 meters per second. What is the maximum height it will reach? If Ronny Heberson kicked the ball instead, how high would it go?

Solution: For any object under the influence of gravity that is thrown upward with a given initial velocity, the formula for the maximum height was derived in Example 3.2:

$$s_{\text{max}} = \frac{1}{2} \cdot \frac{v_0^2}{g}.$$

On the Moon we have $g_{\mathrm{moon}} = 1.6\,\mathrm{m/sec^2}$, and the initial velocity is given to be $v_0 = 15\,\mathrm{m/sec}$. Then we have

$$s_{\mathrm{max}} = \frac{1}{2} \cdot \frac{15^2}{1.6} \simeq 70.3 \text{ m.}$$

So Wallace kicked the ball to a height of about 70 meters.

Now, let us assume that Ronny Heberson would kick the ball at about 60 meters per second, which is four times the initial velocity of Wallace's kick, $v_{\mathrm{Heberson}} = 4v_0$.

Because the maximum height given by the formula is proportional to the square of the velocity, it will increase by a factor of 16:

$$s_{\mathrm{Heberson}} = \frac{1}{2} \cdot \frac{(4v_0)^2}{g_{\mathrm{moon}}} = 16\, s_{\mathrm{max}} \simeq 16 \cdot 70.3 \simeq 1,125 \text{ m.}$$

The ball would go upward more than one kilometer! $\square$

Braking distance on a highway

The same laws and equations of motion that apply to objects falling under gravity hold when you are driving on a freeway and apply the brakes in order to stop. Namely, it is reasonable to assume that the action of the brakes provides a constant negative acceleration $a(t) = -a$, which may depend on the vehicle in question. If we denote the velocity at the instant you start braking as v_0, and measure time starting from that same instant, then the equations describing the position and velocity of the vehicle are:

$$a(t) = -a, \qquad v(t) = v_0 - at, \qquad s(t) = s_0 + v_0 t - \frac{1}{2}at^2.$$

This allows us to perform similar calculations as in the above examples, to model the motion of a car braking on a freeway.

Example 3.14 (Braking distance). Suppose that if you are driving at 30 miles per hour, and applying the brakes hard, your car comes to a stop in 500 feet. If instead the same car is speeding at 60 miles per hour, what is the stopping distance upon applying the brakes?

Solution: To solve this problem, we need to carry out the same analysis as in Example 3.2. However, first we have to convert all quantities into the same physical units. There are $5,280$ feet in a mile and $3,600$ seconds in an hour. Then

$$30 \text{ mph} = 30 \cdot \frac{5,280}{3,600} \text{ ft/sec} = 44 \text{ ft/sec}.$$

Similarly, $60 \text{ mph} = 88 \text{ ft/sec}$.

Now suppose $SD(v_0)$ denotes the stopping distance for the car, from the point where the brakes were applied and the car had velocity v_0 ft/sec. At the moment when the car stops, its velocity is zero:

$$v(t_{\text{stop}}) = v_0 - at_{\text{stop}} = 0 \quad \implies \quad t_{\text{stop}} = \frac{v_0}{a}.$$

Plugging t_{stop} in the equation for $s(t)$, just like in Example 3.2, we obtain the stopping distance:

$$SD(v_0) = \frac{1}{2} \cdot \frac{v_0^2}{a_{\text{car}}},$$

where $-a_{\text{car}}$ is the negative acceleration of the car in feet per second squared, caused by braking. Given the stopping distance at the initial velocity of 30 miles per hour, we can determine a_{car}:

$$500 = SD(44) = \frac{44^2}{2a_{\text{car}}} = \frac{968}{a_{\text{car}}}.$$

This yields: $a_{\text{car}} = 968/500 = 1.936$ ft/sec^2. Now plugging in the initial velocity 60 mph $= 88$ ft/sec, we compute:

$$SD(88) = \frac{88^2}{2a_{\text{car}}} = \frac{7,744}{3.782} = 2,000 \text{ ft.}$$

Thus, just like in Example 3.2, the stopping distance quadruples when the initial speed doubles. $\qquad\square$

The conclusion we can draw from the previous example is: the faster we are driving, the more distance we will need to brake to a complete stop.

EXERCISES

Question 3.1. A passenger in a hot air balloon which is stationary at a height of 80 feet above the ground drops a stone. Compute the time and velocity at which (a) it is 40 feet above the ground, and (b) it hits the ground. Is the time in part (b) twice of the time in part (a), or more than twice, or less? Once you have computed the two times, how do you explain the answer to the previous question?

Question 3.2. A person shoots an arrow vertically up from the ground with some initial velocity, and it reaches its maximum height in 3.5 seconds. Compute (a) the initial velocity, (b) the maximum height attained, and (c) the time(s) in which it reaches half of this height.

Question 3.3. A woman throws a tennis ball down from a height of 5 meters with an initial downward velocity of 5 meters per second. Each time it bounces off the ground, its upward velocity is four-fifths of the downward velocity at the moment of impact. Find the maximum height attained by the ball on the first and the second bounce.

Question 3.4. Venus Williams drops a tennis ball from a height of 6 meters, and at the same time Serena Williams throws another tennis ball upward from the ground with an initial velocity of 9 meters per second. Where and when will the balls meet?

Question 3.5. Solve Example 3.9 with the assumption that the balls meet at seven-eighths of the initial height of the yellow ball.

Question 3.6. Anna throws a small snowball vertically upward from the ground with an initial velocity v_0. At the same time, Elsa throws another snowball from a height $s_0 = 4\,\text{m}$ above the ground with an upward initial velocity $\frac{1}{2}\,v_0$.

1. Find v_0 if the snowballs meet at the height $s_0 = 4\,\text{m}$.

2. Find the velocities of both snowballs at the time they meet.

Question 3.7. A British officer fires a cannonball vertically upward from a cannon located on the ground. It subsequently bursts through a barrier which is placed at a certain height s_0 above the ground. Upon bursting through the barrier, the velocity of the cannonball slows down to 150 feet per second. Eight seconds after bursting through the barrier, the cannonball is 226 feet high. Compute the height of the barrier above the ground. Also compute all time(s) when the cannonball is 226 feet high, and its velocities at those times.

Question 3.8. Suppose you are at a window located 70 feet above the ground, and a pickup truck is coming along the street toward you at a constant speed of 20 miles per hour. You want to drop a small object into the trunk of the truck. At the moment when you drop it, how far should the truck be from the point directly under your window?

Question 3.9. The height of Niagara Falls is about 165 feet. Estimate the vertical velocity of the water stream at the bottom.

Question 3.10. A vehicle driving on a road begins to brake when it is traveling at 50 miles per hour – i.e., 73 feet per second. Suppose the deceleration caused due to the application of the brakes is 10 feet per second squared. How long does the vehicle travel before it comes to a halt, and how far does it travel during this time?

Question 3.11. Two cars are driving along a road, one at 90 feet per second and the other at 80 feet per second. They both start to decelerate at the same time due to application of brakes – the faster car is decelerating at 10 feet per second squared, and the slower car at 7 feet per second squared. Find the stopping distance for each of the cars. Which car will go farther before it stops?

Question 3.12. In the previous question, find the time it takes for each car to stop. Which of the two cars will be the first to stop?

Chapter 4

Irrational: The golden mean and other roots

In this chapter we will focus on a special class of solutions of quadratic equations – the irrational solutions. We will see how these numbers appear naturally in the real world, and why they can be useful.

MATH

Recall from Chapter 1 that for any quadratic equation $ax^2 + bx + c = 0$, where a, b, c are real numbers and x is a real variable, the solutions, if they exist, are given by the formula

$$x = \frac{-b \pm \sqrt{b^2 - 4ac}}{2a}.$$

In this chapter we will be interested in the case when the discriminant $b^2 - 4ac$ is positive, and we have two distinct real solutions. Let us consider some examples.

Example 4.1. Solve the equation $36x^2 - 36x - 7 = 0$.
Answer: Using the formula with $a = 36, b = -36, c = -7$, we get $x_1 = \frac{7}{6}$ and $x_2 = -\frac{1}{6}$. □

Example 4.2. Solve the equation $x^2 - 5 = 0$.
Answer: We have $x^2 = 5$ and $x = \pm\sqrt{5}$. □

Note that in the second case we are unable to express the answer without using the square root sign. The number $\sqrt{5}$ is nothing but the name given to the positive solution of this particular equation.

A real number that can be expressed as a quotient of two integers is called *rational*. All other real numbers are called *irrational*. For example, $\frac{7}{6}$ is rational and $\sqrt{5}$ is irrational. Another example of an irrational number is $\pi = 3.14159265\ldots$. The decimal portion of π is infinitely long and never repeats itself. For the number to be irrational, its decimal expression needs to be infinitely long and non-periodic. We will show in Chapter 10 that any decimal number with a periodically repeating "tail" is rational.

Example* 4.3. How can we be sure that $\sqrt{5}$ is not a quotient of two integers?

Solution: Let us suppose for a moment that there are integers p and q such that $\sqrt{5} = \frac{p}{q}$, and the denominator q is the smallest positive integer with this property. (This means we use the fraction $\frac{7}{3}$ instead of $\frac{14}{6}$). Multiplying by q and taking the square of both sides gives

$$\sqrt{5}q = p \quad \Longrightarrow \quad 5q^2 = p^2.$$

Because p and q are integers, this means that p^2 is divisible by 5. But then p itself is also divisible by 5. An example that satisfies this condition is 100, which is divisible by 5, but then its square root 10 is also divisible by 5. If we take 4 instead of 5, then 36 is a square and divisible by 4, but 6 is not. The reason is that $4 = 2^2$ is a square of an integer, and 5 *is not*. So, in our case, because p^2 is divisible by 5, we can always write $p = 5k$ for some new integer k. Now our equality reads

$$5q^2 = (5k)^2 \quad \Longrightarrow \quad 5q^2 = 25k^2 \quad \Longrightarrow \quad q^2 = 5k^2,$$

where q and k are integers. This is just like the equation we had before for p and q. By the same argument, q^2 is divisible by 5, and so is q. Let $q = 5m$ for an integer m. Then

$$\sqrt{5} = \frac{p}{q} = \frac{5k}{5m} = \frac{k}{m},$$

with m a positive integer smaller than q, which contradicts the assumption that q was the smallest positive denominator of a fraction equal to $\sqrt{5}$. Therefore, the number $\sqrt{5}$ is irrational. $\qquad\square$

Examples of irrational numbers include square roots of integers that are not complete squares, for instance $\sqrt{2}$ or $\sqrt{12}$, cube roots of integers that are not cubes, like $\sqrt[3]{7}$, and so on. Multiplying an irrational number by a rational coefficient or adding a rational number to it produces again an irrational number, as the next example shows.

Example 4.4. Is the number $\frac{7\sqrt{2}}{10} + 3$ rational or irrational?

Solution: Suppose that $a = \frac{7\sqrt{2}}{10} + 3$ is rational and equals $\frac{p}{q}$, where p and q are integers. Then $a - 3 = \frac{p}{q} - 3 = \frac{p-3q}{q}$ is rational. So $\frac{10}{7}(a - 3) = \frac{10}{7} \cdot \frac{p-3q}{q} = \frac{10p-30q}{7q}$ is also rational. But $\frac{10}{7}(a - 3) = \sqrt{2}$, which we know is irrational, a contradiction. Therefore, a is irrational.   □

The golden ratio

Here is a quadratic equation whose irrational solution is quite famous. Suppose you want to divide a segment of a line into two parts, so that the ratio of the larger part (a units long) to the smaller part (b units long) is the same as the ratio of the whole (c units long) to the larger part. This is expressed in the equation:

$$\frac{a}{b} = \frac{c}{a}, \quad \text{or} \quad \frac{a}{b} = \frac{a+b}{a}, \quad \text{or} \quad \frac{a}{b} = 1 + \frac{b}{a}.$$

With the notation $\phi = \frac{a}{b}$, the equation becomes: $\phi = 1 + \frac{1}{\phi}$. Multiplying by ϕ, we get the quadratic equation $\phi^2 = \phi + 1$, or

$$\phi^2 - \phi - 1 = 0, \qquad (4.1)$$

whose solutions are $\phi_1 = \frac{1+\sqrt{5}}{2}$, $\phi_2 = \frac{1-\sqrt{5}}{2}$. We are interested in the positive solution:

$$\frac{a}{b} = \phi = \frac{1 + \sqrt{5}}{2} \simeq 1.6180339887.$$

The number ϕ is irrational by an argument similar to Example 4.4, and is called the *golden ratio*, or the golden mean. It exhibits many amazing properties. We will start by pointing out its relation to the Fibonacci sequence.

The *Fibonacci sequence* $\{f_n\}$ is the sequence of integers starting with $1, 1$, and such that each next element of the sequence is the sum of two previous elements, $f_n = f_{n-1} + f_{n-2}$ for all $n \geq 3$:

$$1, 1, 2, 3, 5, 8, 13, 21, 34, 55, 89, \ldots$$

Thus, $f_1 = 1$, $f_2 = 1$, $f_3 = f_1 + f_2 = 2$, $\quad\ldots,\quad$ $f_6 = f_5 + f_4 = 5 + 3 = 8\ldots$

Let us look at the sequence of ratios of consecutive Fibonacci numbers $\frac{f_{n+1}}{f_n}$:

$$\frac{1}{1} = 1, \ \frac{2}{1} = 2, \ \frac{3}{2} = 1.5, \ \frac{5}{3} \simeq 1.667, \ \frac{8}{5} = 1.6, \ \frac{13}{8} = 1.625, \ \frac{21}{13} \simeq 1.615\ldots$$

As we go further along the Fibonacci sequence, the ratios become closer to each other and seem to approach a certain limit number. This number can only be the golden ratio. Here is why: consider three consecutive numbers $b, a, b + a$ far along in the Fibonacci sequence. Then the ratios $\frac{a}{b}$ and $\frac{b+a}{a}$ should be very close to the limit number, and therefore almost equal. But $\frac{a}{b} = \frac{b+a}{a}$ is the equation for the golden ratio. Indeed, the ratios of the form $\frac{f_{n+1}}{f_n}$ provide a good approximation for ϕ. For example, $\frac{f_{20}}{f_{19}} = \frac{6,765}{4,181} \simeq 1.6180339632$. We will discuss other properties of ϕ in Application 2.

APPLICATION 1

Do irrational numbers have any practical use? One unexpected application is in paper manufacturing. Suppose you want to produce sheets of paper in a variety of sizes (for posters, letters, memos). To minimize production costs, it would be good if a large sheet of paper could be cut up into a number of smaller sheets. Besides, you would like for all sizes to have the same aspect ratio (length to width), for convenient enlargement and reduction of pages. It turns out that the two requirements together lead to a quadratic equation for the aspect ratio, whose solution is an irrational number.

The Lichtenberg conditions

To be more specific, the international paper size standard ISO 216 defines the series An for $n = 0, 1, 2, \ldots 10$ by the following requirements:

1. A landscape-oriented sheet of paper of size An, when cut in half, produces two portrait-oriented sheets of paper of size $A(n + 1)$ for all $n = 0, 1, 2, \ldots 9$.

2. The aspect ratio for all sizes is the same.

3. The area of a sheet of paper of size $A0$ is 1 square meter.

The idea of introducing conditions (1) and (2) to ensure production efficiency and convenient resizing goes back to the German scientist Georg Christoph Lichtenberg, who proposed it at the end of the eighteenth century.

Figure 4.1: The An paper sizes.

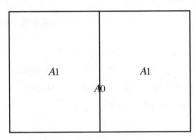

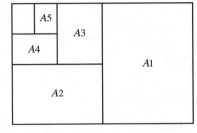

$A0$ divided into two $A1$s $A0$ to $A5$ paper sizes

First let us find the aspect ratio of the An sizes. Let a denote the length and b the width of a size An. Then by (1), b and $\frac{a}{2}$ are, respectively, the length and width of the size $A(n+1)$. To have the same aspect ratio, the parameters should satisfy the equation

$$\frac{b}{a} = \frac{a/2}{b}.$$

Multiplying by b and a, we get

$$b^2 = \frac{a^2}{2} \quad \Longrightarrow \quad 2b^2 = a^2 \quad \Longrightarrow \quad a = \sqrt{2}b.$$

Thus, $\sqrt{2}$ is the aspect ratio common to all An sizes.

Example 4.5. What is the ratio between the length of the $A4$ size and the length of the $A0$ size?

Solution: Let us denote by $l0, l1, l2, l3, l4$ the lengths of the $A0, A1, A2, A3, A4$ sizes, respectively. Then $l1$ is the width of the $A0$ size, which equals $l0/\sqrt{2}$. Similarly, $l2 = l1/\sqrt{2}$, $l3 = l2/\sqrt{2}$, and $l4 = l3/\sqrt{2}$. Taking all this into account, we get

$$l4 = \frac{l3}{\sqrt{2}} = \frac{l2}{\sqrt{2} \cdot \sqrt{2}} = \frac{l2}{2} = \frac{l1}{2 \cdot \sqrt{2}} = \frac{l0}{2 \cdot \sqrt{2} \cdot \sqrt{2}} = \frac{l0}{4}.$$

Finally, $\frac{l4}{l0} = \frac{1}{4}$. $\square$

Now let us find the dimensions of the $A0$ size. If a is the length and b the width, then $a \cdot b = 1m^2$ and $a/b = \sqrt{2}$. We have

$$ab = a\frac{a}{\sqrt{2}} = \frac{a^2}{\sqrt{2}} = 1.$$

Therefore, $a^2 = \sqrt{2}$ and $a = \sqrt{\sqrt{2}}$. What kind of a number is this? Clearly, $\sqrt{\sqrt{2}} \cdot \sqrt{\sqrt{2}} = \sqrt{2}$. Also, $\sqrt{2} \cdot \sqrt{2} = 2$. Therefore, $\sqrt{\sqrt{2}}$ multiplied by itself *four* times gives 2. This irrational number is called the 4th root of 2, and denoted by $\sqrt[4]{2}$. Using your calculator, you can find $\sqrt[4]{2} = \sqrt{\sqrt{2}} \simeq 1.189207$. For the width b we find: $b = \frac{1}{a} = \frac{1}{\sqrt[4]{2}} \simeq 0.840896$. In fact, the dimensions of the $A0$ size are, in millimeters, $1,189\,\mathrm{mm} \times 841\,\mathrm{mm}$.

We will talk more about roots of various degrees in Chapter 5.

Example 4.6. Find the length and the area of the $A5$ size.

Solution: Proceeding just as in Example 4.5, we find that the ratio of the length $l5$ of the size $A5$ to the length $l0$ of the size $A0$ is $l5/l0 = \frac{1}{4\sqrt{2}}$. Therefore, $l5 = l0/(4\sqrt{2}) = \frac{\sqrt[4]{2}}{4\sqrt{2}} \simeq 0.210224...$ In fact, ISO 216 lists the length of $A5$ to be equal to 210 millimeters. To find the area, recall the condition (1): each smaller size has half the area of a larger size. For the area of the $A5$ size we have:

$$A5 = \frac{1}{2}A4 = \frac{1}{4}A3 = \frac{1}{8}A2 = \frac{1}{16}A1 = \frac{1}{32}A0 = \frac{1}{32}\,\mathrm{m}^2 = 0.03125\,\mathrm{m}^2.$$

$\square$

Other standard paper sizes

In addition to the An sizes, there are two more series: Bn sizes and Cn sizes. The Bn sizes are determined by the conditions:

(1b) The length of the Bn size is the *geometric mean*, or the square root, of the product of the lengths of the $A(n-1)$ and An sizes:

$$l(Bn) = \sqrt{l(A(n-1)) \cdot l(An)},$$

and $l(B0) = \sqrt{2} \cdot l(B1)$.

(2b) The aspect ratio of all Bn sizes is $\sqrt{2}$.

For example, the length of the $B1$ size equals

$$l(B1) = \sqrt{l(A0) \cdot l(A1)} = \sqrt{\sqrt[4]{2} \cdot \frac{1}{\sqrt[4]{2}}} = 1\,\text{m}.$$

According to this rule, the Bn sizes lie "in between" the An sizes: $B0$ is larger than $A0$, $B1$ is smaller than $A0$ but larger than $A1$, and so on.

The Cn sizes are determined by the conditions:

(1c) The length of the Cn size is the geometric mean of the lengths of the An and the Bn sizes:

$$l(Cn) = \sqrt{l(An) \cdot l(Bn)}.$$

(2c) The aspect ratio of all Cn sizes is $\sqrt{2}$.

For example, the length of the $C1$ size is

$$l(C1) = \sqrt{l(A1) \cdot l(B1)} = \sqrt{\frac{1}{\sqrt[4]{2}} \cdot 1} = \frac{1}{\sqrt[8]{2}} \simeq 0.917\,\text{m}.$$

This rule makes the Cn sizes perfect to use as envelopes for the An sizes: a sheet of the size An fits perfectly in a slightly larger envelope of the size Cn.

Note that for all series An, Bn, and Cn, the aspect ratio of $\sqrt{2}$ ensures that the Lichtenberg conditions (1) and (2) are satisfied, providing manufacturing efficiency and convenient resizing.

International standard pen nib sizes

The international standard for technical drawing pen nibs upholds the same thickness ratio close to $\sqrt{2}$: the common sizes are 0.70, 0.50, 0.35, 0.25 millimeters, and so on. Then if you draw, say, with a 0.5-millimeter pen nib on $A3$ paper and photocopy the drawing, reducing it to the $A4$ size, you can continue drawing with a 0.35-millimeter pen nib and the thickness of your lines will match the photocopied lines perfectly.

Finally, the paper sizes used in the US, Canada, and Mexico – *letter*, *legal*, *ledger*, and *tabloid* – do not satisfy any mathematical condition. Here tradition seems to be stronger than any rational (in our case, irrational!) arguments.

Practice problem 4.7.   1. Find the reduction factor you have to use
when photocopying from size $A2$ to size $A5$. *Answer:* $\frac{1}{2\sqrt{2}}$.

2. Find the width of a sheet of paper of the $A3$ size.

   Answer: $\frac{\sqrt[4]{2}}{2}$. The international standard value is 594 mm.

APPLICATION 2

Sectio Aurea

In the rest of this chapter, we discuss the role played by the golden ratio ϕ
in the real world. You have probably heard of the presence of the golden
ratio in the design of many works of art and architecture. Here are a few
prominent examples:

1. The Great Pyramid of Giza (2550 BC) is quite close to a *golden
   pyramid* (see Figure 4.4), a pyramid with a square base and isosceles
   triangular faces, such that the ratio of the height of a triangular face
   to half of its base equals ϕ. See Figure 4.2.

2. The Parthenon in Athens, Greece, allegedly exhibits the golden ratio
   in its height-to-width proportion. (See Figure 4.3.) It was built in
   447-432 BC.

3. The proportions of a human body in the works of the Greek sculptor Phidias (490-430 BC) allegedly were based on the golden ratio.
   Although no original work by Phidias survived, the sculptor's name
   gave the golden ratio its current notation, the Greek letter ϕ, or
   "phi."

4. It was Leonardo da Vinci who gave ϕ its present name, *sectio aurea*,
   the golden section. He argued that the golden rectangle (a rectangle
   with aspect ratio equal to ϕ) is present in some proportions of an
   (idealized) human body. It is uncertain, though, if he used the golden
   ratio in his paintings.

5. Some modern artists intentionally base their designs on the golden
   ratio. "The Sacrament of the Last Supper" by Salvador Dali (1955) is
   a well-known example. In this painting, a dodecahedron with twelve

Figure 4.2: Great Pyramid of Giza (photo by Jerome Bon).

Figure 4.3: Parthenon (left, photo by Tilemahos Efthimiadis), and Athena, small replica of a statue by Phidias (right, photo by William Neuheisel).

regular pentagonal faces is the major element of the composition. (For the relation of a regular pentagon and the golden ratio, see Practice Problem 4.9.)

Example 4.8. Find the height of a golden pyramid if its base has sides 2 inches long.

Figure 4.4: A golden pyramid with the $(1, \sqrt{\phi}, \phi)$ triangle.

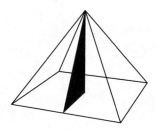

Solution: If a side of the base is 2 inches (in) long, then half the base is 1 inch long, and therefore the height of a triangular face is ϕ inches long. This height is the hypotenuse of the right triangle formed by the height of the pyramid (x) and half of its base (1 inch). Therefore, by the Pythagorean theorem,

$$x^2 = \phi^2 - 1^2 = \phi^2 - 1.$$

But according to the defining equation for ϕ ((4.1)), we have $\phi^2 - 1 = \phi$. Therefore,

$$x = \pm\sqrt{\phi},$$

and because the height should be positive, we obtain $x = \sqrt{\phi} \simeq 1.272$ in. □

Practice problem 4.9. What is the ratio of a diagonal of a regular pentagon to its side? *Answer:* ϕ.
Hint: Take any four of the five vertices of the regular pentagon and apply *Ptolemy's theorem*, which says: for any quadrilateral whose vertices lie on a circle, the product of the two diagonals equals the sum of the products of the opposite sides.

The first mathematical definition and study of the golden ratio can be found in *Elements* by Euclid (c. 300 BC). However, the work that made

Figure 4.5: A regular pentagon inscribed in a circle.

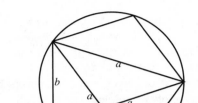

ϕ famous by emphasizing its aesthetic and even "divine" properties was *De Divina Proportione* by the Italian mathematician and Franciscan friar Luca Pacioli. Published in Venice in 1509 and illustrated by Leonardo da Vinci, the book had a considerable influence on the aesthetics of the Renaissance.

Although multiple references to the presence of the golden proportion in nature and art exist, they should be taken with caution. Computing the ratio of dimensions of an object includes measuring the dimensions with a certain precision and then performing the division. If the rational number thus obtained is more or less close to ϕ, what error should be admissible to claim that it is in fact the golden ratio? For example, it is often claimed that a US drivers license or a credit card is designed as a golden rectangle. If you measure the dimensions, you get an aspect ratio of approximately 1.585, which might or might not be an approximation of ϕ. Similar overeagerness sometimes occurs in the analysis of works of art and architecture. However, *not all that glitters is gold*.

Example 4.10. Does the right triangle with sides $3, 4$, and 5 units have anything to do with the golden ratio?

Solution: Consider the ratios of the sides: $\frac{4}{3} \simeq 1.33333$, $\frac{5}{4} = 1.25$, $\frac{5}{3} \simeq 1.66667$. The last ratio is within 3% of ϕ. Of course, this is not surprising because 3 and 5 are two consecutive Fibonacci numbers. However, the use of a triangle with ratios close to $(3, 4, 5)$ in architecture might originate in its simple rational proportions rather than its relation with ϕ. $\qquad\square$

The golden ratio in plants

It is universally accepted that the golden ratio seems to be pleasing to the human eye. The reasons for this must be found in nature, and indeed here is one of the manifestations of ϕ in the real world that can be mathematically justified. Namely, the number ϕ appears to lie in the foundation of *phyllotaxis*, the arrangement of leaves, seeds, or florets in many plants.

A visible manifestation of the golden ratio in phyllotaxis is in the number of the clockwise and counterclockwise (right and left) spirals apparent in the structure of composite flowers, leaf arrangements, and seed heads. In about 85% of plants, these numbers are usually two consecutive Fibonacci numbers: 5 and 8, 8 and 13, and so on. Most remarkably, this fact is independent of the biological species (within the 85%; the remaining 15% of plants use entirely different structural arrangements), manifested as well in a sunflower, as in a pineapple or a pine cone. Which pair of Fibonacci numbers appears in a particular case depends on the relative size of the seed with respect to the size of the seed head.

The same pattern can be obtained by numerical simulation, assuming that the angle between the directions from the stem to any two consecutively sprouting seeds (the *divergence angle*) is always the same and corresponds to the splitting of the complete revolution in the golden ratio 1 to ϕ. This *golden angle* equals $\frac{1}{1+\phi} \cdot 2\pi$, or approximately 137.5078 degrees. As we saw earlier in this chapter, the ratios of consecutive Fibonacci numbers converge to the golden ratio. This implies that a Fibonacci number of seeds arranged at the golden ratio divergence angle will wrap almost precisely in several *complete* revolutions around the center, and thus provide starting points for the spirals. For example, $\frac{1}{1+\phi}$ is approximated by $\frac{1}{1+\frac{8}{5}} = \frac{5}{13}$. Therefore 13 seeds get arranged almost precisely in 5 complete revolutions; the next 13 seeds will get pretty close to these; and so on, forming the 13 spirals. The next best approximation of $\frac{1}{1+\phi}$ is $\frac{8}{21}$, so the 21 seeds get wrapped around almost precisely in 8 revolutions, and can serve as starting points for another family of spirals. The two families of spirals are turning opposite ways because $\frac{5}{13}$ is greater, and $\frac{8}{21}$ smaller, than $\frac{1}{1+\phi}$ (if the divergence angle was a fraction of 2π, we would get rays instead of spirals). In fact, any Fibonacci number of seeds gives rise to a family of spirals, but only two families are easy to discern, depending on the size of the seeds with respect to the size of the seed head.

Thus it is apparent that most plants use the golden ratio to determine their growth algorithm. The most prominent theory explaining this phenomenon is quite sophisticated. It considers a plant as a growing system

Figure 4.6: Numerical simulation with the golden angle: 13 right and 21 left spirals.

where the seeds, leaves, or florets are mutually repelling: the second seed appears as far as possible from the first seed and the stem, the third as far as possible from the two existing seeds and the stem, and so on. The development of such a system is determined by the postulate requiring it to stabilize into a state with minimal energy. It can be shown, both theoretically and experimentally, that such a model achieves its stable state as the divergence angle tends to $\frac{1}{1+\phi} \cdot 2\pi$. An interested reader can find an exposition of this theory in more advanced texts.[1]

We choose here to discuss another argument for the prevalence of the golden ratio phyllotaxis, which is derived from the mathematical properties of ϕ as an irrational number. Plants that follow a simple rule to best use available space for structural elements have an evolutionary advantage, thus assuring the prevalence of a particular phyllotaxis in nature. Therefore, a plant with multiple structural elements of similar size and nature (for example, a pinecone comprising many relatively small seeds) would benefit from a simple and universal algorithm that would allow it to grow indefinitely in size while ensuring the best packing of its structural elements, and their equal exposure to sunlight, dew, and rainwater.

For example, let us consider a pinecone. The seeds originate at the core axis of the pinecone, and grow away from it as the cone grows. To minimize the complexity of the process, let us assume that the pinecone adopts the following rules:

1. The divergence angle between each two consecutively sprouted seeds

---

[1] For example, see M. Livio, *The Golden Ratio: The Story of Phi, the World's Most Astonishing Number*, Broadway Books, 2002.

is the same.

2. The packing of the seeds in the pinecone is optimal at any moment of growth.

3. The exposure to sunlight and rain is optimal for each seed, meaning that the blocking of the light by neighboring seeds is minimal.

4. The size of a fully grown seed is the same in a given pinecone, but may differ among the pinecones and among the trees. However, it would be economical to have the same divergence angle for all pinecones.

Assuming these rules, the best answer for the divergence angle is the golden angle, $\frac{1}{1+\phi}$ of the complete revolution, or approximately 137.5 degrees. In the rest of this chapter, we will try to outline a mathematical reason for this. Essentially it follows from one of the more subtle properties of the golden ratio, namely its *extreme irrationality*.

You would think that any two irrational numbers are irrational to the same extent – they cannot be represented as a fraction of integers. In fact, some numbers are *more irrational* than others. The measure of irrationality is related to how well a number can be approximated by the rationals. For example, $\frac{22}{7} \simeq 3.14286$ is a well-known, quite accurate approximation of $\pi \simeq 3.14159$. Let us see what is the best approximation for $\sqrt{3} \simeq 1.73205$ with denominator less than 10. We quickly find $\frac{7}{4} = 1.75$. There are ways to find best rational approximations for irrational numbers, but we will not worry about them here. In the case of π, we have $\frac{22}{7} - \pi \simeq 0.00126$, while for the $\sqrt{3}$ we have $\frac{7}{4} - \sqrt{3} \simeq 0.01795$. Clearly, π is approximated better than $\sqrt{3}$. Maybe the denominator 4 is just too small? A good candidate for approximating $\sqrt{3}$ with denominator between 10 and 20 is $\frac{26}{15} \simeq 1.73333$, and the error of approximation is $\frac{26}{15} - \sqrt{3} \simeq 0.00128$. This is already closer to what we had for π. To quantify the difference between the approximations, we need to normalize the error with respect to the denominator of the fraction used for the approximation. Indeed, the larger the denominator, the greater the chances of finding a good approximation for any irrational number. The normalization used in number theory is $\frac{E}{M}$, where $E = |\frac{p}{q} - x|$ is the error of the approximation of x with a fraction of integers $\frac{p}{q}$, and $M = \frac{1}{q^2}$. We have the following table:

x	$\frac{p}{q}$	E	M	$\frac{E}{M}$
π	$\frac{22}{7}$	0.00126	$\frac{1}{49}$	0.062
$\sqrt{3}$	$\frac{7}{4}$	0.01795	$\frac{1}{16}$	0.287
$\sqrt{3}$	$\frac{26}{15}$	0.00128	$\frac{1}{225}$	0.289

Of course we tried only a couple of samples of rational approximation with rather small denominators, but the fact is that even with larger denominators, $\sqrt{3}$ never gets approximated nearly as well as π. This means that $\sqrt{3}$ is more irrational than π.

Finally, here is the punch line: the golden ratio ϕ is the *most irrational number* with respect to this kind of comparison. A theorem by German mathematician Adolf Hurwitz, published in 1891, states that any irrational number has infinitely many rational approximations with $\frac{E}{M} < \frac{1}{\sqrt{5}}$. In particular, ϕ is the "limit case" for this theorem: its best approximations have $\frac{E}{M}$ ratios pretty close to $\frac{1}{\sqrt{5}}$. In other words, the rational approximations of ϕ are as bad as they can possibly get.

Figure 4.7: Seed arrangements with divergence angle α.

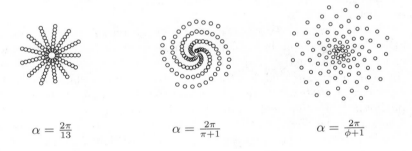

$\alpha = \frac{2\pi}{13}$ $\qquad$ $\alpha = \frac{2\pi}{\pi+1}$ $\qquad$ $\alpha = \frac{2\pi}{\phi+1}$

Now, what does this have to do with the growth of plants? The fraction $\frac{1}{1+\phi}$ of the complete revolution is the best divergence angle a plant can choose. Clearly, any rational fraction of 2π is a bad choice: the seeds line up along radial lines directly in front of each other, leaving empty spaces in between and blocking the light from each other, as in the leftmost pattern in Figure 4.7 (ratio 1 to 12). An irrational splitting of the complete revolution that is easily approximated by the rationals produces an arrangement that has a similar disadvantage: it does not pack well (Figure 4.7 center, ratio 1 to π). The number with the worst possible

Figure 4.8: Sunflower: 34 left and 55 right spirals (photo by Chris Darling).

rational approximations is the *universal* solution to the problem (Figure 4.7 right, ratio 1 to ϕ): it guarantees the optimal arrangement according to the rules (1)–(4) across the different sizes of plants, their seeds, leafs, or florets.

The appearance of two consecutive Fibonacci numbers as the numbers of left and right spirals in plants confirms that the angle of divergence constitutes the fraction of a complete circle corresponding to the 1 : ϕ ratio. This particular seed, leaf, or floret arrangement in plants is called the *Fibonacci phyllotaxis*. Figure 4.7 shows the advantages of the Fibonacci phyllotaxis by numerical simulation: when the golden angle is used for divergence, the packing of the seeds is the best, and alignment is sufficiently random for good access to light. The 13 right and 21 left spirals are prominent in the rightmost pattern.

Counting carefully, you can find exactly 34 left and 55 right spirals in the photo of the sunflower in Figure 4.8.

Practice problem 4.11. Show that ϕ equals the *infinite continued fraction*:

$$\phi = 1 + \cfrac{1}{1 + \cfrac{1}{1 + \cfrac{1}{1 + \cfrac{1}{1 + \cdots}}}}.$$

Solution: Denote the continued fraction (i.e., the second term on the right-hand side) by x and notice that $1 + \frac{1}{x} = x$. Multiplying by x, we get

$x + 1 = x^2$, the same equation that defines the golden ratio. Because the continued fraction is clearly a positive number, it follows that $x = \phi$.

Practice problem 4.12.

1. Check that ϕ satisfies the recursive relation

$$\phi^n = \phi^{n-1} + \phi^{n-2}$$

for any positive integer $n \geq 2$.

Solution: Indeed, if we multiply (4.1) by ϕ, we get $\phi^3 = \phi^2 + \phi$. Multiplying by ϕ again gives $\phi^4 = \phi^3 + \phi^2$, and so on.

2. Using the recurrence above, we can express ϕ^3 in terms of the first power of ϕ and integers. We know that $\phi^2 = 1 \cdot \phi + 1$ by (4.1), so:

$$\phi^3 = \phi^2 + \phi = (\phi + 1) + \phi = 2\phi + 1.$$

Similarly, check that $\phi^4 = 3\phi + 2$. Now do the same for ϕ^6. (*Answer:* $\phi^6 = 8\phi + 5$.) In all these cases, note that the coefficients (1,1), (2,1), (3,2), (8,5) are two consecutive Fibonacci numbers.

EXERCISES

Question 4.1. Which of the following numbers are irrational?

1. $\dfrac{3\sqrt{2} - 2}{5}$.

2. $\dfrac{3\sqrt{\frac{9}{4}} - 2}{5}$.

Question 4.2. Which of the following numbers are irrational?

1. $\sqrt{27} - \sqrt[3]{27}$.

2. $\sqrt{2 + \sqrt{2 + \sqrt{2 + \ldots}}}$.

Question 4.3. Let x and y be two distinct irrational numbers. Can their product $x \cdot y$ be rational? If yes, give an example. If no, explain why.

Question 4.4. A quadratic equation $ax^2 + bx + c = 0$ with a, b, c rational numbers, has only one real solution. Can this solution be irrational?

Question 4.5.   1. What is the length of the $B2$ size?

2. What is the area of the $B3$ size?

3. What is the reduction factor between the lengths of the $A4$ size and the $B6$ size?

Question 4.6. Find the aspect ratio of an imaginary series of paper sizes, say, $X0$, $X1$, $X2$,... such that

1. three portrait-oriented sheets of the $X(n+1)$ size, arranged side by side next to each other, exactly cover two landscape-oriented sheets of the Xn size, arranged one atop the other,

2. for each size, the aspect ratio is the same.

Question 4.7. Suppose we define, based on the An series of international standard paper sizes, a new series of $B'n$ paper sizes by the following conditions:

(1b′) The length of the $B'n$ size is the *arithmetic mean* of the lengths of the $A(n-1)$ and An sizes for all $n = 1, 2, \ldots, 10$:

$$l(B'n) = \frac{l(A(n-1)) + l(An)}{2}.$$

The length of the $B'0$ size is $\frac{\sqrt{2}+1}{2} \cdot l(A0)$.

(2b′) The aspect ratio for all $B'n$ sizes, $n = 0, 1, \ldots 10$, is $\sqrt{2}$.

1. Check that the new $B'n$ paper sizes satisfy Lichtenberg conditions (1) and (2).

2. Is the new paper size $B'4$ larger or smaller than the international standard $B4$ size? *Hint:* express the length of the $B'4$ and $B4$ paper sizes in terms of the length of the $A4$ size.

3. For two positive numbers a and b, which is bigger: $\frac{a+b}{2}$ or $\sqrt{ab}$? Can they be equal? Explain your answer.

Question 4.8. Suppose an interval is divided into two parts, the larger part a units long and the smaller part b units long. In the definition of the golden ratio, we equate the ratio of the larger part to the smaller part with the ratio of the whole to the larger part:

$$\frac{a}{b} = \frac{a+b}{a} \qquad \Longrightarrow \qquad \frac{a}{b} = \phi.$$

Let us tweak this definition to see what kind of answers we can obtain.

1. Suppose that the ratio $\frac{a}{b}$ equals three halves of the ratio of the whole $(a+b)$ to the larger part a:

$$\frac{a}{b} = \frac{3}{2} \cdot \frac{a+b}{a}.$$

   Find the ratio $\frac{a}{b}$.

2. Now suppose that the ratio $\frac{a}{b}$ equals five-sixths of the ratio of the whole $(a+b)$ to the larger part a:

$$\frac{a}{b} = \frac{5}{6} \cdot \frac{a+b}{a}.$$

   Find the ratio $\frac{a}{b}$.

3. Can you guess how the answer for $\frac{a}{b}$ depends on the positive coefficient p in the equation

$$\frac{a}{b} = p \cdot \frac{a+b}{a} \quad ?$$

   Try to solve this equation for the ratio $\frac{a}{b}$ in terms of p.

4. Give an example of a value of p in the equation above such that the number $\frac{a}{b}$ is rational.

Question 4.9. Show that $\phi = \sqrt{1 + \sqrt{1 + \sqrt{1 + \ldots}}}$. *Hint:* Check that the infinitely embedded square root satisfies equation (4.1).

Question 4.10. Consider the partial fractions $\{c_n\}$ of the continued fraction

$$\phi = 1 + \cfrac{1}{1 + \cfrac{1}{1 + \cfrac{1}{1 + \cfrac{1}{1 + \ldots}}}},$$

where $c_1 = 1$, $c_2 = 1 + \frac{1}{1}$, $c_3 = 1 + \frac{1}{1+1}$, and so on. The numbers $\{c_n\}$ are called the *convergents* of ϕ. Check that they are equal to the ratios of the consecutive Fibonacci numbers, $c_n = \frac{f_{n+1}}{f_n}$. Can you explain why?

Remark: In general, the presentation of an irrational number as a continued fraction gives its best rational approximations, and these are precisely the convergents. In number theory, there is a way to make precise what we mean by "best" here.

Question 4.11. Derive the recursive relation

$$\left(\frac{1}{\phi}\right)^n = \left(\frac{1}{\phi}\right)^{n-2} - \left(\frac{1}{\phi}\right)^{n-1}.$$

Hint: Derive the equation $(\frac{1}{\phi})^2 = 1 - \frac{1}{\phi}$ from the defining equation (4.1) for ϕ, and use it to obtain the recursion.

Question* 4.12. Check that for any $n \geq 2$

$$\phi^n = f_n\phi + f_{n-1},$$

where f_n is the nth Fibonacci number.

Hint: Use the recursive relation in Practice problem 4.12 to derive a presentation for ϕ^n as a sum of ϕ with an integer coefficient and another integer. Then observe that these integers are constructed according to the same recursive relation as the defining relation for the Fibonacci sequence.

Question* 4.13.   1. Using the recursion in Question 4.11, and proceeding as in Practice problem 4.12, obtain the following formula (pay attention to the sign!):

$$\left(-\frac{1}{\phi}\right)^n = f_{n-1} - f_n\left(\frac{1}{\phi}\right).$$

2. From the previous part, derive the following equality:

$$\frac{1}{\phi} = \frac{f_{n-1}}{f_n} - \frac{1}{f_n} \cdot \left(-\frac{1}{\phi}\right)^n.$$

Can you conclude that for large n, $\frac{f_{n-1}}{f_n}$ approximates $\frac{1}{\phi}$?

Remark: This is equivalent to the statement we made in the beginning of this chapter: the ratios $\frac{f_{n+1}}{f_n}$ approximate ϕ, and the approximation becomes better as n grows.

Question 4.14. Find a pine cone. Count its left and right spirals. If they are not consecutive Fibonacci numbers, the cone must have experienced a growth disruption. Find another one and repeat!

Chapter 5

Exponents: How much would you pay for the island of Manhattan?

In this chapter, we will study operations with powers. We explain what it means for a number to be raised to a positive, negative, integer, or fractional power, and how to multiply or divide two fractional powers of a number. This technique is indispensable in modern-day banking and finance, and we will apply it to computation of compound interest. What does it mean to compound a thousand dollars monthly at a given annual interest rate? In what follows, we explain these and related concepts.

MATH

Let us start with integer powers. We all know how to take the square of a number, or its cube: $1.6^2 = 1.6 \cdot 1.6 = 2.56$ and $2^3 = 8$. Similarly, if n is a positive integer, then the nth power of a (positive) number is

$$A^n = \underbrace{A \cdot A \cdot A \cdots A}_{n}$$

(multiplied n times). These powers satisfy various properties: for example, $A^3 \cdot A^2 = A^5$ (multiply 3 times, then 2 times – i.e., $3 + 2$ times in all). Similarly, $(A^3)^2 = A^3 \cdot A^3 = A^6$.

If $A > 1$, the numbers A^n grow very fast as n increases. We will talk more about exponential growth in Chapter 8 – for example, Figure 8.1 gives us an idea of the growth of the natural powers of 2.

What if we want to take *fractional powers*? Well, one of these is already very familiar: the power $\frac{1}{2}$ of a positive number is simply its square root. Similarly, the power $\frac{1}{3}$ is the cube root $\sqrt[3]{A}$, and so on.

It is also possible to take negative powers: $A^{-2} = 1/A^2$, for example. In general, given any positive A and any two positive integers n and m,

$$A^{-n} = 1/A^n, \qquad A^{n/m} = \sqrt[m]{A^n}.$$

In fact the same properties as above hold, whether or not the exponent is positive, negative, a power, or a root. In other words, suppose A and B are positive real numbers and c and d are *any* two real numbers (even irrational ones). Then the following properties hold:

$$A^c \cdot A^d = A^{c+d}, \qquad A^c/A^d = A^{c-d}, \qquad (A^c)^d = A^{cd},$$
$$A^c \cdot B^c = (AB)^c, \qquad A^c/B^c = (A/B)^c.$$

These simple rules are called the *laws/properties of exponents*. Note that they imply in particular that the zeroth power of any number is 1: $A^0 = A^{1-1} = A/A = 1$. The properties of exponents are extremely useful in simplifying expressions involving powers. For example,

$$2^3 \cdot \sqrt{2^5} = 2^3 \cdot (2^5)^{1/2} = 2^3 \cdot 2^{5/2} = 2^{3+(5/2)} = 2^{11/2}.$$

The first equality is the definition of a square root. In the second equality, we used the property above that $(A^c)^d = A^{cd}$, with $A = 2, c = 1/2, d = 5$. In the third equality, we used that $A^c \cdot A^d = A^{c+d}$, putting $A = 2, c = 6/2 = 3, d = 5/2$.

Here are some more examples.

EXAMPLES

Example 5.1. Simplify $2^{10} \cdot 5^5/20^4$.

Answer: Using the properties of exponents, we have:

$$\frac{2^{10} \cdot 5^5}{20^4} = \frac{(2^{2 \cdot 5}) \cdot 5^5}{20^4} = \frac{(2^2)^5 \cdot 5^5}{20^4} = \frac{4^5 \cdot 5^5}{20^4} = \frac{20^5}{20^4} = 20^{5-4} = 20.$$

$\square$

It is very instructive to take each equality in the above calculation and see why it is true – and whether you need to use one of the above properties of exponents to justify it.

The next example shows that the exponents need not be concrete real numbers – they can be variables as well.

Example 5.2. Simplify $9^b 4^b - 6^{2b}$. Here, b is a fixed (unknown) real number.
Answer: Using the properties of exponents, we compute:

$$9^b \cdot 4^b - 6^{2b} = (9 \cdot 4)^b - 6^{2b} = 36^b - 6^{2b} = (6^2)^b - 6^{2b} = 6^{2b} - 6^{2b} = 0.$$

☐

Here is a third example. The general philosophy is to try to simplify as much as you can. Do not be surprised if the overall answer is not as simple an expression as "20" or "0" (as in the above examples).

Example 5.3. Simplify $2^a 5^b / 10^{a+b}$, where a and b are real numbers.
Answer: Using the properties of exponents,

$$\frac{2^a \cdot 5^b}{10^{a+b}} = \frac{2^a \cdot 5^b}{10^a \cdot 10^b} = \frac{2^a}{10^a} \cdot \frac{5^b}{10^b} = (2/10)^a \cdot (5/10)^b = (1/5)^a \cdot (1/2)^b$$
$$= (5^{-1})^a \cdot (2^{-1})^b = 2^{-b} 5^{-a}.$$

☐

Example 5.4 (Solving equations involving exponents). Solve the following equations for the unknown real variable:

1. $5(1+x)^3 = 40$.

   Solution: First divide both sides by 5 to get: $(1+x)^3 = 8$. Now take the cube root of both sides. The left-hand side can now be computed using the properties of exponents, to equal:

   $$((1+x)^3)^{1/3} = (1+x)^{3 \cdot 1/3} = 1 + x,$$

   while the right-hand side becomes $8^{1/3} = 2$. So we get: $1 + x = 2$, or $x = 1$.

2. $(3+x)^5 = 32(2x+1)^5$.

Solution: Take the fifth root of both sides – i.e., raise them to the 1/5th power. Using the properties of exponents (do the work!), we can see that

$$(3 + x) = (2^5 \cdot (2x + 1)^5)^{1/5} = 2 \cdot (2x + 1) = 4x + 2.$$

This is a linear equation, so moving all terms involving x to the right-hand side, we get:

$$3 - 2 = 4x - x = 3x.$$

Finally, $x = 1/3$.   □

Practice problem 5.5. Simplify each of the following expressions using the properties of exponents above:

1. $12^6/6^{12}$. *Answer:* $1/3^6 = (1/3)^6$.

2. $(25^3 5^t)^{6-t}$ is what power of 5? *Answer:* 5^{36-t^2}.

3. $\sqrt[4]{2}\sqrt{2}\sqrt[4]{8}$. *Answer:* $2^{\frac{1}{4}} \cdot 2^{\frac{1}{2}} \cdot 8^{\frac{1}{4}} = 2^{\frac{1}{4}+\frac{1}{2}+\frac{3}{4}} = 2^{\frac{3}{2}} = 2\sqrt{2}$.

4. $\dfrac{(\sqrt{RS})^7}{R^3 S^3}$, where R and S are positive numbers. *Answer:* $\sqrt{RS}$.

5. $\dfrac{\sqrt[6]{27} \cdot \sqrt[6]{16}}{\sqrt{6}\sqrt[6]{2}}$.

   Answer: $\dfrac{\sqrt[6]{27} \cdot \sqrt[6]{16}}{\sqrt{6}\sqrt[6]{2}} = \dfrac{3^{\frac{3}{6}} \cdot 2^{\frac{4}{6}}}{2^{\frac{1}{2}} \cdot 3^{\frac{1}{2}} \cdot 2^{\frac{1}{6}}} = 3^{\frac{1}{2}-\frac{1}{2}} \cdot 2^{\frac{2}{3}-\frac{1}{2}-\frac{1}{6}} = 3^0 \cdot 2^0 = 1.$

Practice problem 5.6. Solve the equations:

1. $6^{3x} = 36^{\frac{5}{4}}$. *Answer:* $x = \frac{5}{6}$.

2. $(2y^2 + 1)^5 = 243$. *Answer:* $y^2 = 1$, so $y = \pm 1$.

APPLICATION 1

Computations with exponents are necessary in modeling quickly developing processes, such as chain reactions and avalanches.

Example 5.7. High in the mountains, a comparatively small initial impact (a snowball) can cause a larger snowslide and eventually lead to a huge avalanche.

Suppose that during each second, the volume of the sliding snow grows by a factor of 1/3. What is the volume of the snow in the avalanche after 10 seconds? After 30 seconds?

Solution: Denote the initial volume by V_0. Then after 1 second, we have $V_0 + \frac{1}{3}V_0 = \frac{4}{3}V_0$, after 2 seconds $\frac{4}{3}V_0 + \frac{1}{3}(\frac{4}{3}V_0) = \frac{4}{3}V_0(1 + \frac{1}{3}) = \frac{4}{3} \cdot \frac{4}{3} \cdot V_0$, and so on. After 10 seconds, the volume is

$$V_{10} = \left(\frac{4}{3}\right)^{10} V_0 \simeq 17.76\, V_0.$$

After 30 seconds, we have

$$V_{30} = \left(\frac{4}{3}\right)^{30} V_0 \simeq 5,600\, V_0.$$

For example, if the initial volume V_0 was 1 cubic foot (7.5 gallons), then after 10 seconds we have 133 gallons, and after 30 seconds, almost $42,000$ gallons of sliding snow.

Now suppose we know that the volume of snow in an avalanche grows in such a way that after 30 seconds, it is 100 times ($1,000$ times) the initial volume. How much does the volume grow per second?

Denote by r the factor of volume growth per second. Then we have an equation:

$$(1 + r)^{30} = 100.$$

This implies

$$1 + r = \sqrt[30]{100} \simeq 1.1659,$$

and $r \simeq 0.1659$. In the second case,

$$(1 + r)^{30} = 1,000 \quad \Longrightarrow \quad 1 + r = \sqrt[30]{1,000} \simeq 1.2589,$$

and $r \simeq 0.2589$. □

Another avalanche-type system is provided by a rumor spread.

Example 5.8. Suppose a person obtains some valuable trading information. She probably will not share it massively, but she will tip off her five closest friends. The next day, her friends do the same, and so on. How many *more* people will know the information each day during the next week? We calculate:

$$
\begin{array}{ll}
\text{day 1} & \text{1 person} \\
\text{day 2} & \text{5 more people} \\
\cdots & \cdots \\
\text{day 7} & 5^6 = 15,625 \text{ more people}
\end{array}
$$

At some point in this process saturation will be reached, where all interested people already know the news and others don't care. Saturation is the reason for failure of financial pyramids, which require exponential increases in the number of participants to sustain them. □

Finally, here is an example of a different kind, where proficiency with powers becomes useful.

Example 5.9. Suppose that a store allows three discounts to apply, consecutively:
10% membership discount,
10% seasonal sale discount, and
10% promotional discount.
Another store with the same merchandise offers a flat 28% discount on all sales. Which deal is better?

Solution: Here the main point is that the three discounts in the first store are applied *consecutively* (instead of *simultaneously*). The first discount reduces the price of an item P to $0.9P$, then the second discount reduces it further to $0.9 \cdot 0.9P = (0.9)^2 P$, and the last discount results in a final price of $0.9 \cdot (0.9)^2 P = (0.9)^3 P$. Therefore, a customer would pay $(0.9)^3 P = 0.729P$ in the first store, and only $0.72P$ in the second store. The second deal is better. □

In contrast, if the first store allows the customer to *add* the discounts, then the total discount comes to 30% of the sale price. In this case the first store is clearly offering a better deal than the second.

APPLICATION 2

Raising positive numbers to powers is necessary in finance, in particular in computing *compound interest*. If you take a loan of a thousand dollars from a bank, or deposit a thousand dollars in it, then, after a couple of years, you will either owe or own *more* than that amount. The original amount of money borrowed or deposited is the *principal,* and the extra amount is called the *interest.*

Simple interest

Let us start with *simple interest,* which is, of course, quite simple. If you deposit $1,000 in a bank that has a 5% annual simple interest rate, then the deposited money accumulates interest, $50 per year (which is 5% of $1,000). Thus, after one year you have $1,050, after two years $1,100, and so on.

Here is the "general formula" for simple interest: if we deposit an initial sum of money A in a bank that offers an annual simple interest rate of $r\%$, then each year the principal accumulates the same amount of interest, $A \cdot r/100$. Thus, the interest that we would have accumulated on A after t years is: $A \cdot r \cdot t/100$. The total amount after t years is given by:

$$A(t) = A\left(1 + \frac{rt}{100}\right).$$

Example 5.10. A few years ago, I deposited $2,000 in a bank with a simple interest scheme. The money in my account today is $2,500.

1. If I deposited the money ten years ago, what is the simple interest rate in the bank?

2. If the rate of simple interest is 5%, then how long ago did I deposit the money in the bank?

Solution: In both calculations, we set $A = \$2,000$ and $A(t) = \$2,500$.

1. If the money was deposited ten years ago, then $t = 10$ and r is unknown, so compute:

$$2,500 = 2,000\left(1 + \frac{r \cdot 10}{100}\right).$$

   Solving for r, $500 = 20 \cdot r \cdot 10$, whence $r = 2.5\%$.

2. In this part, $r = 5\%$ and t is unknown, so:

$$2,500 = 2,000\left(1 + \frac{t \cdot 5}{100}\right).$$

   Solving for t, $500 = 20 \cdot t \cdot 5$, whence $t = 5$ yr.   □

Compound interest

This is the model actually used by most banks today. The difference between simple and compound interest is that in the case of simple interest, *only the principal* earns interest. In the case of compound interest, *both the principal and the accumulated interest* earn interest. For example, suppose I deposit $1,000 in a bank that offers a 4% annual rate of compound interest. Then in the first year, I accumulate $40 interest. In the second year, the entire amount of $1,040 is earning interest, so that at the end of the second year 4% of $1,040, or $41.60, is added. Thus, the amount after two years is $1,081.60, and so on.

Let us write down a general formula for this. If the deposit amount is A dollars, and the annual compound interest rate is $r\%$, then after t years, we have:

$$A(t) = A \underbrace{\left(1 + \frac{r}{100}\right)\left(1 + \frac{r}{100}\right)\cdots\left(1 + \frac{r}{100}\right)}_{t \text{ times}} = A\left(1 + \frac{r}{100}\right)^t.$$

Here we have a formula that involves a variable power t.

Some banks compound more often than others. Thus, if the compounding is quarterly as opposed to annually, then we would multiply four times as many factors for each year – one for each quarter – but the corresponding rate of interest would itself be quartered. In other words, the formula for the amount accumulating after t years would be:

$$A(t) \quad = \quad A\left(1 + \frac{r}{400}\right)^{4t} \qquad \text{(quarterly)}.$$

Practice problem 5.11. Using your calculator (or even Google), find all positive numbers r that satisfy:

$$\left(1 + \frac{r}{400}\right)^{20} = 1.4859.$$

Round your answer off to three decimal places. *Hint:* The syntax for raising a number to the 1/20th power, say, in a Google search, is: 5 ^ (1/20). *Answer:* $r = 8$.

Computations such as the last one are very important and commonly show up in studying compound interest, as we will see below.

Example 5.12. Suppose I borrow $10,000 from a bank, and my loan accumulates interest at an annual rate of 6%, compounded quarterly. How much do I owe the bank after three years?

Answer: Use the formula for quarterly compounding, with $A = \$10,000$, $r = 6$, and $t = 3$:

$$A(3) = 10,000 \left(1 + \frac{6}{400}\right)^{4 \cdot 3} = 10,000(1 + 0.015)^{12} = 10,000(1.015)^{12}$$

$$\simeq 11,956.18.$$

Thus, the loan is worth $11,956.18 after three years. □

Similarly, if a bank compounds monthly, the number of factors per year would grow to 12, but the rate of interest would get divided by 12:

$$A(t) \quad = \quad A \left(1 + \frac{r}{1,200}\right)^{12t} \quad \text{(monthly)}.$$

In general, suppose there are n compounding periods per year, and the principal of A dollars is to accrue interest at an *annual* compound interest rate of $r\%$. Then the total amount after t years is

$$A(t) = A \left(1 + \frac{r}{n \cdot 100}\right)^{nt}.$$

Acquisition of Manhattan

To illustrate the compound interest model, let us consider a historical example.

In 1626, the Dutch colonists acquired Manhattan from the Native Americans for some trade goods. According to different estimates, the value of the goods was anywhere between $24 and $1,000 in modern currency. It is not easy to estimate the value of undeveloped Manhattan land now, but one online source estimates a lot of 0.06 acres available for sale for $400,000. This makes an acre of Manhattan land worth $6,666,700. The area of Manhattan is approximately 20 square miles, or 12,800 acres. The total value is then

$$\$6,666,700 \cdot 12,800 \simeq \$85.3 \text{ billion}.$$

This is a huge amount of money, and it is hard to imagine that an investment of just $24 could have grown that much in four centuries. To see how good an investment this was, let us compute the annual compound interest rate that would produce the same growth. We will make a computation for the initial capital investments of $24 and $1,000 separately.

When this book was written, the time elapsed since 1626 was 2014 − 1626 = 388 years. Assuming the annual compounding model, and $24 initial capital, we have the following equation for the annual interest rate r:

$$24 \left(1 + \frac{r}{100}\right)^{388} = 8.53 \cdot 10^{10}.$$

Then

$$\left(1 + \frac{r}{100}\right)^{388} \simeq 3.5 \cdot 10^9 \quad \Longrightarrow \quad \left(1 + \frac{r}{100}\right) \simeq 1.058 \quad \Longrightarrow \quad r \simeq 5.8\%.$$

If we take $1,000 as the initial capital, the annual interest rate r is even lower:

$$1,000 \left(1 + \frac{r}{100}\right)^{388} = 8.53 \cdot 10^{10}.$$

Then

$$\left(1 + \frac{r}{100}\right)^{388} \simeq 8.53 \cdot 10^7 \quad \Longrightarrow \quad \left(1 + \frac{r}{100}\right) \simeq 1.048 \quad \Longrightarrow \quad r \simeq 4.8\%.$$

In both cases, the interest rate is not much different from actual interest rates offered by banks for long-term investments, and is compatible with a realistic economy growth rate. If the people who sold the island had instead invested in a bank offering an annual interest rate of about 5%, they would have accumulated roughly the same amount of money as the value of Manhattan land today. In this sense, the sale/acquisition of Manhattan seems to be a reasonable deal.

Continuous compounding and the number e

What happens if the number of compounding periods per year is allowed to grow indefinitely? To simplify our computations, suppose that the principal $A = \$1$, $t = 1$, and $r = 100\%$. In this case, the formula reads

$$A(1) = \left(1 + \frac{1}{n}\right)^n.$$

Let us see how the result depends on n:

$$
\begin{array}{ll}
n = 1 & A(1) = (1+1)^1 = 2 \\
n = 2 & A(1) = (1+\frac{1}{2})^2 = \frac{9}{4} = 2.25 \\
n = 4 & A(1) = (1+\frac{1}{4})^4 \simeq 2.441 \\
n = 12 & A(1) = (1+\frac{1}{12})^{12} \simeq 2.613 \\
n = 365 & A(1) = (1+\frac{1}{365})^{365} \simeq 2.714 \\
n = 1,000 & A(1) = (1+\frac{1}{1,000})^{1,000} \simeq 2.717
\end{array}
$$

It looks like the values of $A(1)$ become closer together as the number of compounding periods grows. In fact, when n grows indefinitely, this sequence *converges*; that is, it becomes as close as we wish to a certain number, namely to the constant e:

$$
\left(1 + \frac{1}{n}\right)^n \xrightarrow[n\to\infty]{} e = 2.718281828459045\ldots
$$

This number plays an important role in mathematics; for example, it appears as the base of the natural logarithm ln, which we will introduce in Chapter 6. One of the first definitions of e was given in the seventeenth century by the Swiss mathematician Jacob Bernoulli. He defined e as the annual growth factor with 100% compound interest rate *compounded continuously*.

If the compound interest rate is $r\%$, the amount accumulated after t years in the case of continuous compounding is

$$
A(t) = Ae^{\frac{rt}{100}}.
$$

Example 5.13. With the initial capital $A = \$10,000$, how much money will accumulate in $t = 8$ years if the bank compounds continuously with the compound interest rate $r = 3\%$?
Answer: Using the continuous compounding formula, we compute

$$
A(8) = Ae^{\frac{3\cdot8}{100}} = 10,000e^{0.24} \simeq \$12,712.49.
$$

□

Strategy for problem solving

In questions involving compound interest, start with the equation for compound interest (monthly, quarterly, n periods per year). Now plug in the numbers that you are given and identify the unknown quantity. Then solve for the unknown quantity, as is done in the examples below. To make your answer meaningful, you should indicate the units of measurement. For instance, if the amount of money is unknown, the answer should be given in a unit of currency (for example, $). If the interest rate is unknown, the answer should be given as a percentage (%).

When solving the equation, it is a good idea first to isolate the unknown quantity on one side, writing it in terms of the known quantities, and only then use your calculator to compute the answer. This way, even if the numerical answer is wrong, you can get partial credit for the correct expression, and for doing most of the work to solve the problem. If you have to round your answer, then you should do so only at the *end* of your computations, not in the middle. Here is why: compare 1.01^{20} rounded to one decimal place (you get 1.2) with 1.01 rounded to one decimal place *and then* raised to the 20th power (you get $1.0^{20} = 1.0$).

Example 5.14. How much should be deposited today in an account that earns interest at an annual rate of 6%, compounded monthly, so that it will accumulate to $20,000 in 5 years?

Solution: We have $r = 6$, $n = 12$, $t = 5$, and $A(t) = 20,000$. We need to find A.

$$A(t) = 20,000 = A \left(1 + \frac{6}{1,200}\right)^{12 \cdot 5} = A(1 + 0.005)^{60}.$$

Solving for A, we get:

$$A = \frac{20,000}{(1.005)^{60}}.$$

Now use your calculator:

$$A \simeq 14,827.44.$$

Thus, the amount that should be deposited today is $14,827.44.   □

Example 5.15. Suppose I borrow $10,000 from a bank that compounds quarterly. Determine the annual compound interest rate (to four decimal places) if after one year I owe $10,824.32.

Solution: Suppose the annual compound interest rate is r%. Using the formula for interest with quarterly compounding,

$$A(t) = 10,824.32 = 10,000 \left(1 + \frac{r}{400}\right)^{4 \cdot 1} = 10,000 \left(1 + \frac{r}{400}\right)^{4}.$$

To solve for r, we need to divide both sides by $10,000$ and then take the fourth root – i.e., raise both sides to the $1/4$th power. (Do you see why taking the fourth root is the same as taking the square root of the square root? Use properties of exponents.) Thus,

$$(1.082432)^{1/4} = \left(1 + \frac{r}{400}\right)^{4 \cdot (1/4)} = 1 + \frac{r}{400}.$$

In other words, $1 + (r/400) = 1.01999996$. Solving for r, $r = 7.999984 \simeq 8\%$ is the annual compound interest rate (to four decimal places). $\square$

Example 5.16. A bank uses a compound interest model, so that a deposited amount doubles in 18 years. How much should be deposited in the bank today, to obtain \$50,000 in 10 years?

Solution: We need to find the amount of money X such that

$$50,000 = X \left(1 + \frac{r}{n \cdot 100}\right)^{10n} \implies X = \frac{50,000}{(1 + \frac{r}{n \cdot 100})^{10n}}.$$

However, neither the interest rate r nor the period of compounding n is given. All we have is the equation:

$$2A = A \left(1 + \frac{r}{n \cdot 100}\right)^{18n}$$

for any deposit amount A. Dividing by A and taking the 18th power root, we derive

$$\left(1 + \frac{r}{n \cdot 100}\right)^{18n} = 2 \implies \left(1 + \frac{r}{n \cdot 100}\right)^{n} = 2^{\frac{1}{18}}.$$

This is all we need to solve the problem. Using the properties of exponents, we have:

$$X = \frac{50,000}{(1 + \frac{r}{n \cdot 100})^{10n}} = \frac{50,000}{((1 + \frac{r}{n \cdot 100})^{n})^{10}} = \frac{50,000}{(2^{\frac{1}{18}})^{10}}.$$

Now, $(2^{\frac{1}{18}})^{10} = 2^{\frac{10}{18}} = 2^{\frac{5}{9}}$. Finally,

$$X = \frac{50,000}{2^{\frac{5}{9}}} \simeq 34,019.75.$$

The deposit has to be \$34,019.75. $\square$

Remark: The point to understand here is that the number of years enters the compound interest formula as an exponent. Therefore, if a deposit doubles in 18 years, we can tell right away that in one year, *any* deposit will grow by a factor of $2^{\frac{1}{18}} = \sqrt[18]{2}$. Then the growth factor over 10 years is computed by raising $2^{\frac{1}{18}}$ to the 10th power, to obtain $2^{\frac{10}{18}} = 2^{\frac{5}{9}}$.

Effective annual rate

In dealing with banks, you may sometimes encounter the expression *effective annual rate* of compounding. Suppose I borrow $10,000 from a bank with an annual rate of 8%, compounded quarterly. After one year, if the money was compounded only annually, then my loan would be $10,800. But because the compounding is done more frequently, my loan actually increases. See Example 5.15 above for the calculations: the loan is now $10,824.32 – some twenty-four dollars more!

The effective annual rate essentially measures this discrepancy. It is defined as the simple interest rate that produces the same amount of money at the end of one year, as the stated compound annual rate, compounded n times per year (quarterly, monthly, ...). Thus, if $r\%$ is the compound annual rate, and it is compounded n times per year, then the effective annual rate r_{eff} is computed according to the formula

$$1 + \frac{r_{\text{eff}}}{100} = \left(1 + \frac{r}{n \cdot 100}\right)^n.$$

For instance, in the above example, the loan after one year is $10,824.32. Hence the effective annual rate is 8.2432%. (Note that the annual rate was given to be 8%.)

Example 5.17. Find the effective annual rate for the compound annual rate of 6% compounded monthly. Round off your answer to three decimal places.
Answer: Using $r = 6$ and $n = 12$, we compute:

$$1 + \frac{r_{\text{eff}}}{100} = \left(1 + \frac{6}{12 \cdot 100}\right)^{12} = (1 + 0.005)^{12} \simeq 1.0616778.$$

Solving this simple equation, we get $r_{\text{eff}} \simeq 6.168\%$.   □

In case of the continuous compounding, the effective annual rate is determined by the formula

$$1 + \frac{r_{\text{eff}}}{100} = e^{\frac{r}{100}}.$$

Example 5.18. Let $A = \$5,000, r = 4\%$, compounded continuously.

1. How much money will accumulate in $t = 5$ years?

2. Find the effective annual rate.

Solution: Using the formula for continuous compounding, we get

$$A(5) = Ae^{\frac{4 \cdot 5}{100}} = 5,000e^{0.2} \simeq \$6,107.01.$$

For the effective annual rate, we have

$$1 + \frac{r_{\text{eff}}}{100} = e^{\frac{4}{100}} = e^{0.04} \simeq 1.04081.$$

Therefore, $r_{\text{eff}} \simeq 4.081\%$.   □

The effective annual rate is the way to determine which of two (or more) competing banking models yields greater returns/interest over the same length of time. Simply compute which effective annual interest rate is greater! See Question 5.9 below.

EXERCISES

Question 5.1. All variables below are assumed to be positive. Simplify the following expressions:

1. $\dfrac{\sqrt[4]{A}}{\sqrt[5]{A}}$.

2. $\sqrt[3]{56} \cdot 7^{-\frac{1}{3}}$.

3. $2^{45}/32^9$.

4. $3abc - \sqrt{ab}\sqrt{bc}\sqrt{ca}$.

Question 5.2. Suppose a and t are positive numbers. Simplify the following expressions:

1. $\sqrt{16^a}\,\sqrt[3]{8^{2a}}$.

2. $\sqrt{\dfrac{2}{3}} \cdot \dfrac{3^{\frac{1}{3}}}{4^{\frac{1}{4}}}$.

3. $3^{11}2^{55} - 96^{11}$.

4. $(4t^2)^5 - (2t)^9$ (can you take a factor common from both terms?).

Question 5.3. Suppose you borrow \$1,000 from a bank at an annual compound interest rate of 8%. Calculate the interest accumulated over 5 years if the bank compounds (a) monthly, (b) annually. Round all your answers to two decimal places.

Question 5.4.

1. If the amount accumulated at the end of 4 years is $3,600, given an annual simple interest rate of 5%, then what was the principal?

2. If the amount accumulated after 10 years is $2,000, and the principal was $1,500, find the annual simple interest rate.

3. If the amount accumulated after 20 years is exactly twice that of the principal, can you write down an equation for $A(t)$ in terms of A? Use this to compute the interest rate r if the interest is (a) simple; (b) compounded annually; (c) compounded quarterly.

Question 5.5. A person deposits $A = \$20,000$ in a bank that compounds quarterly at an annual compound interest rate of $r\%$. The amount accumulated at the end of 3 years is $A(3) = \$23{,}912.36$.

1. Find the annual compound interest rate r.

2. How much money will accumulate at the end of 6 years after the initial deposit?

3. Compare the amount $A(6)$ obtained in the previous part with the quantity $\frac{A(3)^2}{A}$. Why are they equal?

Question 5.6. A bank uses a compound interest model, so that in 10 years a deposit of $10,000 grows to $15,000. Under the same model,

1. how much will a deposit of $30,000 grow to in 30 years?

2. how much will a deposit of $12,000 grow to in 18 years?

Note: both the period of compounding and the annual compound interest rate are unknown here – but you don't need them to find the answer!

Question 5.7. You have $12,000 to invest. Which investment yields a greater return over 20 years: 7% compounded monthly, or 7.15% compounded annually? Compute the amounts accumulated in both cases. Write the answers rounded to the nearest cent.

Question 5.8. Find the effective annual rate if the bank compounds (a) monthly, and (b) quarterly – both with an annual compound interest rate of 10%. Write the answers to three decimal places. Conclude that the more often the bank compounds (at the same annual compound interest rate), the better it is for the investor.

Question 5.9. Suppose I have \$15,000 to invest, and two banks are offering competing models. Bank A has an annual interest rate of 6.4%, compounded quarterly, while bank B has an annual interest rate of 6%, compounded monthly. Which bank yields greater interest over one year? In other words, which bank has a greater effective annual rate?

Now compute which bank yields greater interest over three years. In fact, the answer remains the same no matter how much money is invested or how many years of compounding one considers. Why is this so? The answer is: compounding monthly or quarterly by a given interest rate is the same as compounding *annually* by its effective rate. Hence to compute the interest for both models, we are basically compounding annually by their effective rates, in both cases. Thus, whichever effective interest rate is greater will *always* yield a greater return.

Chapter 6

Logarithms I: Money grows on trees, but it takes time

To determine the time it takes for an investment to grow to a certain amount according to a compound interest model considered in Chapter 5, we need a new mathematical tool: the logarithm. In mathematics, taking a logarithm is the inverse operation of taking an exponent. In this chapter we will learn to work with the logarithms and discuss their applications, most importantly in finance.

MATH

For positive numbers $a \neq 1$ and B, the *logarithm* $\log_a B$ is defined as the unique solution to the equation:

$$a^x = B \quad \Longleftrightarrow \quad x = \log_a B.$$

By definition, the exponential function a^x and the logarithmic function $\log_a x$ are inverse to each other. This means that if you apply one and then another, you get the same number you started with. More precisely, for any positive numbers $a \neq 1$ and B we have

$$a^{\log_a B} = B,$$

97

and similarly, for any positive $a \neq 1$ and any real x we have

$$\log_a (a^x) = x.$$

The number a is the *base* of $\log_a$. We will denote

$$\log = \log_{10}, \quad \ln = \log_e,$$

where the constant $e = 2.71828\ldots$ was introduced in Chapter 5.

Example 6.1. We compute: $\log 1,000 = \log_{10} 1,000 = 3$, because in order to obtain $1,000$, you need to raise 10 to the third power. $\square$

Here is a quick reminder of the *laws of logarithms*, algebraic rules of dealing with the logarithms. Let a, B, C be positive numbers, $a \neq 1$, and r a real number. The main properties of the logarithms are the following:

$$\log_a B + \log_a C = \log_a(BC), \quad \log_a B - \log_a C = \log_a \left(\frac{B}{C}\right)$$

$$\log_a(B^r) = r \log_a B, \qquad \log_a B = \frac{\ln B}{\ln a},$$

$$\log_a 1 = 0, \qquad \log_a a = 1.$$

These properties follow from the laws of exponents listed in Chapter 5 and the definition of logarithms. Below we prove three of the equations.

Example 6.2. Let a, B, C be positive numbers, $a \neq 1$. Show that

$$\log_a B + \log_a C = \log_a(BC).$$

Solution: Let us introduce the notations: $x = \log_a B$, $y = \log_a C$, and $z = \log_a(BC)$. Then we have by the definition of the logarithm:

$$a^x = B, \quad a^y = C, \quad a^z = BC.$$

By the laws of exponents, this implies

$$a^z = a^x \cdot a^y = a^{x+y}.$$

Hence $z = x + y$. Substituting the values, $\log_a B + \log_a C = \log_a(BC)$. $\square$

Example 6.3. Let a and B be positive numbers, $a \neq 1$, and r a real number. Show that

$$\log_a(B^r) = r \log_a B.$$

Solution: Let $x = \log_a B$. Then by the definition of the logarithm we have:

$$B = a^x.$$

Raising both sides to the power r, we get

$$B^r = (a^x)^r.$$

Taking the logarithm base a of both sides, we have

$$\log_a(B^r) = \log_a((a^x)^r) = \log_a a^{xr} = xr = rx.$$

Plugging in $x = \log_a B$, we get

$$\log_a(B^r) = rx = r \log_a B. \qquad \square$$

Example 6.4. The equality $\log_a 1 = 0$ for any positive $a \neq 1$ is equivalent to the property of exponents: $a^0 = 1$, which holds for any positive a. $\square$

All remaining properties can be proven similarly. (Try it!)

EXAMPLES

In what follows, $\log = \log_{10}$, and $\ln = \log_e$. Each of the following questions can be answered using the definition and laws of logarithms.

Example 6.5. Compute or simplify the following expressions:
(a) $\log \frac{1}{100}$, (b) $(\ln 9 + \ln 4)/\ln(6)$.
Solution: (a) This is the power you need to raise 10 to get $\frac{1}{100}$. We have $10^{-2} = \frac{1}{100}$. Therefore, $\log \frac{1}{100} = -2$.
(b) $(\ln 9 + \ln 4)/\ln 6 = \ln(9 \cdot 4)/\ln 6 = \ln(6^2)/\ln 6 = 2$. $\square$

Practice problem 6.6. Compute or simplify the following expressions:

1. $\log 10^{2t+4}$. *Answer:* $2t + 4$.

2. log of 1 million. *Answer:* 6.

3. $\log 1$. *Answer:* 0.

4. $\log 60 - \log 6$. *Answer:* $\log 60 - \log 6 = \log \frac{60}{6} = \log 10 = 1$.

5. $\log(1 + t) + \log(1 - t) - \log(1 - t^2)$. *Answer:* 0.

Because $\log_a x$ is the inverse function for a^x, one of the most important applications of logarithms is in solving equations where the variable is an exponent.

Example 6.7. Solve the equation using log: $(\frac{1}{10})^{(3x-1)} = 38$.

Solution:

$$\left(\frac{1}{10}\right)^{3x-1} = ((10)^{-1})^{3x-1} = 10^{-3x+1} = 38.$$

The last equality follows from the property of exponents: $(A^a)^b = A^{ab}$. Now we can apply the logarithm (base 10):

$$-3x + 1 = \log 38 \quad \Longrightarrow \quad x = \frac{1}{3}(1 - \log 38).$$

$\square$

Practice problem 6.8. Use suitable logarithms to solve the following equations.

1. Solve the equation using log: $10^x = 563$. *Answer:* $x = \log 563 \simeq 2.7505$.

2. Solve using log and also using ln: $30 \cdot 2^t = 900$. Write down your answer to three decimal places. *Answer:* Dividing by 30, we get $2^t = 30$. Then $t = \log 30 / \log 2 = \ln 30 / \ln 2 \simeq 4.907$.

3. Solve for k in: $50 \cdot 10^{5k} = 5$. *Answer:* Dividing by 50, we get $10^{5k} = \frac{1}{10} = 10^{-1}$. Therefore, $5k = -1$ and $k = -1/5$.

APPLICATIONS

Time-related questions in compound interest models

Now we can use logarithms to answer the following type of questions: With the annual interest rate $r\%$ compounded n times per year, how long does it take for the deposited amount to double? Following the compound interest model in Chapter 5, we have the equation

$$A(t) = A\left(1 + \frac{r}{n \cdot 100}\right)^{nt} = 2A.$$

We need to find t. We can cancel A from both sides of the equation and get

$$\left(1 + \frac{r}{n \cdot 100}\right)^{nt} = 2.$$

Because the time variable appears as an exponent, the only way to solve for it is to take a logarithm. Taking the logarithm with any base would work. As most calculators are equipped with the "ln" button, we choose to take the natural logarithm:

$$\ln\left(1 + \frac{r}{n \cdot 100}\right)^{nt} = \ln 2.$$

Using the law of logarithms $\log_a(B^r) = r \log_a B$, we obtain

$$(nt) \ln\left(1 + \frac{r}{n \cdot 100}\right) = \ln 2.$$

Now we can express the variable t in terms of r and n:

$$t = \frac{\ln 2}{n \cdot \ln\left(1 + \frac{r}{n \cdot 100}\right)}.$$

A similar algorithm works if you need to find how long it takes for the amount to triple, grow by 50%, increase by \$1000, and so on.

In the case of continuous compounding (see Chapter 5), the same question can be asked: given an annual interest rate $r\%$, how long does it take for an investment to double? We need to solve the equation for t:

$$Ae^{\frac{rt}{100}} = 2A.$$

Dividing by A, we have

$$e^{\frac{rt}{100}} = 2.$$

Here taking the natural logarithm is clearly the best choice because the base of the exponent is e.

$$e^{\frac{rt}{100}} = 2 \quad \implies \quad \frac{rt}{100} = \ln 2 \quad \implies \quad t = \frac{100}{r} \ln 2.$$

Example 6.9. How long in years (yr) does it take for your investment to grow by 50% if the annual interest rate is 6% compounded (a) quarterly, (b) continuously?

Solution: Growth by 50% is the same as growth by a factor of 1.5. We can apply the formulas derived above replacing the growth factor 2 by 1.5.

(a)

$$t = \frac{\ln 1.5}{4 \cdot \ln\left(1 + \frac{6}{4 \cdot 100}\right)} \simeq 6.81 \ \text{yr.}$$

(b)

$$t = \frac{100}{6} \ln 1.5 \simeq 6.75 \ \text{yr.}$$

$\qquad\qquad\qquad\qquad\qquad\qquad\qquad\qquad\qquad\qquad\qquad\qquad\qquad\qquad$ □

Example 6.10. How long does it take for your investment to grow from $A = \$10,000$ to $A(t) = \$12,000$, if the annual interest rate is 4%, compounded quarterly? Give your answer to the nearest quarter of a year.

Solution: We have $r = 4\%$, $n = 4$, $A = \$10,000$, $A(t) = \$12,000$.

$$A(t) = 12,000 = 10,000 \left(1 + \frac{4}{4 \cdot 100}\right)^{4t}.$$

Dividing both sides by $1,000$ gives

$$12 = 10(1.01)^{4t} \quad \Longrightarrow \quad \frac{6}{5} = (1.01)^{4t}.$$

Now is a good time to apply the logarithms. You can use the logarithm with any base, because the answer is independent of the base. We choose the natural logarithm because it is a one-touch operation on any calculator:

$$\ln\left(\frac{6}{5}\right) = \ln(1.01)^{4t} = 4t \cdot \ln 1.01 \quad \Longrightarrow \quad t = \frac{\ln 1.2}{4 \ln 1.01}.$$

Finally, use your calculator:

$$t = \frac{\ln 1.2}{4 \ln 1.01} \simeq 4.6 \ \text{yr.}$$

The time needed for the investment to grow to $\$12,000$ is more than 4 years and 2 quarters, but less than 4 years and 3 quarters, because $4.6 > 4.5$ and $4.6 < 4.75$. Therefore, to obtain the desired amount, you will have to wait for 4 years and 3 quarters. □

Strength of an encryption key

Here is an example from a completely different domain. An *encryption key*, or a secret password used in computer security, can be thought of as a sequence of zeros and ones of a certain length. For example, a 56-bit key is a sequence of 56 zeros and ones. It can start like this:

$$(0, 0, 0, 1, 0, 0, 1, 1, 0, 1, 0, 1, 0, 0, 0, 0, 0, 0, 0, 0, 1, 0, 1, 1, 0, 1, \ldots \ldots \ldots \ldots)$$

How many different 56-bit keys are there? There are 2 choices (0 or 1) for the first entry. For each choice of the first entry, there are 2 choices for the second, for each choice of the first and the second entry there are 2 choices for the third, and so on. If the key is 3 bits long, then we will have $2 \cdot 2 \cdot 2 = 8$ different sequences. You can easily list them:

$$(0, 0, 0) \quad (0, 0, 1) \quad (0, 1, 0) \quad (0, 1, 1) \quad (1, 0, 0) \quad (1, 0, 1) \quad (1, 1, 0) \quad (1, 1, 1)$$

Now suppose your encryption key is 56 bits long. Then you have two independent choices for each of the 56 entries. Therefore you have 2^{56} different possible keys.

Now suppose an imaginary intruder – a hacker or an agency – wants to crack the key by brute force, trying all sequences of zeros and ones until the right one is found. Suppose they have a cluster of $1,000$ processors, and each processor can make 10 billion tries per second. How long would it take them to find the key?

With some luck, you don't have to try all 2^{56} combinations before you hit the right one. Let us assume that you have to try half of them. The processor cluster makes $1,000 \cdot 10 \cdot 10^9 = 10^{13}$ tries per second. Let t be the time in seconds required to find the key. Then we have the equation

$$(10^{13}) \cdot t = \frac{1}{2} \cdot 2^{56} \quad \implies \quad t = \frac{2^{55}}{10^{13}} = \frac{2^{42}}{5^{13}} \simeq 3,603 \text{ sec.}$$

It would take the intruder only $3,603$ seconds, or about 1 hour, to crack the key. Even if we assume they need to try *all* possible sequences of length 56, it will take them about 2 hours. If we do not want the key to be vulnerable to brute-force attacks of this kind, we can try to make the key longer. How long should it be to ensure that it takes an adversary decades to crack?

Example 6.11. Assume an adversary has a cluster of $1,000$ processors, each capable of checking 10 billion keys per second. How long should the encryption key be to ensure that it takes more than 30 years for the adversary to crack the key by brute force?

Solution: Let k denote the length of the key. Assume as before that the intruder will have to check half of all possible keys before they hit the right one (actually, taking half or 75% won't make much of a difference for the answer). Then they will have to check $\frac{1}{2} \cdot 2^k$ keys. The time we want them to take is

$$t = 30 \text{ yr} = 10{,}950 \text{ days} = 262{,}800 \text{ h} = 94{,}608 \cdot 10^4 \text{ sec.}$$

So if we want to be safe, we will require that it takes them 10^9 seconds. The computer cluster can make $1{,}000 \cdot 10 \cdot 10^9 = 10^{13}$ tries/sec. We have the following equation for k:

$$10^{13} \cdot 10^9 = \frac{1}{2} \cdot 2^k \quad \Longrightarrow \quad 10^{22} = 2^{k-1}.$$

To solve for k, we need to take a logarithm. For example, we can take $\log_2$:

$$\log_2(10^{22}) = \log_2(2^{k-1}) \quad \Longrightarrow \quad 22 \log_2 10 = k - 1.$$

Then we compute:

$$k = 1 + 22 \log_2 10 \simeq 1 + 73.08 \simeq 74.$$

A 74-bit key would suffice to withstand an attack by this adversary for over 30 years.   □

  As technology progresses, the processing speed of the electronic devices is likely to increase. We can make the model more realistic if we assume that the adversary's processor speed grows, say, by a factor of 2 every two years (this growth is consistent with an observation of the hardware development known as Moore's law). This assumption will result in a more complicated equation for the length of an encryption key, which will involve a *finite geometric series*. This will be considered in Chapter 9 (see Example 9.6).

  We will discuss other mathematical aspects of cryptography and computer security in Chapter 13.

EXERCISES

Question 6.1. In what follows, $\log = \log_{10}$.

  1. Simplify, then compute: $\log\left(\frac{10^{11}}{100^{3.5}}\right)$.

  2. Compute $\log \sqrt[3]{1{,}000}$.

3. Solve for x: $10^{(7x+1)} = \sqrt[3]{100}$.

Question 6.2. In what follows, $\log = \log_{10}$.

1. Simplify, then compute: $\log(15^5) - \log(3^3) - \log(5^5)$.

2. Simplify, then compute: $\log 400 - \log 5 - 3\log 2$.

3. Solve for t: $\sqrt{30} \cdot 100^t = \frac{\sqrt{3}}{1,000}$.

Question 6.3. Solve the equations:

1. $e^{(2t+1)} = 18$.

2. $(0.5)^{(10x-8)} = 3$.

Question 6.4. Solve the equations:

1. $14^{3x} \cdot 7^{(x-3)} = 9 \cdot 2^{(3-x)}$.

2. $5^{(2x-2)} = 17 \cdot 25^{(-2x)}$.

Question 6.5. In what follows, $\log = \log_{10}$. Solve the equations:

1. $\log(x^2 - 16) - \log(x - 4) = 2$.

2. $\log(x^2 + 5x + 7) = 0$.

Question 6.6. Suppose you borrow $4,000$ from a bank at an annual interest rate of 10%. Suppose the interest accumulated over a certain time (which need not be an integer number of years) is 500. Find out how long this takes if the bank compounds (a) monthly, (b) quarterly. Give your answer up to the nearest month.

Question 6.7. How long will it take for an investment to triple at an interest rate of 5% compounded
(a) monthly?
(b) continuously (to the nearest tenth of a year)?

Question 6.8. A bank compounds quarterly at an annual interest rate of $r\%$. With this compounding scheme, the amount of $1,000$ grows to $1,400$ in 10 years.

1. Find the compound interest rate r up to two decimal digits of a percent.

2. With the same compounding scheme, how long will it take for a deposit of $20,000 to grow to $25,000? Give your answer to the nearest quarter of a year.

Question 6.9. Suppose a bank compounds continuously at an annual interest rate of 3.5%. How long will it take for an investment to grow by $10,000 if the initial amount is

1. $20,000?

2. $30,000?

3. $50,000?

Give your answers to the nearest month.

Question 6.10. Suppose a bank compounds monthly. How long will it take for a deposit of $5,000 to grow to $8,000 if the annual compound interest rate is

1. 3%?

2. 4.5%?

3. 5%?

Question 6.11. Suppose the adversary has a cluster of processors capable of making 10^{15} tries per second. You consider an encryption key sufficiently safe if it will take the adversary more than 3 years to crack by brute force. How long should the sequence of zeros and ones be to make a sufficiently safe encryption key? Assume that the adversary will have to check approximately half of all possible sequences before hitting the right one.

Question 6.12. You have received intelligence that currently the adversary has a cluster of 10,000 processors, each capable of making 5 billion tries per second, and that after 5 years the cluster capability will increase twenty-fold and will not change for the next 5 years. You consider an encryption key sufficiently safe if it would take the adversary more than 10 years to crack by brute force. How long should the sequence of zeros and ones be to make a sufficiently safe encryption key? Assume that the adversary will have to check approximately half of all possible sequences before hitting the right one.

Chapter 7

Logarithms II: Rescaling the world

Adders need logs to multiply – the most important property of the logarithms is their ability to convert multiplication into addition. This property lies in the foundation of many applications of logarithms in science, engineering, and everyday life. In particular, it allows us to represent very small and very large quantities together on one scale. Such scales are widely used in various fields – for example, to measure noise levels, earthquake intensity, and the relative apparent magnitude of stars. The same property enables the use of logarithmic scales in the tuning of some musical instruments, which we discuss at the end of this chapter.

MATH

Recall that the logarithm with base $a > 0, a \neq 1$ of a positive number X is defined as follows:

$$\log_a X = Y \iff a^Y = X.$$

The basic *law of logarithms*, derived in Chapter 6, claims that a logarithm (with any base) of a product of two positive numbers equals the sum of the logarithms (with the same base) of the factors:

$$\log_a(BC) = \log_a B + \log_a C,$$

where a, B, C are positive numbers, and $a \neq 1$.

In particular, it implies that the logarithm of a natural power of a number is proportional to the logarithm of the number itself:

$$\log_a(B^n) = n \log_a B.$$

Indeed,

$$\log_a(B^2) = \log_a B + \log_a B = 2 \log_a B,$$
$$\log_a(B^3) = \log_a(B^2) + \log_a B = \log_a B + \log_a B + \log_a B = 3 \log_a B,$$

and so on.

In fact, a more general statement holds: if a, B, C are positive numbers and r is any real number, then

$$\log_a(B^r) = r \log_a B$$

(see Chapter 6).

In particular, for any positive B and any real r we have for the decimal and natural logarithms:

$$\log(B^r) = r \log B \qquad (\log = \log_{10}),$$
$$\ln(B^r) = r \ln B \qquad (\ln = \log_e).$$

Example 7.1. Which is greater: $\log_2 32$ or $\log 10,000$?
Answer: $\log_2 32 = \log_2(2^5) = 5$, while $\log 10,000 = \log(10^4) = 4$. Therefore, $\log_2 32 > \log 10,000$. □

Example 7.2. *Base change formula.* Any calculator has an inbuilt function that computes logarithm base 10. But what if you need to compute a logarithm with an arbitrary positive base $a \neq 1$? Let us derive a formula that expresses $\log_a X$ in terms of $\log X$ for any positive number X. By definition of the logarithm,

$$Y = \log_a X \quad \Longleftrightarrow \quad X = a^Y; \quad \text{and} \quad Z = \log X \quad \Longleftrightarrow \quad X = 10^Z.$$

Then the values Y and Z are related by the equation

$$X = a^Y = 10^Z \quad \Longleftrightarrow \quad \log a^Y = \log 10^Z \quad \Longleftrightarrow \quad Y \log a = Z.$$

Therefore, we find

$$Y = \frac{Z}{\log a}, \quad \text{which answers our question:} \quad \log_a X = \frac{\log X}{\log a}.$$

To find a logarithm base a of a given positive number X, take its logarithm base 10 and divide by $\log a$. □

Taking a logarithm makes the difference between large numbers smaller and between small numbers larger. If two positive numbers differ by a factor of 10, their logs (base 10) differ by one unit, whether it is 0.0001 and 0.001, or $100,000$ and $1,000,000$:

$$\log(10B) = \log 10 + \log B = 1 + \log B.$$

As a consequence, the logarithms provide a tool to deal with data of very large range. For example, if the quantities $B = 100$ and $C = 0.01$ need to be marked on the same scale, then visualizing them together on a linear scale is problematic: the difference $B - C = 99.99$ is too large compared to $C = 0.01$. But on a logarithmic scale they become conveniently spaced: instead of C, we mark $\log C = -2$; instead of B, $\log B = 2$. Then the difference between them is only 4 units.

Figure 7.1: Logarithmic versus linear scale.

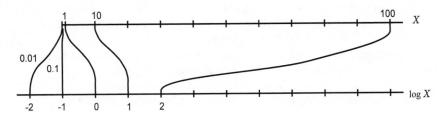

Tiny quantities become easier to discern, and large quantities can fit into the picture. This makes logarithmic scales convenient in applications with a wide range of data.

Practice problem 7.3. What is the difference between the quantities $A = 1$ and $B = 100,000$ on a logarithmic scale (a) base 10, (b) base 100? *Answer:* (a) $\log(100,000) = 5$, (b) $\log_{100}(100,000) = \frac{\log(100,000)}{\log(100)} = \frac{5}{2}$.

APPLICATIONS

Noise levels

Sounds are created by vibrating objects that generate *sound waves* transporting the energy of the vibration through the medium (for example, air).

In physics, the intensity of a sound is defined as the energy transported by the sound wave past a unit of area per unit of time. It is measured in joules per second per square meter ($\frac{\text{J}}{\text{sec} \cdot \text{m}^2}$), or watts per square meter ($\frac{\text{W}}{\text{m}^2}$). In practice the sounds accessible to detection by the human ear range in intensity from $10^{-12} \frac{\text{W}}{\text{m}^2}$ (the *threshold of hearing*) to $10 \frac{\text{W}}{\text{m}^2}$ (pain level). The measurement range of thirteen orders of magnitude suggests the use of a logarithmic scale.

Indeed, the *noise level* of a sound is measured according to the *decibel scale*, which is a logarithmic scale evaluating sound intensity with respect to that of the threshold of hearing (TOH). Given a sound intensity P (in watts per square meter), the decibel (dB) measure of the noise level is given by

$$L_{\text{dB}} = 10 \log \left(\frac{P}{P_0} \right),$$

where P_0 is the TOH intensity in watts per square meter.

The decibel scale is consistent with our perception of noise. For example, the sound intensity of a whisper is about 100 times the sound intensity of TOH: $P_{\text{whisper}} = 100 P_0$, but we do not think of a whisper as being 100 times louder than the tiniest noise we can hear – it is perceived to be just a little louder. Let us find the decibel level of a whisper:

$$L_{\text{whisper}} = 10 \log \left(\frac{P_{\text{whisper}}}{P_0} \right) = 10 \log 100 = 10 \cdot 2 = 20 \text{ dB}.$$

Here is a table of common sounds together with their estimated decibel levels:

Sound	Noise level (dB)
TOH	0
whisper	20
library	40
normal voice	60
lively street	70
nightclub	100
live rock concert	110
pain level	130

The numbers agree with our intuitive perception of the relative intensity of these sounds.

If we know the decibel level of a sound, we can find its intensity with respect to P_0:

$$L = 10 \log \left(\frac{P}{P_0} \right) \quad \Longrightarrow \quad \frac{L}{10} = \log \left(\frac{P}{P_0} \right) \quad \Longrightarrow \quad \frac{P}{P_0} = 10^{L/10}.$$

Example 7.4. Given that the noise level at a nightclub is approximately 100 decibels and that of the normal voice about 60 decibels, find the intensity ratio between the two noise levels.

Solution: Let $L_n = 100$ dB and $L_v = 60$ dB be the decibel noise levels of nightclub music and normal voice, respectively, and P_n and P_v their kilowatt intensities. Then

$$\frac{P_n}{P_0} = 10^{L_n/10} = 10^{10}; \qquad \frac{P_v}{P_0} = 10^{L_v/10} = 10^{6}.$$

Therefore,

$$\frac{P_n}{P_v} = \frac{10^{10}}{10^6} = 10^4 \simeq 10,000.$$

The intensity of a nightclub noise is approximately $10,000$ times that of a normal voice.   □

Apparent magnitude of stars

A logarithmic scale of relative *apparent magnitude of stars* has its origins in the work of an ancient Greek astronomer, Hipparchus (190–120 BC). He suggested dividing the visible stars into six levels of magnitude: 1 (the brightest) through 6 (the faintest), and postulated that the increase in one level of magnitude should correspond to a decrease by half in visible brightness. This way of describing the relative brightness of stars was further popularized by Ptolemy (90–168 AD) in his influential treatise *Almagest*. In the nineteenth century, British astronomer Norman Robert Pogson formalized the system by requiring that the stars of magnitude 1 were 100 times as bright as the stars of magnitude 6. An improved and extended version of this scale is used by astronomers today.[1]

Hipparchus' suggestion that the stars of equidistant magnitudes differ in brightness a certain number of times shows that human vision perceives brightness logarithmically. The same assumption lies in the foundation of the Pogson condition – that an increase in magnitude by 5 units should result in a 100-fold decrease in brightness.

Example 7.5. Given that an increase in magnitude by 5 units corresponds to the brightness ratio of $\frac{1}{100}$, find the base of a logarithmic scale suitable to describe the relative apparent magnitude of the stars.

---

[1] Interested readers can find more details in, for example, *To Measure the Sky: An Introduction to Observational Astronomy*, by Frederick R. Chromey, Cambridge University Press, 2010.

Answer: Let X be the base of the logarithm generating this scale. Then we have:

$$5 = \log_X \frac{1}{100} \quad \Longrightarrow \quad X^5 = \frac{1}{100}$$

$$\Longrightarrow \quad X = \sqrt[5]{\frac{1}{100}} = \frac{1}{\sqrt[5]{100}} = 10^{-\frac{2}{5}} \simeq 0.398.$$

$\square$

Indeed, according to the presently used scale, the apparent magnitude of a star is defined as

$$m = \log_{10^{-\frac{2}{5}}} \left(\frac{F}{F_0} \right),$$

where F is the observed flux (brightness) of the star. The flux is the amount of electromagnetic energy reaching the observer on Earth in unit of time per unit of area, measured in watts per square meter. The quantity F_0 is the flux (brightness) of the star Vega, which is taken as a reference point. The electromagnetic energy comes to the observer in the form of waves of various lengths: X-rays, ultraviolet, visible light, infrared, radio waves, and so on. The ratio of brightness between the sources depends on the wavelengths you take into account: star A can emit more energy in visible light, and star B in the infrared wavelength interval. For the purpose of apparent magnitude, to be consistent with the traditional definition, brightness is understood as the optical broadband flux: the energy emitted across the wavelengths of the visible spectrum.

Example 7.6. Reformulate the definition of the apparent magnitude of stars in terms of log base 10.
Answer: By the base change formula (Example 7.2) we have

$$m = \log_{10^{-\frac{2}{5}}} \left(\frac{F}{F_0} \right) = \frac{\log \left(\frac{F}{F_0} \right)}{\log 10^{-\frac{2}{5}}} = -\frac{5}{2} \log \left(\frac{F}{F_0} \right).$$

$\square$

From now on, we will use the formula for m derived in Example 7.6, which is more convenient for computations.

The scale of apparent magnitude is extended to apply to stars visible only by using a telescope, as well as to certain very bright objects like the planets, the Moon, and the Sun. Because of the negative sign in front of the log, this is an example of a reverse logarithmic scale: the brighter the star, the smaller its magnitude. Very bright objects have negative apparent magnitude. Some well-known celestial bodies and their apparent magnitudes are given in Figure 7.2.

Practice problem 7.7. Compute the apparent magnitude of Sirius, given that its observed flux is 3.87 times that of Vega, and check your answer against the value given in Figure 7.2.

Answer: $m_{\text{Sirius}} = -\frac{5}{2} \log \left(\frac{F_{\text{Sirius}}}{F_0} \right) = -\frac{5}{2} \log 3.87 \simeq -1.47$.

Conversely, given the apparent magnitude of a celestial object, we can find its brightness in the visible spectrum with respect to Vega:

$$m = -\frac{5}{2} \log \left(\frac{F}{F_0} \right) \quad \Longleftrightarrow \quad -\frac{2}{5} m = \log \left(\frac{F}{F_0} \right)$$

$$\Longleftrightarrow \quad \frac{F}{F_0} = 10^{-\frac{2}{5}m} \simeq 0.398^m.$$

Practice problem 7.8. If the apparent magnitude of the faintest stars visible in an urban neighborhood is $m = 4$, find their relative brightness with respect to Vega. *Answer:* $\frac{F}{F_0} = 10^{-\frac{2}{5}m} = 10^{-\frac{8}{5}} \simeq 0.025$. The brightness of these stars is about 2.5% of the brightness of Vega.

Example 7.9. The mean apparent magnitude of the full Moon is $m_{\text{M}} = -12.74$, and the apparent magnitude of Venus at its maximum brightness is $m_{\text{V}} = -4.89$. What is the ratio of brightness between the Moon and Venus?

Solution: We need to find $\frac{F_{\text{M}}}{F_{\text{V}}}$, where F_{M} is the visible brightness of the Moon, and F_{V} that of Venus. The known apparent magnitudes m_{M} and m_{V} allow us to find the ratios $\frac{F_{\text{M}}}{F_0}$ and $\frac{F_{\text{V}}}{F_0}$:

$$\frac{F_{\text{M}}}{F_0} = 10^{-\frac{2}{5}m_{\text{M}}} = 10^{-\frac{2}{5}\cdot(-12.74)} \simeq 124,738,$$

$$\frac{F_{\text{V}}}{F_0} = 10^{-\frac{2}{5}m_{\text{V}}} = 10^{-\frac{2}{5}\cdot(-4.89)} \simeq 90.$$

Then we can find

$$\frac{F_{\text{M}}}{F_{\text{V}}} = \frac{F_{\text{M}}}{F_0} \Big/ \frac{F_{\text{V}}}{F_0} \simeq \frac{124,738}{90} \simeq 1,380.$$

Another solution: We can use the properties of the logarithm to find the required ratio without first computing the brightness of each object with

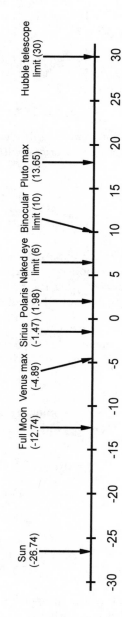

Figure 7.2: Apparent magnitude: stars, planets, and sensitivity limits.

respect to Vega. We have

$$m_M - m_V = -\frac{5}{2}\left(\log\left(\frac{F_M}{F_0}\right) - \log\left(\frac{F_V}{F_0}\right)\right) = -\frac{5}{2}\log\left(\frac{\frac{F_M}{F_0}}{\frac{F_V}{F_0}}\right)$$

$$= -\frac{5}{2}\log\left(\frac{F_M}{F_V}\right)$$

$$\implies -\frac{5}{2}\log\left(\frac{F_M}{F_V}\right) = m_M - m_V = -12.74 + 4.89 = -7.85.$$

Then

$$\log\left(\frac{F_M}{F_V}\right) = 3.14 \implies \frac{F_M}{F_V} = 10^{3.14} \simeq 1,380.$$

The Moon is approximately $1,380$ times as bright as Venus at its brightest.

□

Apparent magnitude gives us an idea of the relative brightness of stars as they appear to an observer on Earth. It does not tell us much about the actual brightness of a star because it does not take into account the distance to the star. The Sun only looks so bright because it is by far the closest star to Earth. To factor out the distance, the scale of *absolute magnitudes* was introduced. The absolute magnitude M of a star is the apparent magnitude the star would have if it is was placed at a distance of 10 parsecs (32.6 light-years) from the observer. The law of inverse squares allows us to calculate the flux (brightness) a star would have at 10 parsecs if its observed flux F and the distance from Earth d, in parsecs (pc), is known:

$$F_{10} = \frac{F}{(\frac{10}{d})^2} = \left(\frac{d}{10}\right)^2 F.$$

Then the absolute magnitude of the star is

$$M = -\frac{5}{2}\log\left(\frac{F_{10}}{F_0}\right) = -\frac{5}{2}\log\left(\frac{F\cdot(\frac{d}{10})^2}{F_0}\right)$$

$$= -\frac{5}{2}\log\left(\frac{F}{F_0}\right) - \frac{5}{2}\log\left(\frac{d}{10}\right)^2 = m - 5\log\frac{d}{10}.$$

Example 7.10. Find the absolute visible magnitude of the Sun if the distance from Earth to the Sun is approximately $4.85 \cdot 10^{-6}$ pc.

Answer: By definition of the absolute magnitude, we have

$$M_{Sun} = m_{Sun} - 5\log\frac{d}{10} = -26.74 - 5\log(4.85 \cdot 10^{-7})$$
$$\simeq -26.74 - 3.43 + 35 = 4.83.$$

If the Sun were placed at the "standard" distance of 10 parsecs from us, it would be just barely visible to the naked eye!   □

Even though the distance is factored out from the definition of the absolute magnitude, this quantity describes only the intensity of light emission by the star in the visible spectrum. It does not allow us to estimate other physical parameters of a star, such as size, surface temperature, or total energy emission. Some smaller stars emit more visible light than larger stars.

Earthquake magnitude

A logarithmic scale to measure the relative magnitude of earthquakes was suggested by the American scientist Charles F. Richter in 1934. An earthquake is a vibration of the ground, associated with seismic waves – elastic waves caused by a sudden break or movement of the Earth's crust. It can be characterized by the maximum amplitude of such vibrations, which is measured in units of length, and compared with the reference amplitude. The rough idea behind the *Richter scale* is to calculate the ratio of the amplitude of vibrations A measured at a distance δ from the epicenter, to a reference amplitude function $A_0(\delta)$, and take the logarithm base 10:

$$M_R = \log\left(\frac{A}{A_0(\delta)}\right).$$

The reference amplitude is chosen to be pretty small, so that an earthquake with maximal amplitude of vibrations $A_0(\delta)$ at distance δ from the epicenter is not felt and can be detected only by sensitive instruments. Observations from many seismic stations are combined to determine the position of the epicenter and the distance δ for each station. For convenience, seismographs (instruments used to measure earthquake magnitudes) have the reference amplitude function inbuilt.

In the 1970s the Richter scale was replaced by a more accurate *moment magnitude scale* (MMS), but the principle of the computation remains the same and the results agree whenever both scales are applicable, so that the MMS magnitude is often referred to as the Richter magnitude.

The use of a logarithm base 10 ensures that the increase by one unit in magnitude corresponds to a tenfold increase of the amplitude of an earthquake. Here is a comparative table of earthquake magnitudes.[2]

Magnitude	Effect	Number/year worldwide	Notable earthquake
< 2		$1,000,000$	
2 to 2.9	Not felt	$100,000$	
3 to 3.9	Minor	$12,000$	
4 to 4.9	Light damage	$2,000$	
5 to 5.9	Moderate damage	200	Long Island, NY, 1884 (5.5)
6 to 6.9	Strong damage; loss of life	20	South Napa, CA, 2014 (6.0)
7 to 7.9	Major; severe damage	3	San Francisco, CA, 1906 (7.9)
≥ 8	Great; total destruction and massive loss of life	< 1	Japan, 2011 (9.0) Chile, 1960 (9.5)

An empirical formula known in geophysics states that the energy of the seismic wave is proportional to the $\frac{3}{2}$ power of the amplitude: $E \sim A^{3/2}$. It follows that an increase by 1 unit in the magnitude corresponds to $10^{3/2} \simeq 31.6$ times the increase of energy released in the earthquake.

Example 7.11. A detonation of about 120 pounds of explosive produces seismic waves of magnitude 2. The magnitude of the 1906 earthquake in San Francisco was 7.9. What is the ratio of the energy released between the San Francisco earthquake and a detonation of 120 pounds of explosive?

Solution: Let A_{SF} and E_{SF} denote the amplitude and energy of the earthquake and A_{exp} and E_{exp} of the explosion. Then

$$\log(A_{\mathrm{SF}}) - \log(A_{\mathrm{exp}}) = \log\left(\frac{A_{\mathrm{SF}}}{A_{\mathrm{exp}}}\right) = 7.9 - 2 = 5.9.$$

$$\frac{A_{\mathrm{SF}}}{A_{\mathrm{exp}}} = 10^{5.9} \quad \Longrightarrow \quad \frac{E_{\mathrm{SF}}}{E_{\mathrm{exp}}} = \left(\frac{A_{\mathrm{SF}}}{A_{\mathrm{exp}}}\right)^{3/2} = (10^{5.9})^{3/2} \simeq 7 \cdot 10^8.$$

There was approximately 700 million times as much energy released in the earthquake as in the explosion. □

---

[2]Source: US Geological Survey.

Egocentric scales

In a lighter vein, we can say that logarithms were invented so that we can keep pretending to be the center of the universe. In his famous illustration for *The New Yorker* magazine titled *A View of the World from 9th Avenue* (1976),[3] Saul Steinberg showed a whimsical map of a large part of the western hemisphere – from 9th Avenue in Manhattan all the way to Japan – as seen by a self-absorbed New Yorker. In particular, the distance from 9th to 10th Avenue is about the same as the distance from 10th avenue to the West Coast of the US. To a mathematician, this perception of distances is clearly logarithmic. Let us compute the base of the logarithm a that could have been used to produce such a scale.

Suppose that the distances are measured westward from a certain reference point, and let x denote the distance from this point to 9th Avenue in Manhattan. Taking x to be the reference distance makes 9th Avenue the zero point of our scale:

$$\log_a \frac{x}{x} = \log_a 1 = 0.$$

Figure 7.3: Logarithmic scale of distances as viewed from 9th Avenue.

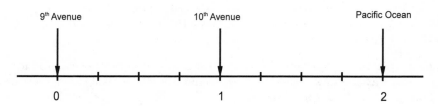

Let Y and Z denote the distances from 9th Avenue to 10th Avenue and the Pacific Ocean, respectively. These distances are known: $Y \simeq 875\,\text{ft} = 0.166\,\text{mi}$, and $Z \simeq 2,700\,\text{mi}$. Then our logarithmic scale should satisfy the following conditions:

$$\log_a \frac{Y + x}{x} = 1, \qquad \log_a \frac{Z + x}{x} = 2.$$

Therefore,

$$\log_a \frac{Z + x}{x} - \log_a \frac{Y + x}{x} = \log_a \frac{Z + x}{Y + x} = 2 - 1 = 1.$$

---

[3] You can find the image, for example, at http://www.saulsteinbergfoundation.org/ gallery_24_viewofworld.html.

From here we find:

$$\frac{Y+x}{x} = \frac{Z+x}{Y+x} = a \qquad \Longrightarrow \qquad \frac{0.166+x}{x} = \frac{2,700+x}{0.166+x} = a.$$

The proportion leads to a quadratic equation for x. In fact, the equation is linear because the x^2 term cancels:

$$0.166^2 + 2 \cdot 0.166 \cdot x + x^2 = 2,700x + x^2$$

$$\Longrightarrow \quad x \simeq \frac{0.0276}{2,699.7} \simeq 0.00001 \, \text{mi} \simeq 0.63 \, \text{in}.$$

So our self-centered Manhattanite measures all distances from a point 0.63 inches to the east of 9th Avenue (we can assume that this is how close his armchair is to the window). The base of the logarithm is

$$a = \frac{2,700 + 0.00001}{0.166 + 0.00001} \simeq 16,264.$$

To find where any given point is situated according to the 9th Avenue worldview scale ($d_{\text{9th Ave}}$), take the distance d from it to the reference point on 9th Avenue in miles, divide it by 0.00001 miles, and take the logarithm base $16,264$.

$$d_{\text{9th Ave}} = \log_{16,264} \frac{d}{0.00001}.$$

Using the base change formula (see Example 7.2), the formula can be rewritten as

$$d_{\text{9th Ave}} = \frac{\log \frac{d}{0.00001}}{\log 16,264} \simeq \frac{5 + \log d}{4.21},$$

where the distance d should be taken in miles.

Example 7.12. How far is Japan from 9th Avenue in New York, according to the 9th Avenue worldview scale?

Solution: The distance from New York to the east coast of Japan is approximately $d = 6,730$ miles. Therefore,

$$d_{\text{9th Ave}}(\text{Japan}) = \frac{5 + \log 6,730}{4.21} \simeq 2.097.$$

Recall that the distance to the West Coast according to this scale is 2. This is why the Pacific Ocean looks like a narrow strip of blue in Saul Steinberg's picture.   □

The illustration, initially used as the cover for the March 29, 1976, issue of *The New Yorker*, gave rise to multiple imitations and parodies.

Equal temperament scale in music

In all examples we have encountered so far, logarithmic scales were used mainly for the purpose of accommodating data with a very large range on a single scale. Now we will discuss an application where a logarithmic scale is used primarily for its *multiplicative uniformity*: the property of transforming equal ratios into equal distances.

Traditional European music is based on a seven-pitch octave scale with ratios between the frequencies of the pitches given by ratios of small integers. For example, the octave interval has the ratio of frequencies 2 : 1; perfect fifth 3 : 2; perfect fourth 4 : 3; major third 5 : 4; minor third 6 : 5. It is believed that our preference for such intervals originates from their presence in nature, which can be explained by the physics of basic natural sounds. The simplest natural drums and whistles that ancient people could have heard (a stick hitting on dry fallen wood, wind blowing in a wooden pipe) produce exactly these kinds of intervals. As early as the third millennium BC, the Babylonian lyre is believed to have had its first seven strings tuned to represent some of the perfect intervals. In classical antiquity there is evidence of the presence of a perfect fourth and major third in the tuning of the strings of the Greek lyre. The heptatonic (seven-pitch) major scale $C - D - E - F - G - A - B - C^2$ and various versions of heptatonic minor scales gradually gained popularity in European music mainly because of the richness of the harmonic intervals they included.

Figure 7.4: Heptatonic major scale and some of its perfect intervals.

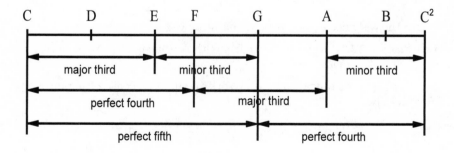

Example 7.13. If the ratio of frequencies between the fourth and the first pitch in the heptatonic major scale is a perfect fourth $\frac{f_F}{f_C} = \frac{4}{3}$, and between the sixth and the fourth – a major third $\frac{f_A}{f_F} = \frac{5}{4}$, then what

should be the ratio of frequencies between the eighth and the sixth pitch to make the eighth pitch an octave up from the first?

Solution: We need to find the ratio of frequencies $\frac{f_{C^2}}{f_A}$. The ratios should satisfy the equation:

$$2 = \frac{f_{C^2}}{f_C} = \frac{f_{C^2}}{f_A} \cdot \frac{f_A}{f_F} \cdot \frac{f_F}{f_C} = \frac{f_{C^2}}{f_A} \cdot \frac{4}{3} \cdot \frac{5}{4} = \frac{f_{C^2}}{f_A} \cdot \frac{5}{3}.$$

From here we find

$$\frac{f_{C^2}}{f_A} = 2 \cdot \frac{3}{5} = \frac{6}{5}.$$

Therefore, the interval between the sixth and the eighth pitch has to be the minor third. ☐

But the heptatonic major scale has a serious drawback: it does not behave well under translations. If you start playing a tune with a pitch several steps up or down from the original starting pitch and then try to repeat the same sequence of pitches with respect to the new starting pitch, you will hear a totally different song.

Example 7.14. Using the heptatonic major scale, try to reproduce the major third interval starting from the third pitch up.

Solution: The major third interval is achieved by skipping one pitch between C and E. If we do the same starting from the third pitch E, we arrive at the fifth pitch G. Let us compute the ratio of frequencies $\frac{f_G}{f_E}$. Because the ratio between the first and the fifth pitch is $\frac{f_G}{f_C} = \frac{3}{2}$, and between the first and the third $\frac{f_E}{f_C} = \frac{5}{4}$, then the ratio between the third and the fifth pitch is

$$\frac{f_G}{f_E} = \frac{3}{2} \cdot \frac{4}{5} = \frac{6}{5}.$$

Therefore, the third and the fifth pitch form a minor third, while the first and the third pitch form a major third, a different interval, even though there is exactly one skipped pitch in between in both cases. It is easy to check that the interval E-A will fall even farther away from the desired ratio of frequencies than E-G. The heptatonic major scale has no pitch to form a major third interval up from E. ☐

In other words, in the heptatonic major scale, it is impossible to translate a song a few pitches up or down; at best, it would require retuning the instrument. In the case of a pipe organ, shortening and lengthening of pipes would be needed because each pipe is built to produce a particular pitch. To overcome the difficulty, the first idea was to increase the number of pitches per octave to accommodate more intervals. However, if the intervals between the old seven pitches are kept perfect (this way of tuning is called *just intonation*), some other intervals are bound to be dissonant. An attempt to tune an instrument to produce perfect intervals in several keys (i.e., starting pitches) often resulted in a disaster if a performer deviated from the list of the allowed keys. One of the particularly dissonant chords produced this way was called the *wolf howl*. The scale is not translationally invariant because a just intonation tuning is not multiplicatively uniform: the ratios between each two consecutive frequencies are different from pitch to pitch.

It was due to the efforts of several musicians and mathematicians, notably Vincenzo Galilei, the father of Galileo Galilei (sixteenth century) and Simon Stevin (early seventeenth century), that a solution was found. The new scale satisfied the following requirements:

1. The scale is multiplicatively uniform, or logarithmically equidistant: the ratios between each two consecutive pitches are equal. This assures a translation invariance: any song can be translated a few pitches up or down, with the ratios of frequencies staying exactly the same. This way of tuning is called *equal temperament*.

2. Among the pitches in the new scale, the seven old ones can be identified such that their relative frequencies are as close as possible to the original heptatonic scale.

3. The number of pitches per octave should not be too large.

The perfection of intervals was traded for multiplicative uniformity. The resulting scale, the *twelve-tone equal temperament scale*, was popularized by J.-S. Bach in his *Well-Tempered Clavier*, a collection of pieces in each of the twenty-four new major and minor keys defined by the twelve new pitches. It became a standard tuning scale in European music in the seventeenth century and is still in use now, notably for the organ and the piano.

Example 7.15. Find the ratio of frequencies between two adjacent pitches in the twelve-tone equal temperament scale.

Solution: Let x be the required ratio. The twelve intervals should have equal ratios of frequencies and span the octave; therefore

$$x^{12} = 2, \quad \implies \quad x = \sqrt[12]{2} = 2^{\frac{1}{12}}.$$

The ratio of frequencies between each two consecutive pitches is $2^{\frac{1}{12}}$. Taking the logarithm base 2, we obtain a uniform logarithmic scale with step $\frac{1}{12}$. □

Thus, the octave contains twelve equal intervals, and the ratio between the first and the thirteenth pitch is exactly 2. Other perfect intervals suffer, but not substantially. The heptatonic major scale pitches C-D-E-F-G-A-B are approximated in the twelve-pitch scale by the first, third, fifth, sixth, eighth, tenth, and twelfth pitch, respectively.

Figure 7.5: Twelve-tone equal temperament scale: the white keys correspond to the pitches of the heptatonic major scale.

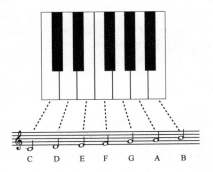

Example 7.16. Find the ratio of frequencies between the fifth and the first pitch of the twelve-tone equal temperament scale. Which of the perfect intervals does it approximate?

Solution: The ratio of frequencies between the kth and the lth pitch is

$$\frac{f_k}{f_l} = (\sqrt[12]{2})^{k-l}.$$

We compute:

$$\frac{f_{5th}}{f_{1st}} = (\sqrt[12]{2})^4 = 2^{\frac{1}{3}} \simeq 1.2599.$$

The closest perfect interval is the major third $\frac{5}{4} = 1.25$. The error of approximation is relatively small:

$$\frac{1.2599 - 1.25}{1.25} \simeq 0.0079, \quad \text{or} \quad 0.79\%.$$

□

The number 12 in the equal temperament scale is chosen because among the multiples of the 12th root of 2 one can find numbers that are fairly close to the desired ratios representing perfect intervals. Clearly, if we took a one-million-tone equal temperament scale, we would be able to approximate any perfect interval with much better precision than the twelve-tone scale allows. A natural question to ask is: what is the minimal number of equidistant pitches that would produce a better approximation to the heptatonic major scale than the twelve-tone one provides?

Here we need to specify what we mean by a better approximation. It can happen, for example, that the major third is approximated much better in a new scale, but the perfect fifth – not so well. There are methods developed in statistical analysis that provide a way to compare the quality of approximations. For instance, one fair approach is to list all important perfect intervals as represented in both scales, and take the sum of the squares of the errors with respect to the perfect ratios. The squares are taken because the errors may be positive or negative, but we don't want them to cancel in the sum. If the obtained number is smaller for the new scale than for the twelve-tone scale, we can claim that the new approximation is better.

In 1961, Joel Mandelbaum calculated that the next number of pitches per octave providing a better approximation to the perfect intervals is nineteen. In this equal temperament scale the ratio of frequencies of any two adjacent pitches is $\sqrt[19]{2}$. To get still further improvement, you have to take thirty-one pitches per octave![4] In fact, it turns out that a nineteen-tone scale was known and music was composed for it already in the sixteenth century. In the nineteenth century, mathematician and musical theorist Wesley Woolhouse advocated for the use of an equal temperament nineteen-tone scale for its precision.

Example 7.17. Find the pitches approximating the perfect fifth and the major third in the 19-tone equal temperament scale, and their relative precision.

Solution: We need to find the power of $\sqrt[19]{2}$ closest to $\frac{3}{2}$ and $\frac{5}{4}$. We have:

$$2^{k/19} = \frac{3}{2} = 1.5 \quad \Longrightarrow \quad k \ln 2 = 19 \ln 1.5 \quad \Longrightarrow \quad k \simeq 11.11.$$

---

[4]Unaware of this work, I tried to answer the same question in 1987, during my senior year in high school, and obtained the numbers 19 and 31. This was my first attempt at mathematical research – and I was happy to learn, years later, that my conclusions agreed with a well-known result. (*A.L.*)

The perfect fifth is realized by the twelfth pitch of the scale (11 steps from the first pitch). The relative error is

$$\left| \frac{2^{\frac{11}{19}} - 1.5}{1.5} \right| \simeq 0.0042 = 0.42\%.$$

For the major third, we have

$$2^{l/19} = \frac{5}{4} = 1.25 \quad \Longrightarrow \quad l \ln 2 = 19 \ln 1.25 \quad \Longrightarrow \quad l \simeq 6.12.$$

The major third is realized by the seventh pitch of the scale (6 steps from the first pitch). The relative error is

$$\left| \frac{2^{\frac{6}{19}} - 1.25}{1.25} \right| \simeq 0.0042 = 0.42\%.$$

$\square$

Music is not necessarily a competition in precision. Perfect intervals are pleasing, but other combinations of sounds can open new attractive possibilities for musicians. Many non-equal temperament scales are still being used in different parts of the world. Music is being written for the twelve-tone, fifteen-tone, seventeen-tone, as well as nineteen-tone scales. Mathematics can answer a question only within a given model – whether or not this model represents the best desirable result is up to us to decide.

EXERCISES

Question 7.1. Use $\log = \log_{10}$ or $\ln = \log_e$, the base change formula, and your calculator to solve the following equations or compute the following expressions.

1. Solve for t: $10^{3t+1} = 80$.

2. $\log_{13.5} 7,300$.

3. $\log_{\sqrt{e}} 80$.

Question 7.2. Same instructions as for the previous question.

1. Solve for y: $18 \cdot 2^{3y} = 9$.

2. $\log_{10^{15}} 10^{23}$.

3. $\log_{8^{-8}} 1,024$.

Question 7.3. The sound intensity of a vacuum cleaner is about 31.6 times higher than that of a normal voice. If the decibel level of a normal voice is 60, find the decibel level of the vacuum cleaner.

Question 7.4. Rustling leaves make a faint noise of about 15 decibels. How many times more intense is the sound of a noisy urban environment with a noise level of about 80 decibels?

Question 7.5. The apparent magnitudes of the Sun and the Moon are $m_S = -26.74$ and $m_M = -12.74$. What is the ratio of brightness between the Sun and the Moon?

Question 7.6. The ratio of brightness of Aldebaran to that of Vega is 0.46.

1. Find the apparent magnitude of Aldebaran.

2. Find the absolute magnitude of Aldebaran if its distance from Earth is 18.4 parsecs.

Question 7.7. Sirius has an apparent magnitude $m_S = -1.46$, Rigel $m_R = 0.12$, and Canopus $m_C = -0.72$. Rank the stars according to their absolute magnitude if the distance from Earth to Sirius is 2.6 parsecs, to Rigel 430 parsecs, and to Canopus 22.7 parsecs.

Question 7.8. The maximum brightness of Neptune is 0.076% that of Vega. Is it visible

1. to the naked eye (apparent magnitude sensitivity limit 6)?

2. with common 7×50 binoculars (apparent magnitude sensitivity limit 10)?

Question 7.9. The absolute magnitude of the Hale-Bopp comet, also known as the Great Comet of 1997, is -2.7. What is the maximal distance from Earth at which this comet can be visible to the naked eye (apparent magnitude sensitivity limit 6)?

Question 7.10. The magnitude of the 2013 earthquake on Santa Cruz Island was 8.0, and the magnitude of the 2005 earthquake in northwestern Tennessee was 3.9. Find the ratio of the energies released in the Santa Cruz and the Tennessee earthquakes.

Question 7.11. Find the base of the logarithmic timescale in which 0 corresponds to the present moment, 1 to the moment 3 days in the future, and 2 to the moment 1 year in the future.

Question 7.12. The major sixth is the interval between the first and the sixth pitches (C-A) in the heptatonic major scale.

1. If the interval A-C^2 in the heptatonic major scale is a minor third with a ratio of frequencies $\frac{6}{5}$, what is the ratio of frequencies corresponding to the major sixth interval C-A?

2. Find the pitch that approximates A in the twelve-tone equal temperament scale. What is the relative error of approximation?

Question 7.13.

1. Compare the approximations of a perfect fourth interval (ratio of frequencies $\frac{4}{3}$) provided by each of the twelve-tone, fifteen-tone, and nineteen-tone equal temperament scales.

2. Compare the approximations of a minor third interval (ratio of frequencies $\frac{6}{5}$) provided by each of the twelve-tone, seventeen-tone, and nineteen-tone equal temperament scales.

Question 7.14. Which pitches represent the pitches E, F, and G in the 31-tone equal temperament scale? Find the relative errors of approximation for the intervals C-E $\left(\frac{5}{4}\right)$, C-F $\left(\frac{4}{3}\right)$ and C-G $\left(\frac{3}{2}\right)$.

Chapter 8

e : The queen of growth and decay

Exponential growth and decay are among the simplest and most commonly used mathematical models in science. The models are based on the properties of exponents and can be used to describe many natural phenomena – in particular, population growth and radioactive decay. The decay of a radioactive carbon isotope ^{14}C lies at the foundation of the radiocarbon dating method, a valuable tool in history, archaeology, and paleontology.

MATH

The processes of *exponential growth and decay* are described by a single formula that shows the dependence of a variable $A(t)$ on time:

$$A(t) = A(0)e^{kt}.$$

Here $A(t)$ is the quantity at time t, $A(0)$ the initial quantity, and k a nonzero constant, called the *relative growth rate* (if $k > 0$), or the *relative rate of decay* (if $k < 0$). The number e is the universal constant approximately equal to $2.718281828459045\ldots$ We introduced it in Chapter 5 as the base of the exponent in the continuous compounding model. This remarkable number is also the base of the natural logarithm $\ln = \log_e$ (see Chapter 6).

To see how one formula can describe both growth and decay, we need to understand the following property of exponents: if base a is greater than 1, then

$$a^x > 1, \qquad x > 0,$$
$$a^x = 1, \qquad x = 0,$$
$$a^x < 1, \qquad x < 0.$$

Start with the first line. If $a > 1$ and n is a natural number, then $a^n > 1$ is a product of several numbers that are greater than 1. For the same reason, it is easy to observe that $\sqrt[m]{a} > 1$ for any natural m (see Question 8.1 at the end of this chapter). It follows that $a^{\frac{n}{m}} > 1$ for any natural n and m, or equivalently $a^x > 1$ for all rational $x > 0$. The fact that this inequality holds for all positive x is intuitively clear, because the rational numbers are spread densely everywhere on the number line. A rigorous proof requires calculus.

The second line follows from the laws of exponents (see Chapter 5): for any number y, we have $a^0 = a^{y-y} = \frac{a^y}{a^y} = 1$.

Finally, the third line follows from the first. If $x < 0$, then $-x > 0$ and $a^{-x} = \frac{1}{a^x} > 1$, which implies $a^x < 1$. $\qquad\qquad\square$

Note that if $a > 1$, the function a^x grows very fast as x increases. For example, Figure 8.1 shows the comparative growth of the exponential 2^x and linear function $2x$ for $x \geq 0$.

Figure 8.1: Exponential versus linear growth.

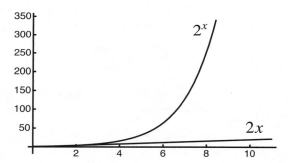

Now we can examine the behavior of the exponential model $A(t) = A(0)e^{kt}$ with respect to the sign of k.

If $k > 0$, then for $t_2 > t_1$ we have $A(t_2)/A(t_1) = \frac{A(0)e^{kt_2}}{A(0)e^{kt_1}} = e^{k(t_2-t_1)} > 1$. This is a growth model, where k determines the rate of growth.

On the other hand, if $k < 0$, then for $t_2 > t_1$ we have $A(t_2)/A(t_1) = e^{k(t_2-t_1)} < 1$, and we observe a decay, where k determines the rate of decay. $\qquad\qquad\square$

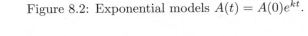

Figure 8.2: Exponential models $A(t) = A(0)e^{kt}$.

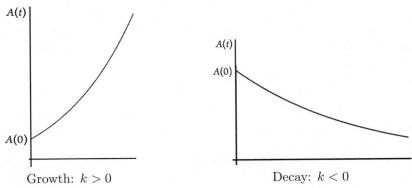

Growth: $k > 0$ Decay: $k < 0$

What kind of processes can be described by an exponential model? If a quantity grows or decreases by a fixed factor in equal intervals of time, the process is exponential and the exponential model can be applied. Let us see how this works.

Suppose the quantity $A(t)$ is such that it grows (or decreases) by a factor p in each interval of time T. Then we have

$$
\begin{array}{ll}
t = 0 & A(0) \\
t = T & pA(0) \\
t = 2T & p^2 A(0) \\
t = 3T & p^3 A(0) \\
\cdots & \cdots \\
t = nT & p^n A(0)
\end{array}
$$

We claim that the same process can be described by the exponential model. Let us find its parameters. The initial quantity is given to be $A(0)$. To find k, we have the equation

$$A(nT) = p^n A(0) = A(0)e^{knT} \implies p^n = e^{knT} \implies$$

$$p^n = (e^{kT})^n \implies p = e^{kT}.$$

Taking the natural logarithm of both sides (see Chapter 6), we get

$$\ln p = kT \iff k = \frac{\ln p}{T}.$$

Therefore, knowing the parameters p and T, we can uniquely determine k and use the exponential formula to model the process:

$$A(t) = A(0)e^{kt} = A(0)e^{\frac{\ln p}{T}t}.$$

Conversely, if a process is described by an exponential model, we can find how often the quantity grows or decreases by any given factor p:

$$T = \frac{\ln p}{k}.$$

Two important special cases of this formula are $p = 2$ (in this case the time interval is called the *doubling time*), and $p = \frac{1}{2}$ (in this case the time interval is called the *half-life*).

The formula shows that the units of k are inverse to the units of time: if the time is measured in minutes, then k is per minute; if the time is in years, then k is per year. It is important to remember that the value of k, which characterizes an exponential process, depends on the units of measurement.

Example 8.1. A quantity decreases exponentially with the relative rate of decay $k = -\frac{1}{100}$ per day. Find the interval of time needed for the quantity to get reduced by a factor of 5.

Solution: In this case, $k = -\frac{1}{100}$ and $p = \frac{1}{5}$. Then we have

$$T = \frac{\ln \frac{1}{5}}{-\frac{1}{100}} = -100(\ln 1 - \ln 5) = -100(-\ln 5) = 100\ln 5 \simeq 161 \text{ days}.$$

In the computation, we used the property of the logarithm to convert a quotient into a difference, and the fact that $\ln 1 = 0$ (see Chapter 6). $\square$

APPLICATIONS

Population growth models

Most populations develop according to an exponential model. The parameter commonly used to characterize the process is the *doubling time* – the time needed for a population to double. If the doubling time is given, we know that the process is exponential and can determine the parameters of the exponential model.

Suppose that a population of a certain species doubles every T_d years (i.e., has the doubling time T_d). Then

$$A(n \cdot T_d) = 2^n A(0),$$

where $A(0)$ is the initial population. We know by the above analysis that the process can be modeled as exponential growth:

$$A(t) = A(0)e^{kt}, \quad \text{where} \quad k = \frac{\ln 2}{T_d}.$$

This formula allows us to determine the population $A(t)$ at any moment of time t in the future. The parameter k shows the growth rate of a population relative to the population size.

Conversely, if the relative growth rate k is given (say, per year), then the doubling time T_d (in years) is determined by the formula

$$T_d = \frac{\ln 2}{k}.$$

The units of measurement of k, T_d, and t are related. In the above example, T_d was given in years; therefore k shows the relative growth per year, and the time t in the exponential model should be taken in years. If T_d was given in minutes, then k would have been per minute, and t should be taken in minutes.

Example 8.2. A bacteria culture is observed to grow exponentially in time, with a relative growth rate $k = 0.0462/\text{h}$. If there are $2,000$ bacteria at the start of the experiment, we would like to

1. find out how many hours it takes for the bacteria to reach $10,000$.

2. compute the doubling time, i.e., how long it takes to reach $4,000$.

Solution: Because the growth model is exponential, we have the equation $A(t) = A_0 e^{kt}$, where $A_0 = 2,000$ and $k = 0.0462$.

To answer the first question, we write:

$$10,000 = 2,000e^{0.0462t} \quad \Longrightarrow \quad e^{0.0462t} = 5.$$

Taking the natural logarithm of both sides and solving for t, we get the answer:

$$0.0462t = \ln(5) = 1.609 \quad \Longrightarrow \quad t = \ln(5)/0.0462 = 34.86 \text{ h}.$$

To answer the second question, we use the formula that relates T_d with k:

$$T_d = \frac{\ln 2}{k} = \frac{\ln 2}{0.0462} \simeq \frac{0.69315}{0.0462} \simeq 15.003,$$

which means that the doubling time is roughly 15 hours.   □

Practice problem 8.3. In the conditions of Example 8.2, find

1. how long it takes for the bacteria to grow by half, i.e., to reach $3,000$.
   Answer: 8.78 hours.

2. how long it takes for the bacteria to grow by a factor of 8.
   Answer: 45.01 hours.

Let us apply the exponential model to the historically recorded population surge of rabbits in Australia in the absence of natural predators.

Example 8.4. Twenty-four rabbits were brought from England to western Australia for hunting purposes in 1859. In 1920, there were estimated to be ten billion rabbits in the area.[1]

1. Find the relative growth rate per year for this population.

2. Find the growth factor of the population per year.

Solution: Let k_{Aus} denote the required relative growth rate of the population. There are 61 years between the known population records $A(0) = 24$ and $A(61) = 10^{10}$. The exponential model provides the equation for k_{Aus}:

$$A(61) = A(0)e^{k_{Aus} \cdot t} \quad \implies \quad 10^{10} = 24e^{61k_{Aus}}$$
$$\implies \quad 61k_{Aus} = 10\ln 10 - \ln 24 \simeq 19.85.$$

From here we find that

$$k_{Aus} \simeq \frac{19.85}{61} = 0.3254 \,/\text{yr}.$$

The growth factor per year is determined by the formula $p = e^{k_{Aus}T}$, where T should be taken to be 1 year. Then we have

$$p = e^{k_{Aus} \cdot 1} = e^{0.3254} \simeq 1.3846.$$

The population was growing each year by approximately 38%.   □

---

[1]Source: Geography.About.com.

The growth factor of 38% per year is pretty high, and the damages inflicted by the rabbits to the crops in Australia were growing at the same rate. Various measures were taken in the twentieth century to control the bunny invasion, and the relative growth rate dropped significantly. However, keeping the rabbit threat under control is a persistent concern of Australian farmers.

In the next example, we will try to apply the exponential model to the human population.

Example 8.5. The world population was 2,560 million people in 1950 and 3,040 million in 1960. Assuming that the population grows exponentially, find the relative growth rate per year, and calculate the world population in 2010.

Solution: Set $t = 0$ in 1950. Then $A(0) = 2,560 \cdot 10^6 = 2.56 \cdot 10^9$, and $A(10) = 3.04 \cdot 10^9$. According to the exponential growth model, we have

$$A(10) = A(0)e^{10k} \quad \Longrightarrow \quad 3.04 \cdot 10^9 = 2.56 \cdot 10^9 \cdot e^{10k} \quad \Longrightarrow \quad \frac{3.04}{2.56} = e^{10k}.$$

Applying the natural logarithm, we find

$$\ln\left(\frac{3.04}{2.56}\right) = 10k \quad \Longrightarrow \quad k = \frac{1}{10}\ln\left(\frac{3.04}{2.56}\right) \simeq 0.017185.$$

Therefore, the relative growth rate per year of the world population, as determined by the data for 1950 and 1960, is $k = 0.017185$.

Now, the time elapsed between 1950 and 2010 is 60 years. The exponential growth model gives

$$A(2010) = A(1950)e^{0.017185 \cdot 60} = 2.56 \cdot 10^9 \cdot e^{0.017185 \cdot 60} \simeq 7.18 \cdot 10^9.$$

We can obtain the same answer based on the population data for 1960 and the same k. Then the time elapsed is 50 years, and the initial population is $3.04 \cdot 10^9$.

$$A(2010) = A(1960)e^{0.017185 \cdot 50} = 3.04 \cdot 10^9 \cdot e^{0.017185 \cdot 50} \simeq 7.18 \cdot 10^9.$$

$\square$

In fact, the world population in 2010 was approximately 6.896 billion people. The difference between the actual and the predicted number might have occurred for many reasons. First, we based our calculation on only two data points, from 1950 and 1960. It would have been wiser to find relative growth rates based on several pairs of data points and take the average. We also assumed that the relative growth rate for the world population was constant, while, in the reality, it can change. The relative growth rate is the coefficient of proportionality between the rate of growth of a population per unit of time and the size of the population. For example, the advances of medical care in many developing countries could have reduced the birthrate, and therefore the relative growth rate in the world.

"Essentially, all models are wrong, but some are useful," said the famous statistician George E.P. Box.[2] It is important to remember that any mathematical model is just a model; it gives correct answers only if all the assumptions of the model are satisfied in reality. You should decide the question of applicability in each particular case separately. In general, the more complicated the system, the harder it is to formulate an adequate model. For example, the exponential growth model describes reproduction of bacteria in a lab fairly adequately, but when dealing with more complex systems such as human populations, multiple factors should be taken into account. For example, changing conditions of life – advances of medicine, wars, natural catastrophes – can contribute to birthrate and life expectation and make the system more complicated. Still, simple models like exponential growth are useful because they quickly provide a rough estimate of the actual process.

Fibonacci rabbits*

Recall the Fibonacci sequence that we discussed in Chapter 4:

$$1, \ 1, \ 2, \ 3, \ 5, \ 8, \ 13, \ 21, \ 34, \ \ldots, \ f_n = f_{n-1} + f_{n-2}, \ \ldots$$

This sequence was introduced by Fibonacci in his book *Liber Abaci* (*The Book of Calculation*) in the context of a population growth model. Namely, the numbers in the sequence were defined as the numbers of couples of rabbits in a population satisfying the following assumptions:

1. A couple of rabbits produces another couple of rabbits each month, starting from the age of two months.

2. The rabbits live forever.

Clearly, immortal rabbits are biologically unrealistic, but they define a new growth model that leads to the Fibonacci sequence. We want to

---

[2]G.E.P. Box and N.R. Draper, *Empirical Model Building and Response Surfaces*, John Wiley & Sons, New York, 1987.

know if this model is exponential. If we can show that the population growth of the Fibonacci rabbits is exponential, we should be able to find their relative growth rate.

First, let us see why the assumptions listed above lead to the Fibonacci sequence in the numbers of couples of rabbits. The table below shows these numbers at the beginning of each month.

Time (mo)	New couples	Old couples	Total couples
0	1	0	1
1	0	1	1
2	1	1	2
3	1	2	3
4	2	3	5
5	3	5	8
...	...	...	...
$n-1$	f_{n-2}	f_{n-1}	$f_n = f_{n-2} + f_{n-1}$
n	f_{n-1}	$f_n = f_{n-2} + f_{n-1}$	$f_{n+1} = f_{n-1} + f_n$

A couple is new if it is younger than one month. After one month, all new couples become old and get moved to the next column (the number in "Old couples" is the sum of those in "Old couples" and "New couples" one line above). Each old couple produces a new couple at the beginning of the next month (the number in "New couples" is copied from that in "Old couples" one line above). The total number of couples at any time is the sum of the new and the old couples. There is just one couple of rabbits at the beginning. The table shows how these rules lead to the Fibonacci law:

$$f_1 = 1, \quad f_2 = 1, \quad f_n = f_{n-2} + f_{n-1}.$$

Is this an exponential growth? To answer this question, we need a formula that expresses, or at least approximates, the Fibonacci number f_n as it depends on n. Luckily, such a formula exists:

$$f_n = \frac{\phi^n - \left(-\frac{1}{\phi}\right)^n}{\sqrt{5}},$$

where $\phi = \frac{1+\sqrt{5}}{2}$ is the golden ratio (see Chapter 4). This formula is usually referred to as Binet's formula, in honor of the French mathematician J.P.M. Binet (1786–1856).

Example 8.6. Derive Binet's formula for f_n from the following two relations:

$$\phi^n = f_n \phi + f_{n-1}$$

(see Question 4.12), and

$$\left(-\frac{1}{\phi}\right)^n = f_{n-1} - f_n \frac{1}{\phi}$$

(see Question 4.13).

Solution: Subtracting the second equality from the first, we can get rid of the term f_{n-1}:

$$\phi^n - \left(-\frac{1}{\phi}\right)^n = f_n \left(\phi + \frac{1}{\phi}\right).$$

The coefficient $(\phi + \frac{1}{\phi})$ can be computed using the definition of ϕ and the properties of square roots:

$$\phi + \frac{1}{\phi} = \frac{\sqrt{5}+1}{2} + \frac{2}{\sqrt{5}+1} = \frac{(5+2\sqrt{5}+1)+4}{2(\sqrt{5}+1)} = \frac{2(5+\sqrt{5})}{2(1+\sqrt{5})}$$

$$= \frac{\sqrt{5}(\sqrt{5}+1)}{1+\sqrt{5}} = \sqrt{5}.$$

Then

$$\phi^n - \left(-\frac{1}{\phi}\right)^n = \sqrt{5} f_n.$$

Dividing both sides by the square root of 5, we obtain Binet's formula for f_n. $\qquad\square$

Let t be the time in months, and $R(t)$ the number of Fibonacci rabbits in the population. If we recall that f_n counts the couples of rabbits, then for the number of individuals we have

$$R(t) = \frac{2}{\sqrt{5}} \left(\phi^t - \left(-\frac{1}{\phi}\right)^t\right).$$

We are interested in an estimate for the population growth when t is large. For this purpose, the term $\left(-\frac{1}{\phi}\right)^t$ can be discarded because it becomes negligibly small as t grows, because $0 < \frac{1}{\phi} < 1$. Thus we have

$$R(t) \simeq \frac{2}{\sqrt{5}} \phi^t, \quad \text{for large } t.$$

Now it is clear that the Fibonacci rabbit population grows exponentially. To find the relative growth rate k_F, we write

$$A(t) = A(0)e^{k_F t} = R(t) = \frac{2}{\sqrt{5}} \phi^t,$$

where t is measured in months (mo). Now we have $A(0) = \frac{2}{\sqrt{5}} \simeq 1$, and

$$e^{k_F t} = \phi^t \quad \Longrightarrow \quad e^{k_F} = \phi \quad \Longrightarrow \quad k_F = \ln \phi \simeq 0.4812 \,/\text{mo}.$$

The relative growth rate of the Fibonacci rabbit population is approximately 0.4812 per month. To see how grotesquely high this number is, let us compare the Fibonacci model with the actual growth of the population of rabbits in Australia (see Example 8.4).

Example 8.7. Find out how much the initial population of twenty-four rabbits would grow between 1859 and 1920 if it was developing according to the exponential model with the Fibonacci growth rate $k_F = \ln \phi \simeq 0.4812 \,/\text{mo}$.

Solution: By the exponential model we have $A(t) = 24e^{k_F t}$, where $t = 61 \cdot 12 = 732$ mo. Then

$$A(732) = 24e^{0.4812 \cdot 732} \simeq 24e^{352}.$$

If we try to type this into a calculator, most probably it will return an error message. The number is too big. To overcome this difficulty, we can rewrite the number using the properties of exponents (see Chapter 5):

$$A(732) \simeq 24(e^{35.2})^{10} \simeq 24(1.9 \cdot 10^{15})^{10} \simeq 24 \cdot 1.9^{10} \cdot 10^{150}$$
$$= 1.5 \cdot 10^4 \cdot 10^{150} = 1.5 \cdot 10^{154}.$$

Another solution: In the Fibonacci model, we have f_n couples of rabbits after n months. There were 12 couples of rabbits to start with. The Fibonacci numbers are known at least up to $n = 1,000$, in the form of their factorizations into products of prime factors.[3] In our case,

$A(732) = 12f_{732}$
$= 12 \cdot 2^4 \cdot 3^2 \cdot 1,097 \cdot 4,513 \cdot 19,763 \cdot 102,481 \cdot 10,225,307 \cdot$
$\quad \cdot 21,291,929 \cdot 21,791,641 \cdot 555,003,497 \cdot 14,686,239,709 \cdot$
$\quad \cdot 5,600,748,293,801 \cdot 24,848,660,119,363 \cdot 533,975,715,909,289 \cdot$
$\quad \cdot 14,297,347,971,975,757,800,833 \cdot$
$\quad \cdot 7,181,634,929,637,355,776,701,081,412,683,$

---

[3] Source: mersennus.net.

a number of the same order of magnitude.   □

This is an enormous number – about $(10\,\text{billion})^{15}$, compared to the estimated 10 billion in 1920 in Australia. To a large extent, this result is a
consequence of the fantastic assumption that the Fibonacci rabbits are immortal. The Fibonacci population model works better for small t (within a
rabbit's life span). More interestingly, it provides a tangible illustration of
the fact that the Fibonacci numbers grow exponentially, with the relative
growth rate equal to the natural logarithm of the golden ratio.

Radioactive decay

The decay of a radioactive substance is a consequence of the instability
of its atoms, which tend to collapse into a more stable state with the
emission of radiation. Radioactivity was discovered at the turn of the
twentieth century independently by Henri Becquerel (who worked with
samples of uranium), Ernest Rutherford, Paul Villard, and Pierre and
Marie Curie. Since this time, many chemical elements have been found to
have radioactive isotopes. The decay process of a radioactive substance is
reliably described by an exponential model with the negative parameter k
– the relative rate of decay:

$$A(t) = A(0)e^{kt}.$$

Here $A(t)$ is the quantity of a radioactive substance at time t, and $A(0)$ is
the initial quantity. The units of k are inverse to the units of t: if the time
is measured in hours, than k is per hour; if the time is in years, then k is
per year. With the parameter k negative, the quantity $A(t)$ after time t is
smaller than the initial quantity $A(0)$.

The *half-life* $T_{1/2}$ plays the role of the doubling time for the decay
model. It is defined as the time needed for a quantity of a radioactive
substance to decrease by half, and is often used as the parameter charac-
terizing the process. The relation between the relative rate of decay k and
the half-life $T_{1/2}$ is given by the formulas

$$T_{1/2} = -\frac{\ln 2}{k}, \qquad k = -\frac{\ln 2}{T_{1/2}}.$$

The half-lives of radioactive isotopes range from microseconds (e.g., for
polonium-213) to billions of years (e.g., for rubidium-87).

Practice problem 8.8. The half-life of the radioactive substance A is three times the half-life of substance B. How much longer will it take for A, than for B, to decay to $1/5$ of its initial mass? *Answer:* Three times as long.

Example 8.9. The mass of a sample of radioactive radium-226 reduced from 100 milligrams (mg) to 95.76 milligrams in 100 years.

1. Find the half-life of radium-226.

2. How long will it take for the mass of the sample to be reduced to 30 milligrams?

Solution:

1. We will first find k and then use the formula $T_{1/2} = -\frac{\ln 2}{k}$ to find the half-life. To find k in the formula $A(t) = A(0)e^{kt}$ we plug $A(t) = 95.76$, $A(0) = 100$, $t = 100$ yr. Then

   $$95.76 = 100e^{100k} \quad \Longrightarrow \quad 0.9576 = e^{100k} \quad \Longrightarrow \quad \ln 0.9576 = 100k,$$

   so that
   $$k \simeq -0.0004332 \, /\text{yr}.$$

   Then the half-life of radium-226 is
   $$T_{1/2} = -\frac{\ln 2}{-0.0004332} \simeq 1,600 \, \text{yr}.$$

2. To solve the equation with respect to t, we write
   $$A(t) = 100e^{-0.0004332t} = 30.$$

   Dividing by 100 and applying the natural logarithm, we have
   $$e^{-0.0004332t} = 0.3 \quad \Longrightarrow \quad -0.0004332 \cdot t = \ln 0.3$$
   $$\Longrightarrow \quad t = \frac{\ln 0.3}{-0.0004332} \simeq 2,779.$$

   Therefore, it would take approximately $2,779$ years. $\qquad\square$

Radiocarbon dating

The method of *radiocarbon dating* of objects containing organic materials relies on the exponential decay model for the radioactive isotope carbon-14. The method was developed in 1949 by the American physicist Willard Libby (University of Chicago), who was eventually rewarded with a Nobel Prize in chemistry. Roughly, the idea is the following. Carbon, which is contained in all organic materials, has various isotopes, and the one with six protons and eight neutrons (^{14}C) is radioactive – hence, it decays according to the exponential model. While an animal or a plant is alive, it keeps replenishing its ^{14}C supply to balance with the atmospheric level, which in the first approximation is assumed to be stable. As soon as the plant or animal dies, the amount of carbon-14 in its body starts to decrease at an exponential rate.

Thus, if $A(t)$ denotes the amount of carbon-14 in a fossil t years after death, then

$$A(t) = A(0)e^{kt},$$

where k is the relative rate of decay per year for carbon-14. Knowing the amount of carbon-14 in a fossil and the amount at the time of death (assumed to be equal to the atmospheric level), the equation can be solved for time t.

Because the behavior of carbon-14 atoms is well understood (when the energy is not too high or too low), the exponential decay model gives remarkably precise results. Modern methods of radiocarbon dating use calibration curves, which take into account the variations of the level of ^{14}C in the atmosphere. The precision of the method depends on the age of the object. For example, medieval documents can be dated within a range of 30 years. The method produces good results for objects up to 60,000 years old. For older samples, the relative amount of ^{14}C is so small that it can be obscured by possible random variations, which makes the method unreliable.

Example 8.10.

1. Carbon-14 has a half-life of 5,730 years. Find its relative rate of decay.

2. A parchment document has about 74% as much ^{14}C as does the atmosphere. Estimate its age.

Solution: The relative rate of decay is determined according to the formula

$$k = -\frac{\ln 2}{T_{1/2}} = -\frac{\ln 2}{5,730} \simeq -1.20968 \cdot 10^{-4} = -0.000120968\,/\text{yr}.$$

To find the age of the document, we write

$$A(t) = 0.74A(0) = A(0)e^{kt} = A(0)e^{-0.000120968t}.$$

Eliminating $A(0)$ from both sides and taking the natural logarithm, we get

$$\ln 0.74 = -0.000120968t \quad \implies \quad t \simeq 2,489.$$

Therefore, the parchment is about $2,500$ years old.   □

Strictly speaking, in the last example the time calculated is the time elapsed since the animal whose skin was used for the parchment was slaughtered. Similarly, if the object in question is a wooden tool, the time obtained by the radiocarbon dating method is the time elapsed since the tree was cut. In most cases, however, this information is enough to give historians and archaeologists an idea of the object's age.

Radiocarbon dating is only one in a range of radiometric dating methods, based on measuring levels of radioactive substances with known relative rates of decay in various natural and manufactured materials. For example, radioactive isotopes of uranium and lead are used to determine the timescale of geological formations. If the amount of a radioactive substance in the material at the time of formation is known, and both the substance and the decay product do not enter or leave the material after formation, then we can compute the time of the formation. This method is also applied in paleontology, where fossils that are too old for the radiocarbon method can be dated according to the geological age of the surrounding formations.

The Voynich manuscript

The Beinecke Rare Book and Manuscript Library at Yale University holds one of the most mysterious medieval documents in the world, the Voynich manuscript. It is a vellum manuscript containing over 200 pages of text written in ink in a flowing cursive script, and hundreds of color illustrations. Judging by its appearance, the book was intended as a scientific or magical treatise. Here are some of the characteristics that earned the manuscript its reputation as the ultimate enigma in the world of ancient texts:

1. The title, author, precise date, and place of creation are unknown.

2. The script and language of the text are unknown. Similarly to many European languages, the text seems to be composed of twenty to

thirty "letters" grouped into "words"; the word distribution follows natural patterns. However, unlike in most European languages, there are almost no one- or two-letter words, some words are repeated up to three times in a row, and some differ by a single letter. Despite the efforts of many professional cryptographers, including teams of counter-intelligence code breakers, and hundreds of amateurs, the text remains undeciphered. The hypotheses range from "an eastern Asian language written in an invented script" to "a very elaborate cipher involving multiple alphabets" to the claim that the text, mostly meaningless, contains information hidden in minute details, the tiny pen strokes, which can be interpreted as an ancient Greek shorthand. Some evidence can be cited both in favor of and against each of the hypotheses. High-definition images of all pages of the manuscript are available online from the Beinecke library website for anyone wishing to try their luck with the text.

3. Judging by the illustrations, the text can be divided into six chapters: botanical, astronomical, biological or medical, cosmological, pharmaceutical, and "a list of recipes." Color illustrations are plentiful, but mysterious. For instance, although many identifications have been suggested, none of the botanical pictures is conclusively identified with a known plant.

4. The history of the manuscript is mysterious. The earliest mention comes from Prague, where it was allegedly in the possession of Emperor Rudolf II (1552–1612). From there it was eventually sent to Athanasius Kircher (1602–1680), a Roman Jesuit scholar and polymath. It is likely that the manuscript stayed in the library of Collegio Romano until the twentieth century. The first well-documented mention dates from 1912, when the Polish-American book dealer Wilfrid Voynich bought it from a Jesuit college near Rome in Italy. Subsequently, the book acquired his name. Later it was purchased from Voynich's heirs and presented to the Beinecke library.

In this situation, any scientifically confirmed information can be a big help. In 2009, one piece of the puzzle was set in place: the approximate date of manufacture of the vellum on which the manuscript is written. A group of researchers from the University of Arizona, led by assistant research scientist Greg Hodgins, performed a radiocarbon analysis of the vellum of four different pages of the manuscript. All four pieces dated between 1404 and 1438.[4]

Figure 8.3: The Voynich manuscript, page 14*v* (Beinecke Rare Book and Manuscript Library, Yale University).

Because different pages were dated within the same narrow range, it is reasonable to assume that this was the time when the manuscript was created. This puts the creation of the manuscript in the early fifteenth century, about 100 years earlier than was assumed before; in particular it almost certainly rules out the English astrologer John Dee (1527–1608)

[4]Daniel Stolte, "University of Arizona experts determine age of book 'nobody can read'," *PhysOrg*, February 10, 2011.

as well as the philosopher Roger Bacon (1214–1294) as possible authors. More importantly, the dating of the vellum also allows researchers to discard the nineteenth- or twentieth-century hoax hypothesis. It is very unlikely that someone who lived centuries later would be able to procure a substantial quantity (requiring dozens of calf hides) of high-quality ancient vellum. A chemical analysis of the paints used in the illustrations, performed in 2009 at the McCrone Research Institute in Chicago, revealed no traces of industrial manufacture and confirmed that the paints could have been produced by medieval methods around the same time.

An obvious next step would be to perform a radiocarbon analysis of the ink used in the manuscript. If the ink turned out to be about as old as the vellum, this would lock the date of the creation of the manuscript in the early fifteenth century. Unfortunately, the amount of ink needed for the analysis would require the destruction of several pages, which is probably too high a price to pay for an answer. Another possible next move would be to try to determine the DNA of some of the cattle whose hides were used. Comparing it with the DNA of modern cattle, one could determine the region where the animals most probably lived, and hence narrow down geographical regions where the manuscript could have been created.

In 2014, Stephen Bax, a professor of applied linguistics at the University of Bedfordshire, England, announced that he has tentatively deciphered ten words in the manuscript and identified about fourteen symbols that occur in the text.[5] This is an exciting development. In particular, if the words are deciphered correctly, they are the names of the plants and the constellation depicted on the same page, which confirms that the manuscript was intended as a book of scientific knowledge, and not a Renaissance-era hoax or a nonsense text. However, the language of the manuscript remains unknown, because the ten deciphered words have approximate counterparts in about a dozen Indo-European and Semitic languages. Not all specialists are convinced that the words are deciphered correctly, and it will take the time and efforts of many dedicated linguists and scientists to pursue the quest.

As of today, the Voynich manuscript mystery remains unsolved: stored under the number MS 408 in the Medieval and Renaissance Manuscript section in the Beinecke library, the book is waiting for its secrets to be revealed. Figure 8.3 shows the image of page 14v of the botanical section of the manuscript.

---

[5] See www.stephenbax.net for the research article.

EXERCISES

Question 8.1. Suppose that $a > 1$ and m is a natural number. Use an argument by contradiction to show that $\sqrt[m]{a} > 1$: assume, to the contrary, that $\sqrt[m]{a} \leq 1$ and deduce that $a \leq 1$. This implies the required statement.

Question 8.2. A quantity grows exponentially with relative growth rate $k = 3/\text{h}$. How much time is needed for the quantity to grow 100-fold?

Question 8.3. A quantity decreases by a factor of 7 every 20 minutes. Find its relative rate of decay

1. per minute.

2. per hour.

Question 8.4. A furry species population develops according to an exponential growth model with a relative growth rate of $k = 0.0126/\text{yr}$. The initial population of the species is $200,000$.

1. Compute the population (to the nearest integer) after 15 years.

2. How long does it take for the furry species population to double? (Find the doubling time.)

3. How long does it take for the furry species population to reach one million?

Question 8.5. If a population of a certain bacteria culture doubles every 3 minutes, and at 11 am there is only one bacterium, find the approximate number of bacteria at 12 noon. (Bacteria reproduce asexually.)

Question 8.6. Find data for the world population in 1990 and 2000. Assuming an exponential growth model, use this data to compute the relative growth rate k and estimate the world population in 2010. Compare the result with the actual number.

Question 8.7. Suppose the population growth of a species is exponential, with growth rate $k = 0.00693/\text{yr}$. If the population today is $45,000$, find

1. how many years it takes to reach $100,000$.

2. the doubling time.

Question 8.8. A population of green bugs grows exponentially, with the relative growth rate $k_g = 0.09$/week. A population of yellow bugs grows exponentially, with the relative growth rate $k_y = 0.02$/day. Both populations count 100 bugs at time $t = 0$. For each population, find

1. how many green and yellow bugs there are after 4 weeks (28 days).

2. the doubling time for each of the species.

3. the time when the population of the yellow bugs is 10 times that of the green bugs.

Question 8.9. In 1,063 days, the amount of radioactive curium-242 in a sample dropped by a factor of 100. Find the half-life of curium-242. Give your answer up to a day.

Question 8.10. The radioactive isotope plutonium-239 has a half-life of 24,000 years.

1. Find the relative rate of decay.

2. 100 grams of plutonium-239 was buried in a container at the turn of this millennium (2,000 AD). If it were to be removed in 10,000 AD, how much Plutonium would be left in the container?

3. How much plutonium-239 would need to be buried in 2,000 AD so that 100 grams would remain in 10,000 AD?

Question 8.11. The half-life of carbon-11 is about one-fourth of the half-life of barium-139. In 10 minutes, a sample of barium-139 reduces to 92% of its initial mass.

1. How much would a sample of carbon-11 reduce in the same time?

2. Find the half-lives for both isotopes.

Question 8.12. A sample of uranium-232 reduces to four-fifths of the initial amount in about 23 years, while a sample of strontium-90 does the same in about 9 years and 4 months. Find the relative rates of decay and half-lives for both elements.

Question 8.13. The relative rate of decay of the radioactive substance A is five times that of the substance B.

1. What is the relation between the half-lives of substances A and B?

2. If a sample of substance A reduces to 98% of its initial mass in 1 year, how much will a sample of substance B reduce in the same time?

Chapter 9

Finite series: Summing up your mortgage, geometrically

The idea of a *geometric series* comes from the necessity of summing up quantities that increase (or decrease) exponentially. For example, if we pour half a cup of water, then add a half of a half (a quarter), then a half of a half of a half (one-eighth), and so on, will we ever be able to fill the cup? How many steps will it take for the water to reach a given mark?

In this chapter, we will introduce the notion of a *finite geometric series*, where the sum has finitely many terms, and derive the formula for the sum of such series. Finite geometric series find applications in mortgage computation, in finding lengths and areas of geometric patterns, and in many other real-world settings.

MATH

A *finite geometric series* is a finite sum of consecutive powers of a fixed number:

$$(1 + r + r^2 + \cdots + r^n) = \sum_{k=0}^{n} r^k.$$

Here r is a real number and n is a natural number. The right-hand side is a notation for the sum, where the index k takes all integer values between

149

0 and n. For example, if $r = \frac{1}{2}$ and $n = 5$, we will have

$$1 + \frac{1}{2} + \left(\frac{1}{2}\right)^2 + \left(\frac{1}{2}\right)^3 + \left(\frac{1}{2}\right)^4 + \left(\frac{1}{2}\right)^5.$$

This may look tedious to compute, but in fact, it turns out that we can derive a formula for such sums for arbitrary r and n.

First, if r is 1, then the sum equals $n + 1$.

To compute the sum for $r \neq 1$, let us multiply it by $(1 - r)$:

$$(1 - r)(1 + r + r^2 + \cdots + r^n) = 1 - r + r - r^2 + r^2 - \cdots - r^n + r^n - r^{n+1}.$$

It is clear that all the middle terms occur twice with the opposite signs: $-r$ and r, $-r^2$ and r^2, and so on. These terms cancel, and we have:

$$(1 - r)(1 + r + r^2 + \cdots + r^n) = 1 - r^{n+1}.$$

To obtain the answer for the original sum, we divide both sides by $(1 - r)$:

$$(1 + r + r^2 + \cdots + r^n) = \frac{(1 - r^{n+1})}{1 - r}.$$

This formula holds for all integers $n \geq 1$ and all real numbers $r \neq 1$.

Example 9.1. In the example we considered above, the formula gives

$$1 + \frac{1}{2} + \left(\frac{1}{2}\right)^2 + \left(\frac{1}{2}\right)^3 + \left(\frac{1}{2}\right)^4 + \left(\frac{1}{2}\right)^5 = \frac{1 - (\frac{1}{2})^6}{1 - \frac{1}{2}} = \frac{1 - \frac{1}{64}}{\frac{1}{2}} = \frac{2 \cdot 63}{64} = \frac{63}{32}.$$

$\square$

A more general form of the geometric series occurs when we allow the quantity to be multiplied by a fixed number a:

$$a + ar + ar^2 + \cdots + ar^n = \sum_{k=0}^{n} a \cdot r^k.$$

By taking the common factor out, we have:

$$a + ar + ar^2 + \cdots + ar^n = a(1 + r + r^2 + \cdots + r^n) = \frac{a(1 - r^{n+1})}{1 - r}.$$

This is the *finite geometric series formula* that holds for any real a, any natural n, and a real $r \neq 1$. Using the sum notation, we can rewrite it as

$$\sum_{k=0}^{n} a \cdot r^k = \frac{a(1 - r^{n+1})}{1 - r}.$$

EXAMPLES

Practice problem 9.2. Compute the following sums:

1. $3\left(1 + \frac{1}{2} + \left(\frac{1}{2}\right)^2 + \cdots + \left(\frac{1}{2}\right)^{30}\right)$.

   Answer: $\frac{3(1-(\frac{1}{2})^{31})}{1-\frac{1}{2}} = 6 \cdot \left(1 - \left(\frac{1}{2}\right)^{31}\right) \simeq 5.999999997 \simeq 6$.

2. $1 + 3 + 3^2 + 3^4 + \cdots + 3^{12}$. *Answer:* $\frac{1-3^{13}}{1-3} = -\frac{1}{2}(1 - 3^{13}) = 797,161$.

3. $\sum_{k=0}^{8} \left(\frac{1}{e}\right)^k$.

   Answer: $\displaystyle\sum_{k=0}^{8} \left(\frac{1}{e}\right)^k = 1 + \frac{1}{e} + \left(\frac{1}{e}\right)^2 + \cdots + \left(\frac{1}{e}\right)^8 = \frac{1 - (\frac{1}{e})^9}{1 - \frac{1}{e}} \simeq$
   $\frac{0.99987659}{0.6321206} \simeq 1.5818$.

Example 9.3. Compute the sum $0.3 + (0.3)^3 + (0.3)^5 + (0.3)^7 + \cdots + (0.3)^{11}$.

Solution: Here we will need to do some algebra before we can use the formula for the geometric series. First, we can take the common factor (0.3) out:

$$0.3 + (0.3)^3 + (0.3)^5 + (0.3)^7 + \cdots + (0.3)^{11}$$
$$= 0.3(1 + (0.3)^2 + (0.3)^4 + (0.3)^6 + \cdots + (0.3)^{10}).$$

Next, notice that the sum is taken over the consecutive powers of $(0.3)^2$:

$$0.3(1 + (0.3)^2 + (0.3)^4 + (0.3)^6 + \cdots + (0.3)^{10})$$
$$= 0.3(1 + (0.3)^2 + ((0.3)^2)^2 + ((0.3)^2)^3 + \cdots + ((0.3)^2)^5).$$

Now the geometric series formula with $a = 0.3$ and $r = (0.3)^2$ applies:

$$0.3(1 + (0.3)^2 + ((0.3)^2)^2 + ((0.3)^2)^3 + \cdots + ((0.3)^2)^5)$$
$$= 0.3\frac{1 - ((0.3)^2)^6}{1 - (0.3)^2} = 0.3\frac{1 - (0.3)^{12}}{1 - (0.3)^2} \simeq 0.3 \cdot \frac{0.9999995}{0.91} \simeq 0.32967.$$

$\square$

APPLICATION 1

Grains of wheat on a chessboard

According to legend, when the game of chess was first presented to the king of India, he was so delighted with it that he asked the inventor, an astrologer and polymath named Sessa, to name his own reward. The inventor, a shrewd mathematician, requested that his award was to be calculated as follows:

1 grain of wheat on the first square of the chessboard, plus

2 grains on the second square, plus

4 grains on the third square, plus

8 grains on the fourth square, and so on,

doubling the number of grains on each next square until all 64 squares were filled. Figure 9.1 shows only the first row of the chessboard, because the grains in the second row would have covered the entire board and spilled around.

Figure 9.1: First row of the chessboard with the grains of wheat.

How many grains would this make in total? We have the following geometric series:

$$1 + 2 + 2^2 + 2^3 + \cdots + 2^{63}.$$

According to the geometric series formula, this sums to

$$\frac{1 - 2^{64}}{1 - 2} = 2^{64} - 1 \simeq 1.85 \cdot 10^{19} \quad \text{grains.}$$

Assuming that the volume of one grain of wheat is 0.03 cm^3, or $3 \cdot 10^{-8}$m^3, we get the total volume of wheat

$$V_{\text{wheat}} \simeq 1.85 \cdot 10^{19} \cdot 3 \cdot 10^{-8} \simeq 5 \cdot 10^{11} \text{ m}^3.$$

This amount of wheat would cover the total area of India approximately six inches deep, which was well beyond the means of the king of India.

More examples

Here is another setting where the geometric series comes in handy. Unlike the previous story, it is not based on a well-known legend, but is of a similar flavor.

Example 9.4. There are eighty participants in a race. The prizes are awarded according to the following rule: the first prize is $1,000$; the second prize is $3/4$ of the first prize, the third prize $3/4$ of the second prize, and so on, till the eightieth prize for the last arriving athlete. How much money do the organizers need to have for the prize fund? (Assume that no two athletes will arrive at the same time.)

Solution: The prizes are distributed as follows:

$$
\begin{array}{ll}
\text{1st} & 1,000 \\
\text{2nd} & \frac{3}{4} \cdot 1,000 \\
\text{3rd} & \left(\frac{3}{4}\right)^2 \cdot 1,000 \\
\cdots & \cdots \\
\text{80th} & \left(\frac{3}{4}\right)^{79} \cdot 1,000.
\end{array}
$$

To find the total amount of all prizes, we write

$$
1,000 + 1,000 \cdot \frac{3}{4} + 1,000 \cdot \left(\frac{3}{4}\right)^2 + \cdots + 1,000 \cdot \left(\frac{3}{4}\right)^{79}
$$

$$
= 1,000 \left(1 + \frac{3}{4} + \left(\frac{3}{4}\right)^2 + \cdots + \left(\frac{3}{4}\right)^{79}\right)
$$

$$
= 1,000 \cdot \frac{1 - \left(\frac{3}{4}\right)^{80}}{1 - \frac{3}{4}} \simeq 1,000 \cdot \frac{1 - 10^{-10}}{\frac{1}{4}} \simeq 4,000.
$$

Therefore, the organizers will need to have a total of $4,000$ for the prize fund. $\qquad\square$

Now let us consider the example mentioned in the introduction to this chapter. Putting aside the question of whether we can ever fill the cup (we will return to it in Chapter 10, where we consider *infinite geometric series*), let us find the number of steps needed to fill the cup to a given level.

Example 9.5. Take an empty cup. Pour half a cup of water into it, then add a quarter of a cup, then one-eighth, and so on. How many steps will it take to fill nine-tenths of the cup?

Solution: Suppose we poured the water k times, starting from half a cup and reducing the amount of water each time by half. Then the total amount of water in the cup is

$$\frac{1}{2} + \left(\frac{1}{2}\right)^2 + \cdots + \left(\frac{1}{2}\right)^k.$$

This sum is a finite geometric series with $a = \frac{1}{2}$, $r = \frac{1}{2}$, and $n = k - 1$:

$$\frac{1}{2} + \left(\frac{1}{2}\right)^2 + \cdots + \left(\frac{1}{2}\right)^k = \frac{1}{2}\left(1 + \frac{1}{2} + \cdots + \left(\frac{1}{2}\right)^{k-1}\right).$$

Using the formula for the sum, we get

$$\frac{1}{2}\left(1 + \frac{1}{2} + \cdots + \left(\frac{1}{2}\right)^{k-1}\right) = \frac{1}{2} \cdot \frac{1 - (\frac{1}{2})^k}{1 - \frac{1}{2}} = 1 - \left(\frac{1}{2}\right)^k.$$

This is the amount of water in the cup after k steps, and we have to make it greater or equal to nine-tenths. This leads to the equation

$$1 - \left(\frac{1}{2}\right)^k = \frac{9}{10} \quad \Longrightarrow \quad \left(\frac{1}{2}\right)^k = \frac{1}{10}.$$

Taking the logarithm of both sides, we get

$$k \ln\left(\frac{1}{2}\right) = \ln\left(\frac{1}{10}\right) \quad \Longrightarrow \quad k \simeq 3.32.$$

Therefore, it will take four steps. Indeed,

$$\frac{1}{2} + \frac{1}{4} + \frac{1}{8} + \frac{1}{16} = \frac{8 + 4 + 2 + 1}{16} = \frac{15}{16} = 0.9375 > \frac{9}{10}.$$

$\square$

Let us return to the question of the strength of an encryption key against a brute-force attack, introduced in Chapter 6 (see Example 6.11).

Example 9.6. An encryption key, or a secret password used in computer security, is a sequence of zeros and ones of a certain length. Cracking the key by brute-force attack consists of trying all sequences of zeros and ones until the right one is found. Assume the adversary has a cluster of $1,000$ processors capable of checking 10 billion keys per second. According to

Moore's law of hardware development, you expect the processing speed to double every two years. How long should the encryption key be to ensure that it takes more than thirty years for the adversary to crack the key by brute force? Assume that the adversary will have to check half of all possible keys before the right one is found.

Solution: Let k denote the length of the key. There are 2^k different sequences of zeros and ones of length k. The adversary will have to check $\frac{1}{2} \cdot 2^k = 2^{k-1}$ keys.

To compute the number of tries the adversary can perform in thirty years, we will have to divide the time into two-year stretches and take into account the doubling of the processing speed. Two years contain roughly

$$2 \cdot 365 \text{ days} = 17,520 \, \text{h} = 63,072,000 \, \text{sec}.$$

It is safe to take $6.4 \cdot 10^7$ seconds as a two-year stretch. During the first two years, the $1,000$ processors working at 10 billion tries per second will make at most

$$1,000 \cdot 10 \cdot 10^9 \cdot 6.4 \cdot 10^7 = 6.4 \cdot 10^{20}$$

tries. In the next two years, they will make at most twice as many, and so on to the last two-year stretch, when the processor speed will be 2^{14} times as high as in the beginning. Totally, the number of tries is the sum of the geometric series:

$$6.4 \cdot 10^{20} \cdot \left(1 + 2 + 2^2 + 2^3 + \cdots + 2^{14}\right)$$
$$= 6.4 \cdot 10^{20} \cdot \frac{1 - 2^{15}}{1 - 2} = 6.4 \cdot 10^{20} \cdot 32,767 \simeq 2.1 \cdot 10^{25}.$$

This is an upper bound for the number of tries the adversary will be able to make in thirty years. We have the equation for k:

$$2.1 \cdot 10^{25} = 2^{k-1}.$$

Taking the logarithm base 2 of both sides, we get

$$\log_2(2.1 \cdot 10^{25}) = \log_2(2^{k-1}) \quad \Longrightarrow \quad \log_2 2.1 + 25 \log_2 10 = k - 1.$$

Then we find

$$k = 1 + \log_2 2.1 + 25 \log_2 10 \simeq 1 + 84.1 \simeq 86.$$

A key of length 86 should withstand a brute-force attack by this adversary for thirty years. $\square$

Here is an example where the factor r of the geometric series has to be carefully computed before the series is summed.

Example 9.7. A snow fortress is built with fifty round snowballs, whose radii (in meters) are as follows:

$$1\,\text{m}, \quad \frac{5}{7}\,\text{m}, \quad \left(\frac{5}{7}\right)^2\,\text{m}, \quad \cdots \left(\frac{5}{7}\right)^{49}\,\text{m}.$$

If the volume of a ball with radius R is $V = \frac{4}{3}\pi R^3$, find the total volume of the fortress.

Figure 9.2: Wall built of snowballs of decreasing sizes.

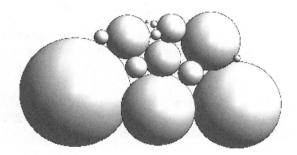

Solution: The total volume of the snow fortress is the sum of the volumes of all the balls:

$$V_{\text{total}} = \frac{4}{3}\pi\left(1^3 + \left(\frac{5}{7}\right)^3 + \left(\left(\frac{5}{7}\right)^2\right)^3 + \cdots + \left(\left(\frac{5}{7}\right)^{49}\right)^3\right) =$$

$$= \frac{4}{3}\pi\left(1^3 + \left(\frac{5}{7}\right)^3 + \left(\frac{5}{7}\right)^6 + \cdots + \left(\frac{5}{7}\right)^{147}\right).$$

This is a geometric series with $a = \frac{4}{3}\pi$ and $r = \left(\frac{5}{7}\right)^3$. Applying the formula, we get

$$V_{\text{total}} = \frac{4}{3}\pi\frac{1 - \left(\frac{5}{7}\right)^{150}}{1 - \left(\frac{5}{7}\right)^3} \simeq 6.59\,\text{m}^3.$$

$\square$

APPLICATION 2

Future value of money

The most prominent application of geometric series in daily life is in computing mortgage payments. First we need to know how to compute the future value of money: when you take a loan or a mortgage, you will have to pay it back with all the interest accumulated by the time you pay it in full. In particular, we need to know the value of a dollar at a given point of time in the future (given a compounding scheme). You probably have heard the expression "A dollar in hand is worth less tomorrow." This is true about a dollar you keep in hand, i.e., not invested. It does not accumulate interest, and thus it loses value with respect to an invested dollar. To see how the value of the invested money changes with time, consider the following example.

Example 9.8. Suppose a bank compounds quarterly, at an annual interest rate of 8%. How much will a dollar deposited in the bank in 2012 be worth in 2022?

Solution: The answer is computed using the compound interest formula: in ten years, the 2022-value of a dollar in 2012 is:

$$A(10) = 1 \cdot \left(1 + \frac{8}{4 \cdot 100}\right)^{4 \cdot 10} = 1.02^{40} \simeq \$2.21.$$

Conversely, what was the value in 2012 of a dollar in 2022? Suppose the value was A, then

$$1 = A\left(1 + \frac{8}{4 \cdot 100}\right)^{4 \cdot 10} = A \cdot 1.02^{40} = 2.21A \quad \Longrightarrow \quad A = \frac{1}{2.21} \simeq \$0.45.$$

This is what happens with your dollar if you keep it in your pocket for ten years – it loses more than half of its value! □

Practice problem 9.9. Given a dollar in 2012, find its value under each of the following compounding schemes:

1. In 2022, compounded quarterly at an annual rate of 12%.

    Answer: $3.26.

2. In 2002, compounded quarterly at an annual rate of 16%.

   Answer: $0.21.

3. In 2052, compounded quarterly at an annual rate of 10%.

   Answer: $51.98.

4. In 2021, compounded quarterly at an annual rate of 20%.

   Answer: $5.79.

Mortgage payments calculation

Now we can use geometric series to compute mortgage payments. Here is our first example.

Example 9.10. If someone takes out a loan of $10,000 from a bank that compounds quarterly at an 8% annual interest rate, what quarterly installment must they pay (starting the next quarter) for the next five years, to pay off the loan?

Solution: Suppose the quarterly installment is A. We then equate the value of all payments made in five years, with the value in five years of the loan. The latter is easy to compute: five years means twenty quarters of compounding, so the value of the loan in five years is:

$$A(5) = 10,000 \cdot \left(1 + \frac{8}{4 \cdot 100}\right)^{4 \cdot 5} = 10,000 \cdot 1.02^{20} = \$14,859.474.$$

Computing the value of the payments in five years is not too hard either: the first installment of A gets compounded 19 times, because it is paid one quarter from now. Thus, it is worth $A \cdot 1.02^{19}$ in five years. The second is compounded 18 times, hence worth $A \cdot 1.02^{18}$, and so on. Thus, the value of our total payment in five years is

$$A \left(1.02^{19} + 1.02^{18} + \cdots + 1\right),$$

where the last payment is made at the very end and so does not get compounded; hence it is worth $A = A \cdot 1$ on the closing date.

Now we can add up this expression using the geometric series formula. We have

$$A(1 + r + r^2 + \cdots + r^n) = \frac{A(1 - r^{n+1})}{1 - r}$$

for every real number $r \neq 1$. Applying this to our example, $r = 1.02$ and $n = 19$, the value on the closing date of all our payments is:

$$\frac{A(1 - 1.02^{20})}{1 - 1.02} = A \cdot (-0.4859/ - 0.02) = 24.295 \cdot A.$$

Finally, this equals the value on the closing date of our loan, if we are to pay it off entirely. Thus,

$$A = 14,859.474/24.295 \simeq \$611.63.$$

This is the quarterly installment that is needed to pay off the loan in five years.   □

Now let us try to derive a general strategy for solving mortgage problems. Suppose you take a mortgage of M dollars. The bank uses a compounding scheme with $r\%$ annual interest rate, compounded n times per year. You have to pay the mortgage back in t years by paying equal installments n times per year. Find the installment amount.

First, we compute the value of the mortgage in t years:

$$M \left(1 + \frac{r}{n \cdot 100}\right)^{n \cdot t} = M p^{nt},$$

where we denoted the compounding factor $\left(1 + \frac{r}{n \cdot 100}\right)$ by p for convenience.

Now, suppose each installment is A dollars. The first installment is paid *one compounding period after the mortgage is taken*, and therefore it gets compounded $(nt - 1)$ times. The next installment gets compounded $(nt - 2)$ times, and so on until the last installment, which does not get compounded at all. Thus the net worth of all installments equals the following sum, which is calculated using the geometric series formula:

$$A \left(1 + p + p^2 + \ldots + p^{nt-1}\right) = \frac{A(1 - p^{nt})}{1 - p}.$$

We now equate the two amounts:

$$M p^{nt} = \frac{A(1 - p^{nt})}{1 - p}.$$

Finally,

$$A = M \cdot \frac{p^{nt}(1 - p)}{(1 - p^{nt})}.$$

In practice, it is convenient to first calculate the compounding factor $p = \left(1 + \frac{r}{n \cdot 100}\right)$, and then plug it in the above formula for A. The compounding periods are usually months ($n = 12$) or quarters ($n = 4$).

The next question is a variation on a mortgage calculation.

Example 9.11. Suppose you want to establish a monthly deposit to a bank that compounds monthly at an interest rate of 6%, so that in fifteen years you can have $100,000. Find the monthly payment.

Solution: In this case, the final amount in fifteen years is already known: $100,000. Only the monthly payments get compounded. The compounding factor per month is

$$p = \left(1 + \frac{r}{n \cdot 100}\right) = \left(1 + \frac{6}{1,200}\right) = 1.005.$$

Here is an important difference from the mortgage calculation: the first payment of A dollars is made *right now*; therefore it gets compounded $n \cdot t = 12 \cdot 15 = 180$ times, and not $n \cdot t - 1 = 179$ times. The next payment is compounded $nt - 1$ times, and so on, until the last payment, which is made a month before 15 years are over, and so *it gets compounded once*. Therefore, the total value of all installments is

$$A\left(p + p^2 + p^3 + \cdots + p^{nt}\right).$$

Now, the sum in the brackets is not exactly the geometric series as it was first defined, because it starts with p instead of 1. To apply the geometric series formula, we need to take out the common factor p. Then the sum in the brackets starts with 1 and the geometric series formula can be applied:

$$A\left(p + p^2 + p^3 + \cdots + p^{nt}\right) = A \cdot p \cdot \left(1 + p + p^2 + \cdots + p^{nt-1}\right)$$
$$= \frac{A \cdot p \cdot \left(1 - p^{nt}\right)}{1 - p}.$$

Plugging in the numbers $n = 12$, $t = 15$, and $p = 1.005$, and equating the amount to $100,000, we have

$$100,000 = A \cdot 1.005 \cdot \frac{1 - 1.005^{12 \cdot 15}}{1 - 1.005} \simeq A \cdot 1.005 \cdot \frac{-1.45409}{-0.005} \simeq 292.27A.$$

Therefore,

$$A = \frac{100,000}{292.27} \simeq \$342.15.$$

$\square$

To make the difference between the two cases clear, here is an example of a mortgage payment calculation for the same amount and under the same compounding scheme, but with different payment schedules: (1) if the first payment is made one compounding period from now, and (2) it is made right now.

Example 9.12. A person borrows $500,000 from a bank, to be repaid over the next twenty years in eighty quarterly installments. Suppose the bank compounds quarterly at an annual rate of 16%. Find the value of the quarterly installment, if the payments start (1) a quarter from now; (2) today. *Hint:* Think how many times the installments are compounded in both of these situations.

Solution: The crucial point in solving mortgage-type questions is to understand that *each payment* is compounded a different number of times and contributes a different value to the total accumulated payment.

The amount borrowed from the bank is $500,000 today – so its value in twenty years, at the annual rate of 16% compounded quarterly, is:

$$500,000 \cdot \left(1 + \frac{16/4}{100}\right)^{20 \cdot 4} = 500,000 \cdot 1.04^{80} = \$11,524,899.60.$$

This is to equal the amount paid in installments (and compounded over twenty years). We start with question (1) and suppose each installment is A. There are eighty total payments. We can denote them A_1, A_2, up to A_{80}. Now let us compute the value of each installment in twenty years. The first is A_1, which is paid a quarter from today, compounded seventy-nine times. Hence, $A_1 = A(1.04)^{79}$. The second payment of A is compounded seventy-eight times, so $A_2 = A(1.04)^{78}$, and so on. The last (eightieth) payment is not compounded at all, so $A_{80} = A$.

Now it remains to add them up, and equate to the above amount. We have:

$$11,524,899.6 = A\left(1.04^{79} + 1.04^{78} + \cdots + 1.04 + 1\right).$$

The geometric series in the parentheses adds up to $\dfrac{1 - 1.04^{80}}{1 - 1.04}$, so we have

$$11,524,899.60 = A \cdot \frac{1 - 1.04^{80}}{1 - 1.04}.$$

And finally,

$$A = \$20,907.04.$$

Now solve question (2). Denote each installment by B. There are again eighty installments. The first B_1, which is paid today, *is compounded eighty, not seventy-nine, times.* Hence, $B_1 = B(1.04)^{80}$. The second payment of B is compounded seventy-nine times, so $B_2 = B(1.04)^{79}$, and so on. The last (eightieth) payment is made a quarter before the closing of the mortgage, and hence *it is compounded once,* so $B_{80} = B(1.04)$. As before, we need to add them up and equate with the value of the loan in twenty years. We have:

$$11,524,899.6 = B \left(1.04^{80} + 1.04^{79} + \cdots + (1.04)^2 + 1.04\right).$$

Here it is important to remember that the formula for summing a finite geometric series works only when the starting term is 1. In this case, the starting term is 1.04, so we cannot apply the formula to the equation as it stands.

Instead, take 1.04 out as a common factor from every term, and write:

$$11,524,899.6 = B \cdot 1.04 \left(1.04^{79} + 1.04^{78} + \cdots + 1.04^1 + 1\right).$$

Now we have a geometric series to which the formula can be applied. Hence, we get

$$11,524,899.6 = B \cdot 1.04 \cdot \frac{1 - 1.04^{80}}{1 - 1.04}.$$

So this sum results in the exponent 80, *and not* 81 – but now there is an extra factor of 1.04 outside, which was not there in question (1). From here, you can solve for B:

$$B = \$20,102.92.$$

(See also Question 9.8 in the Exercises.)

Moral: the sooner you start paying back your debt, the less you will have to pay.   □

EXERCISES

Question 9.1. Find the sums of the following finite geometric series:

1. $1 + 5 + 5^2 + 5^3 + \ldots + 5^{16}$.

2. $\sum_{k=0}^{10} (-\frac{1}{2})^k$.

3. $1 + 1.25^4 + 1.25^8 + 1.25^{12} + \ldots + 1.25^{40}$.

Question 9.2. Find the sums of the following finite geometric series:

1. $1 - 0.4 + 0.4^2 - 0.4^3 + \ldots + 0.4^{24}$.

2. $\frac{1}{7} + \frac{1}{7^2} + \frac{1}{7^3} + \ldots + \frac{1}{7^{15}}$.

3. $\sum_{k=0}^{12} \left(\frac{8}{5}\right)^k$.

Question 9.3. Find the values in 2013 of a dollar at the same time of year in 2023, 2003, and 2035, if it is compounded quarterly at an annual interest rate of $r = 12\%$.

Question 9.4. Find the present values of a dollar at the same time of year in 2005, 2020, and 2039, if it is compounded monthly at an annual interest rate of $r = 12\%$.

Question 9.5. A person takes a mortgage of $150,000 from a bank that compounds monthly at an annual interest rate of 4%. The mortgage has to be paid back in equal monthly installments in 20 years. The first payment is made a month from now. Find the monthly payments.

Question 9.6. Solve Question 9.5 if the bank compounds quarterly and the mortgage has to be paid back in equal quarterly installments in twenty years. The first payment is made a quarter from now.

Question 9.7. Suppose you make a monthly deposit (of the same amount) each month for six years to a bank that compounds monthly at an annual interest rate of 10%. How much should you deposit every month so that at the end of six years the amount in the bank equals $100,000? (In this case, the first payment is made *today*, and hence compounded seventy-two times.)

Question 9.8. Consider the two payment schemes that were described in Example 9.12. They are identical in all aspects except for the starting and ending dates for the payments. Now compute the ratio between the two answers for the monthly installment. We obtain, to seven decimal places:

$$\frac{20,907.04}{20,102.92} = 1.04,$$

which is exactly the compounding factor in the example. Is this a coincidence, or do you think that there is a reason for this equality?

Question* 9.9. Recall the formula for the per-period installment A to be paid back on a mortgage or loan M to a bank that uses the compounding factor $p = 1 + \frac{r}{n \cdot 100}$:

$$A = M \cdot \frac{p^{nt}(1-p)}{1-p^{nt}}.$$

This formula applies if the first installment is paid one compounding period from the time of taking the loan.

Derive a similar formula for the same values of M, A, and p, with the assumption that the first installment is paid at the time of taking the loan.

1. How much is the new monthly payment?

2. Divide the monthly payment in the first scheme by the one obtained in part (1). What is the ratio?

Question 9.10. Suppose a person borrows $\$25,000$ for a new car from a bank that compounds monthly at an annual interest rate of 10%. If he has to repay the loan in four years, how much should he pay each month? (The loan is taken today, but the first payment is made a month from now, and hence compounded 71 times.)

Question 9.11. In a 10-mile race with twenty registered competitors, the prize fund is divided as follows: the winner gets $\$1,000$, the runner arriving second gets $\frac{4}{5}$ of the first prize, the runner arriving third gets $\frac{4}{5}$ of the second prize, and so on to the last arriving athlete. Assume that no two athletes arrive at the same time. How much money should the prize fund have?

Question 9.12. A snail covers 1 foot in the first ten minutes, $\frac{8}{9}$ of a foot in the next ten minutes, and each subsequent ten minutes, $\frac{8}{9}$ of the distance it covered in the previous ten minutes. What distance will the snail cover

1. in the first thirty minutes?

2. in the first seventy minutes?

Give your answer to the nearest inch.

Question 9.13. A decoration on a jewelry box consists of a golden wire bent into a rectangular spiral in the following way: the first 1 inch of it is laid vertically down, the next $\frac{7}{8}$ of an inch is laid horizontally to the right, the next piece up, the next piece to the left, and so on. Each next piece is $\frac{7}{8}$ of the previous piece by length. Figure 9.3 shows such a spiral made of fifteen pieces. Find the total length of the golden wire needed to make the decoration, if it consists of

1. ten pieces,

2. thirty pieces.

Figure 9.3: Rectangular spiral.

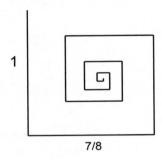

Chapter 10

Infinite series: Fractals and the myth of forever

In this chapter, we will discuss infinite geometric series and their applications in studying infinitely subdivided processes and patterns. In particular, infinite geometric series play an important role in the description of fractals – patterns that look the same no matter how much we zoom in on the picture. These amazing self-repeating shapes are present in nature and everyday life. At the end of the chapter we describe a recent example of an application of fractal geometry in archaeology.

MATH

An *infinite geometric series* is the sum of infinitely many consecutive powers of a fixed real number r, times a fixed real coefficient a:

$$a(1 + r + r^2 + r^3 + \cdots) = \sum_{k=0}^{\infty} ar^k.$$

Example 10.1. Here are a few examples of infinite geometric series:

1. $\sum_{k=0}^{\infty} \frac{2^k}{10} = \frac{1}{10}(1 + 2 + 4 + 8 + 2^4 + 2^5 + \cdots)$.

2. $\sum_{n=0}^{\infty} (-\frac{1}{3})^n = 1 - \frac{1}{3} + \frac{1}{9} - \frac{1}{27} + \frac{1}{3^4} - \frac{1}{3^5} + \cdots$.

3. $\sum_{k=0}^{\infty} 5(0.1)^k = 5(1 + 0.1 + 0.01 + 0.001 + 0.0001 + \cdots)$.

$\square$

To see how the behavior of an infinite geometric series depends on the value of r, recall the formula for the sum of the finite geometric series that holds for any real a, any $r \neq 1$, and any natural n (Chapter 9):

$$a(1 + r + r^2 + r^3 + \cdots + r^n) = \sum_{k=0}^{n} ar^k = \frac{a(1 - r^{n+1})}{1 - r}.$$

What happens if we allow n to grow indefinitely?

1. If $r > 1$, then as n grows, r^{n+1} becomes infinitely large. Then the numerator $1 - r^{n+1}$ becomes infinitely large negative; because $1 - r$ is also negative, but finite, the result is an infinitely large number whose sign is determined by the sign of a.

2. If $r < -1$, then as n grows, r^{n+1} becomes an infinitely large number that changes sign depending on the parity of n. The sum of the finite geometric series oscillates between ever growing positive and negative values and does not converge to any real number.

3. If $r = 1$, we cannot use the formula for the finite geometric series, but the expression $a(1 + 1 + 1 + 1 + \cdots)$ is just a sum of infinitely many copies of a, which is not a finite number.

4. If $r = -1$, the formula for the finite series gives 0 or a depending on the parity of n. When n grows indefinitely, the expression oscillates between these two values and does not converge to any real number.

5. Finally, if $-1 < r < 1$ while n grows indefinitely, the expression r^{n+1} becomes arbitrarily small. Note that this is true independent of the sign of r: for both positive and negative r, the term r^{n+1} becomes infinitely close to 0 when n grows. Therefore, the expression

$$\frac{a(1 - r^{n+1})}{1 - r} \quad \text{tends to} \quad \frac{a}{1 - r},$$

which is the sum of the infinite geometric series in this case.

Cases (1)–(4) are not interesting to us because in these cases the infinite sum does not make any sense. The only meaningful case is the last one, where $-1 < r < 1$. In this case, the sum is a finite number and is given by the formula we just derived:

$$a(1 + r + r^2 + r^3 + \cdots) = \sum_{k=0}^{\infty} ar^k = \frac{a}{1 - r}.$$

The formula holds for any real r such that $-1 < r < 1$ and any real a. It comes in handy in many applications, as we will see below.

In the particular case when $a = 1$ and $-1 < r < 1$, the sum of the infinite geometric series is given by the formula

$$\sum_{k=0}^{\infty} r^k = \frac{1}{1-r}.$$

EXAMPLES

Example 10.2. Find the sums of the infinite geometric series listed in Example 10.1, whenever they make sense.

Solution:

1. The first sum $\sum_{k=0}^{\infty} \frac{2^k}{10}$ is infinite because $r = 2 > 1$.

2.
$$\sum_{n=0}^{\infty} \left(-\frac{1}{3}\right)^n = \frac{1}{1-(-\frac{1}{3})} = \frac{1}{\frac{4}{3}} = \frac{3}{4}.$$

3.
$$\sum_{k=0}^{\infty} 5(0.1)^k = \frac{5}{1-0.1} = \frac{5}{0.9} = \frac{50}{9} \simeq 5.55556.$$

$\square$

Practice problem 10.3. In each of the examples below, find the sum of the infinite geometric series, if it is finite.

1. $\sum_{k=0}^{\infty} 88 \cdot 0.12^k$. *Answer:* $\frac{88}{1-0.12} = \frac{88}{0.88} = 100$.

2. $\sum_{m=0}^{\infty} \left(\frac{3}{5}\right)^m$. *Answer:* $\frac{1}{1-\frac{3}{5}} = \frac{1}{\frac{2}{5}} = \frac{5}{2} = 2.5$.

3. $\sum_{k=0}^{\infty} \frac{(-2.8)^k}{100,000}$. *Answer:* This is not a finite number because $r = -2.8 < -1$.

4. $\sum_{n=0}^{\infty} \frac{0.99^n}{15}$. *Answer:* $\frac{1}{15}\frac{1}{1-0.99} = \frac{100}{15} = \frac{20}{3}$.

5. $\sum_{m=0}^{\infty} \frac{1}{8}\left(\frac{2}{17}\right)^m$. *Answer:* $\frac{1}{8}\frac{1}{1-\frac{2}{17}} = \frac{1}{8}\frac{1}{\frac{15}{17}} = \frac{1}{8}\frac{17}{15} = \frac{17}{120} \simeq 0.1416667$.

6. $\sum_{p=0}^{\infty} 63(-\frac{1}{5})^p$. *Answer:* $\frac{63}{1-(-\frac{1}{5})} = \frac{63}{\frac{6}{5}} = \frac{63 \cdot 5}{6} = \frac{105}{2} = 52.5$.

7. $\sum_{s=0}^{\infty} -6(\frac{5}{4})^s$. *Answer:* This is not a number because $r = \frac{5}{4} > 1$.

APPLICATION 1

The fact that an infinite geometric series can sum up to give a finite number shows us that adding up infinitely many parts (for example, of distances or time intervals) can make sense and result in a finite value. What kind of a process can be divided into infinitely many steps? A true infinity is hard to come by in the real world. There is a finite number of hairs on your head, and there is a finite number of grains of sand on any given beach. The observable universe is finite in both space and time. To illustrate the difference between the mathematical concept of infinity and the almost-infinities we encounter in reality, consider the following game. Two people take turns inserting a needle in a sheet of paper along a straight line. The needle cannot be placed at the same point twice. The player who has nowhere to place the needle loses. In reality, there is always a loser, as the needle has a nonzero size and after a certain very large number of moves, there is no place on the line to put it. But if we ask the players to give the coordinate of the point where the needle should be placed, instead of placing it, the game becomes infinite because there are infinitely many numbers in any given interval.

Most of our mental experiments involving infinity actually model real processes that are finite. However, it is often convenient to think of an *unknown very large number* as infinity. It is especially efficient if the process can be modeled by infinitely many ever-decreasing steps that can be described by a convergent geometric series. Perhaps the most famous example of such a process is the paradox of Achilles and the tortoise, described in Question 10.1, at the end of this chapter. Several other examples are discussed below.

Filling a cup with water

Let us return to the question formulated at the beginning of Chapter 9. Suppose we have an empty cup. We pour half a cup of water into it, then add a quarter of a cup, then one-eighth, and so on. Will we ever be able to fill the cup?

The amount of water in the cup after infinitely many steps is

$$\frac{1}{2} + \left(\frac{1}{2}\right)^2 + \left(\frac{1}{2}\right)^3 + \cdots = \frac{1}{2}\left(1 + \frac{1}{2} + \left(\frac{1}{2}\right)^2 + \cdots\right) = \frac{1}{2} \cdot \sum_{k=0}^{\infty} \left(\frac{1}{2}\right)^k.$$

This infinite geometric series sums up to

$$\frac{1}{2} \cdot \sum_{k=0}^{\infty} \left(\frac{1}{2}\right)^k = \frac{1}{2} \cdot \frac{1}{1 - \frac{1}{2}} = \frac{1}{2} \cdot 2 = 1.$$

After infinitely many steps, the cup will be full of water. What shall we conclude? This depends on the meaning of "ever" in the question. Following the described procedure, it is impossible to fill the entire cup with water in finite time, but it is possible to almost fill it. Indeed, after n steps the amount of water in the cup is

$$\frac{1}{2} \cdot \sum_{k=0}^{n-1} \left(\frac{1}{2}\right)^k = \frac{1}{2} \cdot \frac{1 - \left(\frac{1}{2}\right)^n}{1 - \frac{1}{2}} = \frac{1 - \left(\frac{1}{2}\right)^n}{2} \cdot 2 = 1 - \left(\frac{1}{2}\right)^n.$$

As n grows, the quantity $\left(\frac{1}{2}\right)^n$ decreases exponentially and the water will eventually reach any mark below the top of the cup (see Example 9.5 for the nine-tenths mark). We cannot fill the cup to the top, but we can get as close as we like to it in finitely many steps.

Two trains and a bee

Recall the question we considered earlier in Chapter 2 – Example 2.5 – when we discussed linear constant speed motion.

Two trains, A from the west and B from the east, are approaching along the same straight train track, each moving at a constant speed of 5 miles per hour. When the distance between the trains is 10 miles, a bee starts flying east from the head of train A. The bee moves at a constant speed of 6 miles per hour until it hits train B, then it turns around and flies west at the same constant speed until it hits train A, and so on until the two trains collide. Question: find the total distance covered by the bee before the crash.

In Chapter 2 we found a quick, but tricky, solution: because the trains take 1 hour to reach the meeting point, and the bee always moves at the same constant speed of 6 miles per hour, the total distance covered by the bee is 6 miles.

Figure 10.1: Bee flying between two trains.

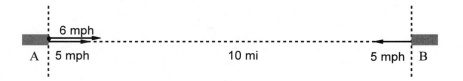

A longer, but more straightforward, way to solve this problem is to model the process by an infinite geometric series. We can consider the distance covered by the bee as the sum of infinitely many segments between the consecutive changes of direction. In Chapter 2, we already computed the length of the first two segments:

$$s_1 = \frac{60}{11} \text{ mi}, \text{ and } s_2 = \frac{60}{121} \text{ mi}.$$

Let us recall how we obtained these numbers. For s_1, we had to divide the initial distance between the bee and train B, $d_1 = 10$ mi, by the sum of their speeds, $5 + 6 = 11$ mph, and multiply by the speed of the bee, 6 mph. Thus $s_1 = 10 \cdot \frac{1}{11} \cdot 6 = \frac{60}{11}$ mi. To find s_2, we performed the same operations starting with the initial distance $d_2 = \frac{60}{11} - \frac{50}{11} = \frac{10}{11}$ mi, where $\frac{50}{11}$ miles is the distance covered by train A by the time the bee meets train B and turns around. The obtained result is $s_2 = \frac{10}{11} \frac{1}{11} \cdot 6 = \frac{60}{11^2}$ mi. Now the point is to realize that the problem of finding s_2, and all subsequent segments, is exactly the same as the problem of finding s_1, the only difference being in the initial distance between the bee and the train it is bound to meet. For the computation of s_1, this distance was $d_1 = 10$ mi, and for s_2, it was $d_2 = \frac{10}{11}$ mi. Each time, the initial distance gets divided by 11, and consequently the length of each new segment covered by the bee is one-eleventh of the length of the previous segment.

If the bee is assumed to be just a point, this process is infinite. Then the total distance covered by the bee before the trains collide is given by the infinite geometric series:

$$D = \frac{60}{11} + \frac{60}{11^2} + \frac{60}{11^3} + \cdots = \frac{60}{11} \left(1 + \frac{1}{11} + \frac{1}{11^2} + \cdots \right).$$

Applying the formula for the sum with $a = \frac{60}{11}$ and $r = \frac{1}{11}$, we get

$$D = \frac{60}{11} \cdot \frac{1}{1 - \frac{1}{11}} = \frac{60}{11} \cdot \frac{11}{10} = \frac{60}{10} = 6 \text{ mi}.$$

This longer way to get the same answer serves two purposes: first, it illustrates the statement that a finite distance can be presented as a sum of infinitely many ever-decreasing pieces. Second, it confirms the idea that replacing a large unknown number of steps by infinity works just fine – actually, the number of the segments the bee flew was finite, because its nonzero size will prevent it from turning around or moving after the trains get close enough – but the size of the bee is very small compared with the distance between the trains, and the answer we obtained is close enough to reality for all practical purposes.

This same question was supposedly asked of the famous mathematician and computer scientist John von Neumann, who gave the correct answer in seconds. "Ah, you found the shortcut," the questioner said. "Most people start summing up a geometric series." "That's what I did," von Neumann said. "Is there a shortcut?" [1]

A train slowing down

Here is another example of a nearly infinite process successfully modeled by an infinite geometric series.

Example 10.4. As a train approaches a station, it slows down according to the following rule: in the first second, it covers thirty meters, and in each subsequent second, $\frac{4}{7}$ of the distance covered in the previous second. What is the total distance covered by the train before it stops?

Solution: The distance covered by the train in the first second is 30 m, in the second second $\frac{4}{7} \cdot 30$ m, in the third second $(\frac{4}{7})^2 \cdot 30$ m, and so on to infinity. The total distance covered by the train before it stops is given by the infinite geometric series

$$D = 30 \left(1 + \frac{4}{7} + \left(\frac{4}{7} \right)^2 + \left(\frac{4}{7} \right)^3 + \cdots \right) = \frac{30}{1 - \frac{4}{7}} = \frac{30}{\frac{3}{7}} = \frac{30 \cdot 7}{3} = 70 \text{ m.}$$

$\square$

In reality the train does not take infinitely long to stop. The model implies infinite stopping time because the conditions of the problem mean that the speed of the train is never zero. In fact, at some point, the speed becomes so slow it is negligible, and the train effectively stops. Let us find when this happens. We will have to decide which speed to consider as negligible.

---

[1]Source: http://www.primepuzzle.com/leeslightest/howfar.html.

Example 10.5. In the conditions of Example 10.4, when does the speed of the train become slower than 10^{-6} meters per second (one thousandth of a millimeter per second)? Find the distance covered by the train by this time.

Solution: The distance covered by the train in the nth second is $30 \cdot (\frac{4}{7})^{n-1}$ m. To find n such that this distance is less than 10^{-6} meters, we have to solve the equation for n:

$$30 \cdot \left(\frac{4}{7}\right)^{n-1} = 10^{-6}.$$

Dividing by 30 and taking the natural logarithm of both sides, we get

$$(n-1)\ln \frac{4}{7} = \ln \frac{10^{-6}}{30} = -6\ln 10 - \ln 30 \simeq -17.2.$$

Then we obtain

$$n = 1 + \frac{-6\ln 10 - \ln 30}{\ln \frac{4}{7}} \simeq 1 + \frac{-17.2}{-0.5596} \simeq 1 + 30.74 \simeq 32.$$

It will take 32 seconds for the speed of the train to drop below 10^{-6} meters per second, or for the train to effectively stop.

To find the distance D_{32} covered by the train in 32 seconds, we can use the formula for the finite geometric series (Chapter 9):

$$D_{32} = 30 \left(1 + \frac{4}{7} + \left(\frac{4}{7}\right)^2 + \ldots + \left(\frac{4}{7}\right)^{31}\right) = \frac{30(1 - (\frac{4}{7})^{32})}{1 - \frac{4}{7}}$$

$$\simeq \frac{30 \cdot 0.9999999833}{\frac{3}{7}} = 70 \cdot 0.9999999833 \simeq 69.99999883 \text{ m}.$$

$\square$

The distance D_{32} covered in the first 32 seconds of motion differs from the distance D that would have been covered in infinite time only by about 10^{-6} of a meter, and the computation of D was considerably easier. For all practical purposes, we could have used the infinite geometric series to obtain the answer.

$$\boxed{\text{APPLICATION 2}}$$

Repeating decimals

Everyone knows that 0.333333333... is equal to $\frac{1}{3}$ (the dots mean that the digit 3 is repeated infinitely many times in the decimal form of this number). In general, a *periodic decimal* is a number whose decimal form has an infinite "tail" of repeating digits. For example, 3.42561561561... is a periodic decimal, where the digits 561 are repeated infinitely many times, while $\pi \simeq 3.1415926...$ is not, because no periodic repetition is observed in its digits. Often the repeated sequence in a periodic decimal is denoted by parentheses: $3.42561561561... = 3.42(561)$.

The general statement is that *any periodical decimal number is rational*, i.e., it is equal to a ratio of two integers. The infinite geometric series allows us to prove this statement by calculating the two integers whose quotient gives the exact value of the periodic decimal in each particular case.

Let us consider a couple of examples.

Example 10.6. Find the exact value of the number $1.(2) = 1.2222\ldots$.

Solution: First, note that we have the non-repeating part – namely, 1 – and the repeating part $0.2222\ldots = 0.(2)$. To compute the exact value of the repeating part, we add up the digits taking into account their place values:

$$0.2222\cdots = (2/10) + (2/100) + (2/1000) + \cdots$$
$$= \frac{2}{10^1} + \frac{2}{10^2} + \cdots + \frac{2}{10^n} + \cdots = \frac{2}{10}\left(1 + \frac{1}{10^1} + \frac{1}{10^2} + \cdots\right).$$

This is an infinite geometric series with $a = \frac{2}{10}$ and $r = \frac{1}{10} < 1$, and the sum is given by the formula

$$\frac{2}{10}\left(1 + \frac{1}{10^1} + \frac{1}{10^2} + \cdots\right) = \frac{\frac{2}{10}}{1 - \frac{1}{10}} = \frac{\frac{2}{10}}{\frac{9}{10}} = \frac{2}{9}.$$

Thus, $0.2222\ldots = 2/9$, so adding 1 to both sides, we obtain $1.2222\ldots = 1 + (2/9) = 11/9$. $\qquad\square$

Example 10.7. Find the exact value of 3.42(561).

Solution: Here the non-repeating part is 3.42. The repeating part, which we will represent in the form of an infinite geometric series, is 0.00561561... This part is equal to

$$0.00561561561\ldots = \frac{561}{10^5} + \frac{561}{10^8} + \frac{561}{10^{11}} + \cdots = \frac{561}{10^5}\left(1 + \frac{1}{10^3} + \frac{1}{10^6} + \cdots\right).$$

This is an infinite geometric series with $a = \frac{561}{10^5}$, and $r = \frac{1}{10^3} < 1$. The sum is given by the formula

$$\frac{561}{10^5}\left(1 + \frac{1}{10^3} + \frac{1}{10^6} + \cdots\right) = \frac{561}{10^5} \cdot \frac{1}{1 - \frac{1}{10^3}} = \frac{561}{10^5} \cdot \frac{10^3}{999} = \frac{187}{10^2 \cdot 333}$$

$$= \frac{187}{33,300}.$$

Now, for the non-repeating part we have

$$3 + \frac{42}{100} = 3 + \frac{21}{50} = \frac{150 + 21}{50} = \frac{171}{50}.$$

Finally, we obtain the exact value of the required number:

$$3.42(561) = \frac{171}{50} + \frac{187}{33,300} = \frac{171 \cdot 666 + 187}{33,300} = \frac{114,073}{33,300}.$$

$\square$

The algorithm used in Examples 10.6 and 10.7 can be generalized to produce a proof that any periodic decimal is rational. Indeed, the repeating part of the number is given by the sum of a convergent infinite geometric series with rational parameters a and $0 < r < 1$. The non-repeating part is finite in its decimal form and therefore is rational as well. The same algorithm allows us to compute the exact value of any given repeating decimal.

Note that the inverse operation, for example converting $\frac{114,073}{33,300}$ into 3.42(561), is done easily by any calculator; but calculators are not designed to find the *exact* value of a repeating decimal. Instead, it can be found with the help of an infinite geometric series as in Examples 10.6 and 10.7.

Practice problem 10.8. Find the exact value of each of the following repeating decimals:

1. 0.7777.... *Answer:* 7/9.

2. $1.121212\ldots$. *Answer:* $37/33$.

3. $1.2545454\ldots$. *Answer:* $69/55$.

APPLICATION 3

Drug levels in the body

Geometric series can help compute the amount of a drug found in a patient's body. If you take pills at regular intervals, the concentration of the drug in your body increases with each dose. Typically the body retains a certain percentage of the drug at the time the next dosage is ingested. Thus, the minimum drug levels in the body tend to go up with time. We will show how these drug levels can be computed using geometric series.

Example 10.9. Suppose a patient takes a 500-milligram tablet of a drug every day. In twenty-four hours, the patient's body excretes 90% of the existing drug in the body. What is the minimum drug level in the body of the patient after three days (just before the patient takes the next dose)? After ten days?

Solution: Let D_n be the drug levels in the body at the end of n days (measured in milligrams). We start with $D_0 = 0$ and $D_1 = 50$, because 90% of the 500 milligrams was excreted the first day. To compute D_2, we add the second-day intake, 500 milligrams, to D_1, and then rescale by 10%, because 90% of the entire drug amount is excreted by the end of the second day. Thus,

$$D_0 = 0, \qquad D_1 = (D_0 + 500) \cdot 0.1 = 500 \cdot 0.1,$$
$$D_2 = (D_1 + 500) \cdot 0.1 = 50 \cdot 0.1 + 500 \cdot 0.1 = 500 \cdot 0.1^2 + 500 \cdot 0.1.$$

Similarly, for all $n > 0$, the amount at the end of the nth day just before the next dose is taken is computed as follows:

$$D_n = (D_{n-1} + 500) \cdot 0.1 = \cdots = 500 \cdot 0.1^n + 500 \cdot 0.1^{n-1} + \cdots + 500 \cdot 0.1.$$

Using the geometric series formula, we can find the sum of all terms:

$$D_n = 500 \cdot 0.1(1 + 0.1 + \cdots + 0.1^{n-1}) = 50 \cdot \frac{1 - 0.1^n}{1 - 0.1} = 50 \cdot \frac{1 - 0.1^n}{9/10}$$

$$= \frac{500}{9}(1 - (1/10)^n).$$

Thus, after 3 days the minimum amount is

$$D_3 = \frac{500}{9}(1 - 0.001) = 55.5 \text{ mg},$$

while after 10 days it is $D_{10} = \frac{500}{9}\left(1 - (1/10)^{10}\right) = 55.55555555$ mg.   □

Example 10.9 shows that as time goes on, the minimum drug levels in the body stabilize to a limit. This limit is precisely the sum of the infinite geometric series above. Let us call this amount the *long-term minimum level* of the drug. In the previous example, the long-term minimum level of the drug is 55.(5) milligrams. Similarly to our earlier examples on recurring decimals, you can verify that this equals $55 + \frac{5}{9} = 500/9$ mg.

In general, suppose a patient takes D amount of a drug in each installment of time, and over the same time the body excretes f fraction of the drug (in Example 10.9, $D = 500$ mg and $f = 0.9$). What is the minimum amount D_n in the body after the nth day, and what is the long-term minimum level? As in the above analysis, we can compute D_n:

$D_0 = 0,$

$D_1 = (D_0 + D)(1 - f) = D(1 - f),$

$D_2 = (D_1 + D)(1 - f) = (D(1 - f) + D)(1 - f) = D(1 - f)^2 + D(1 - f),$

and so on; therefore

$$D_n = D(1 - f) + D(1 - f)^2 + \cdots + D(1 - f)^n.$$

This is a finite geometric series with initial term $D(1 - f)$ and common ratio $(1 - f)$. Using the geometric series formula, we obtain:

$$D_n = D(1 - f) \cdot \frac{1 - (1 - f)^n}{1 - (1 - f)} = D(1 - f) \cdot \frac{1 - (1 - f)^n}{f}.$$

The long-term minimum amount in the body D_∞ can be computed using the formula for the corresponding infinite geometric series. The formula makes sense because f is a fraction between 0 and 1. We obtain

$$D_\infty = \frac{D(1 - f)}{f}.$$

The total amount of the drug in the patient's body on the nth day drops from $D_{n-1} + D$ the moment after taking a dose, to D_n just before taking the subsequent dose. The long-term level of the drug is between D_1 (the minimum level on the first day) and $D_\infty + D$ (the limiting maximum level).

Note that the above example deals with drug levels in a patient's body. The mathematics is the same if applied to other settings. In the next example, we will call the *biological half-life* of caffeine the time it takes for the level of caffeine in the body to reduce by half.

Example 10.10. The biological half-life of caffeine for a given brand of coffee is approximately six hours. Joe likes this coffee and has a cup every morning. If each cup of this coffee contains forty milligrams of caffeine, what is the minimum amount of caffeine in Joe's body after three days? After a week? Longer?

Figure 10.2: Caffeine level in Joe's body.

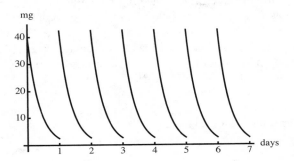

Solution: We can use the above formulas for D_n and for the long-term minimum amount. In these formulas $D = 40$, and we need to compute f. Because the amount of caffeine in the body is halved every six hours, it reduces to $(1/2)^4$ in 24 hours – i.e., in a day it becomes one-sixteenth of its original amount. Thus $1 - f = \frac{1}{16}$, or $f = \frac{15}{16}$. Now we compute:

$$D_n = 40 \cdot \frac{1}{16} \cdot \frac{1 - (1/16)^n}{15/16} = \frac{40}{15}(1 - (1/16)^n).$$

Figure 10.2 shows how the levels of caffeine in Joe's body change with time in the first seven days. After three days, the minimum level is $D_3 = 2.666015625$ mg; after seven days, we have $D_7 \simeq 2.666666$ mg; and the long-term minimum amount is $D_\infty = D(1 - f)/f = 40/15 = 2\frac{2}{3}$ mg. During each day, the amount of caffeine in Joe's body changes from $(D_{n-1} + 40)$ to D_n. The long-term caffeine level will always be between

$D_1 = 2.5$ mg and $D_\infty + 40 = 42\frac{2}{3}$ mg, with the lower bound gradually increasing and approaching $D_\infty = 2\frac{2}{3}$ mg.   □

APPLICATION 4

Geometric applications: fractals

Geometric series can be used to compute areas or other parameters of many recursively defined self-similar geometric shapes, or *fractals*. Self-similarity means that the object has infinitely repeating identical structure at all scales. One of the simplest visually engaging examples is the Koch snowflake. It is constructed as follows: add three equilateral triangles to the middle third of the three sides of a larger equilateral triangle. The resulting star has the boundary of twelve segments. Next, place an equilateral triangle on the middle third of each of these segments. Repeat an infinite number of times (see Figure 10.3).

Figure 10.3: The Koch snowflake – the first four steps. (Licensed under CC BY-SA 3.0 via Wikimedia Commons).

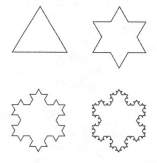

The area of the snowflake is clearly finite (for example, it fits in a circle with a radius equal to the side of the original triangle). But its perimeter is infinite: at each step, the length of each side increases by one-third, making each side and the perimeter grow by a factor of four-thirds. If the side of the original triangle is 1 inch, then after the first step the perimeter of the six-pointed star is $4/3 \cdot 1$ inches. After n steps, the resulting shape has the perimeter $(\frac{4}{3})^{n-1} \cdot 1$ inches, which becomes infinitely large as n grows.

Example 10.11. Let us compute the area of the Koch snowflake. Suppose the area of the initial triangle is 1 square inch. The area of the star obtained after the first step (upper right in the picture) is

$$1 + 3 \cdot \frac{1}{9} = 1 + \frac{1}{3},$$

because the area of each of the three added small triangles is one-ninth that of the original one. At the next step, twelve more triangles are added, each with an area one-ninth that of the small triangles, or $1/9^2$ that of the initial triangle. The area increases by

$$\cdots + 12 \cdot \frac{1}{9^2} = \cdots + \frac{4}{3^3}.$$

During each subsequent step, four times as many triangles are added as at the previous step, because each side of the snowflake gets replaced with four smaller sides. The area of each added triangle is one-ninth that of a triangle added at the previous step.

Therefore, the total area is given by the infinite geometric series:

$$1 + \frac{1}{3} + \frac{4}{3^3} + \frac{4^2}{3^5} + \frac{4^3}{3^7} + \cdots = 1 + \frac{1}{3}\left(1 + \frac{4}{9} + \frac{4^2}{9^2} + \frac{4^3}{9^3} + \cdots\right).$$

The geometric series in the parentheses can be summed using the general formula with $a = \frac{1}{3}$ and $r = \frac{4}{9}$:

$$1 + \frac{1}{3}\left(1 + \frac{4}{9} + \frac{4^2}{9^2} + \frac{4^3}{9^3} + \cdots\right) = 1 + \frac{1}{3} \cdot \frac{1}{1 - \frac{4}{9}} = 1 + \frac{1}{3} \cdot \frac{9}{5} = 1 + \frac{3}{5} = \frac{8}{5}.$$

Therefore, the area of the Koch snowflake is eight-fifths the area of the initial equilateral triangle. Even though the object is formed by infinitely many tiny shapes put together, its area is finite and can be computed using the geometric series.   □

Here is another example. Figure 10.4 shows the Sierpinski triangle – an equilateral triangle with infinitely many shaded inverted equilateral triangles inside. The first shaded triangle is constructed by joining the middles of the sides of the original triangle. Next, the same operation is performed with the three white triangles to the left, to the right, and above the shaded triangle, and so on to infinity (the picture shows the result of the first three steps, but it should be clear how to continue the process).

Example 10.12. Suppose that the area of the original white triangle is 1 square inch. What is the total area of all shaded triangles? The first shaded triangle has an area of 1/4 because it is one of the four equal triangles forming the original one. Next, three triangles of area 1/16 are added. Next, nine triangles of area 1/64 are added, and so on. We have the infinite geometric series:

$$\frac{1}{4} + \frac{3}{16} + \frac{9}{64} + \cdots .$$

Figure 10.4: The Sierpinski triangle, the first three steps.

One quarter is the common factor in all terms:

$$\frac{1}{4}\left(1 + \frac{3}{4} + \frac{3^2}{4^2} + \frac{3^3}{4^3} + \cdots\right) = \frac{1}{4}\sum_{k=0}^{\infty}\left(\frac{3}{4}\right)^k .$$

This is an infinite geometric series with $a = \frac{1}{4}$ and $r = \frac{3}{4}$. Because $r < 1$, the sum is given by the formula:

$$\frac{1}{4}\sum_{k=0}^{\infty}\left(\frac{3}{4}\right)^k = \frac{1}{4}\cdot\frac{1}{1-\frac{3}{4}} = \frac{1}{4}\cdot\frac{1}{\frac{1}{4}} = \frac{1}{4}\cdot\frac{4}{1} = 1.$$

After infinitely many steps, the shaded triangles will fill all of the white triangle! This means that we can get the total area of the shaded triangles as close as we like to the area of the white triangle by repeating the process sufficiently many times. Consequently, the area of all the white parts of the Sierpinski triangle (after infinitely many iterations) is zero.   □

Fractal dimension

What is the dimension of a fractal, for example, of the Koch snowflake?

First, let us try to formalize the idea of dimension. Everyone knows that a line is one-dimensional and a flat square is two-dimensional. If you try to think about it, you will probably come up with some version of a definition of the dimension as the number of *linearly independent* directions in the object. To formulate this notion rigorously, you will need methods of linear algebra. Besides, this definition is not suitable for a fractal, whose lacy structure makes the notion of direction inapplicable.

On the other hand, we can think of a line segment and a square as fractals, or self-similar objects. Indeed, if we divide a line segment into two equal pieces, each piece is similar to the original segment, meaning that it can be obtained by shrinking the original segment (see Figure 10.5). Each of the smaller pieces can be stretched back to become the original line segment if we multiply its length by 2. Therefore, the *magnification factor* of each of the smaller line segments in this case is 2. In the same way, if we divide a line segment into any given number of equal parts, then the magnification factor will be equal to the number of parts.

Figure 10.5: Line segment as a self-similar object.

Now consider the square. It is also a self-similar object. We can divide the square into four equal smaller squares, with half the sides of the original square (see Figure 10.6). The side of a smaller square has to be stretched by a factor of 2 to become equal to the side of the bigger square. Therefore, the magnification factor is 2. If we divided the square into, say, twenty-five smaller squares, then the magnification factor for each smaller square would be 5.

We observe a clear difference between the line segment and the square: in the first case, the number of self-similar parts is equal to the magnification factor, while in the second case, it is the square of the magnification factor. It turns out that this difference is exactly the difference between the one-dimensional and two-dimensional objects.

So we can assume that the dimension of a self-similar shape is the exponent of the magnification factor, needed to obtain the number of self-similar pieces that constitute the shape: $2^1 = 2$ pieces for the line, and $2^2 = 4$ pieces for the square. Taking the logarithm allows us to write down

Figure 10.6: Square as a self-similar object.

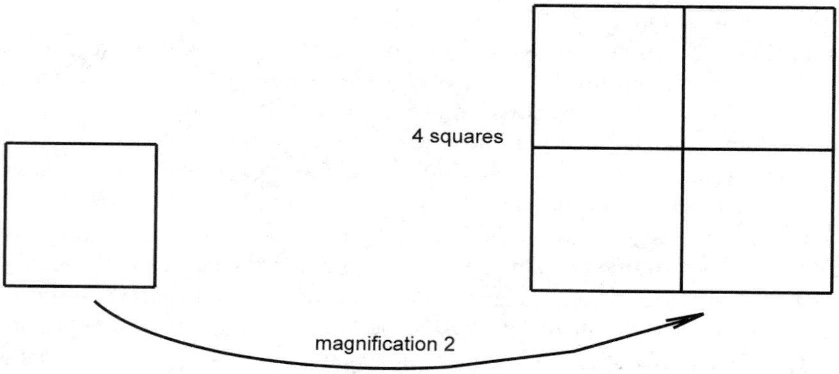

4 squares

magnification 2

an explicit formula for the dimension:

$$\text{dim} = \frac{\log(\text{number of pieces})}{\log(\text{magnification factor})}.$$

In our examples, the line segment L and the square S, we have

$$\text{dim L} = \frac{\log 2}{\log 2} = 1, \quad \text{dim S} = \frac{\log 4}{\log 2} = \frac{\log 25}{\log 5} = 2.$$

This definition of dimension can be applied to fractals. The result will often be a fractional dimension, as in the examples below.

Example 10.13. Find the fractal dimension of the shape consisting of the white parts of the Sierpinski triangle (ST).

Solution: ST contains three smaller white triangles that are similar to the original triangle. The sides of the smaller triangles are one-half of the sides of the original triangle; therefore, the magnification factor is 2. There are three self-similar pieces of that size in the Sierpinski triangle. Then the pattern repeats: each small white triangle contains three even smaller white triangles with sides half the size of the bigger one, and so on. For the dimension, we find

$$\text{dim ST} = \frac{\log 3}{\log 2} \simeq 1.58.$$

$\square$

Example 10.14. Let us compute the fractal dimension of the Koch snowflake (KS).

Answer: Here each side of the snowflake is similar to one of the four sides built on it, so the number of self-similar pieces is four. As the smaller side is one-third the size of the bigger side, the magnification factor is 3. The fractal dimension of KS is

$$\dim \text{KS} = \frac{\log 4}{\log 3} \simeq 1.26.$$

□

The Koch snowflake consists of infinitely many infinitely small line segments, while the white part of the Sierpinski triangle consists of infinitely many infinitely small triangles. It is intuitively clear that the Sierpinski triangle is more substantial than the Koch snowflake – and the fractal dimension reflects this fact: 1.58 for ST is bigger than 1.26 for KS.

The algorithm we developed to calculate the fractal dimension works only in the simplest cases. In mathematics, a more elaborate definition is used that allows one to compute the fractal dimension of any self-similar pattern, and even obtain estimates for some real-life objects such as channel networks, lightning strikes, and the tracheobronchial system in humans. The fractal dimension is used to measure the complexity and space-filling capacity of a pattern.

Many real-world phenomena show (at least statistically) some self-similar behavior. For example, any coastline is a self-similar curve: the closer you look at it, the more complex it becomes, and the patterns encountered on a large scale are bound to appear again on a smaller scale. In fact, the fractal dimension of a given coastline can be estimated, and the smoother the coastline, the closer to 1 is its dimension. For example, the fractal dimension of the coastline of Great Britain is estimated to be 1.25; that of South Africa, only 1.05; and that of Norway, 1.52 (they have the fjords! See Figure 10.7).[2] The coastline example is famous for its appearance in the foundational work of modern fractal geometry, as presented in an article by the French-American mathematician Benoit Mandelbrot titled "How Long is the Coast of Britain? Statistical Self-Similarity and Fractional Dimension".[3] The answer to the question depends on how closely you look: the shorter your measuring stick, the bigger the length of the coastline. If you could measure with infinite precision, the coastline would be infinitely long, for the same reason that the perimeter of the

---

[2]Source: Fractalfoundation.org.

[3]*Science*, May 5, 1967, Volume 156, no. 3775, pp. 636-638.

Figure 10.7: Norwegian coastline (Bergen harbor, photo by Bernt Rostad).

Koch snowflake is infinite. Here, again, we cannot observe infinity because our measuring stick will always have a nonzero length. But we can observe that the length we are trying to measure grows beyond any bound as the precision increases. This may be as close as we can get to infinity in the real world. The fact that this effect can be detected in reality is one of the miracles that makes mathematics worth studying.

The concept of a fractal dimension is widely applied in image analysis, in particular in diagnostic medicine[4] and in neuroscience. Below we choose to describe a fascinating and quite unexpected application of fractal dimension in archaeology.

An application of fractal geometry in archaeology

Methods of fractal geometry were applied by a group of German scientists to demonstrate the existence of human activity on a large scale at some point in the past in an area of Egypt known as the necropolis of Dahshur.[5]

---

[4]A.L. Goldberg and B.J. West, "Fractals in physiology and medicine," *Yale Journal of Biology and Medicine*, 60 (5), 1987, pp. 421–435.

[5]Arne Ramischa, Wiebke Bebermeiera, Kai Hartmanna, Brigitta Schütta and Nicole Alexanian, "Fractals in topography: Application to geoarchaeological studies in the surroundings of the necropolis of Dahshur, Egypt," *Quaternary International*, Volume 266, July 17, 2012, pp. 34–46.

Figure 10.8: The necropolis of Dahshur with the Bent Pyramid (photo by Arian Zwegers).

The royal necropolis of Dahshur is situated in a desert about 20 miles south of Cairo and close to the Nile's floodplain. Big rivers are surrounded by branching networks of channels that play a major role in the formation of landscape topography, if this formation is defined only by natural factors. The channels carve fractal patterns in the land. If the landscape of Dahshur was formed naturally, it should have followed the fractal pattern similar to those shown in Figure 10.9. Mathematically, this means the following. Suppose we have a pretty precise topographic map of the area. Moving a square frame from left to right on the map, for each position of the frame we compute the fractal dimension of the channel network inside the frame (this should be a number between 1 and 2), and the fractal dimension of the surface topography within the same frame (a number between 2 and 3). If the landscape developed naturally, there should be a strong correlation – clear dependence – between the two obtained sets of data: roughly, the more complicated the channel network, the more uneven the surface. Indeed, this is the case in the areas *surrounding* the necropolis of Dahshur.

But this is not the case in a large area of the necropolis itself. The correlation between the fractal dimension data of the channel network and the surface topography data is low or insignificant in an area of at least 2.5 square miles. This indicates that the creation of the landscape was

Figure 10.9: Natural floodplain formation: aerial view of Crystal River, Florida (left, photo by Visit Citrus), and view of Columbia River, Washington (right, photo by Paul Schultz).

disrupted at one point by a human impact on a large scale. The scientists concluded that the whole area once experienced a major construction or landscape architecture project, which took place most probably around the time the Bent Pyramid and the Red Pyramid were built (Old Kingdom, around 2600 BC). The channel beds might have been used for transportation of the building materials, or more directly for irrigation or landscape architecture. Traditional methods of archaeological research did not bring clear evidence of human landscape-changing activity on such a scale, and so the discovery is entirely due to the mathematical methods of fractal geometry applied to archaeology.

EXERCISES

Question 10.1. The most famous example of dividing a distance into infinitely many pieces is the Zeno paradox of Achilles and the tortoise. The tortoise offers to run a race against Achilles, if the tortoise can have a 100-foot head start. The speed of Achilles is assumed to be 10 times that of the tortoise. While Achilles covers the 100 feet, the tortoise advances by 10 feet. While Achilles runs that distance, the tortoise goes 1 foot more, and so on to infinity. Formulate a mathematical model for this problem, and persuade yourself that Achilles will be able to overtake the tortoise in finite time. To do this, you will have to sum up an infinite geometric series.

Question 10.2. In Question 10.1, what if the tortoise had a 200-foot head start? Find the distance covered by Achilles before he overtakes the tortoise in either of two ways: computing the time at which Achilles overtakes the tortoise if he runs with a given speed, or computing directly the distances. In the second case you will have to sum an infinite geometric series.

Question 10.3. After a tropical rainfall, the rainwater discharge from the roof of a building amounts to 1 gallon in the first minute, and in each consecutive minute, four-fifths of the volume discharged in the previous minute.

1. How much water will be eventually discharged from the roof?

2. How soon will the rate of discharge drop below one coffee spoon (0.0093 gallon) of water per minute?

3. How much water is discharged from the roof by that time?

Question 10.4. Find the exact value of each of the following repeating decimals:

1. 0.575757....

2. 1.864864864....

3. 0.2376 2376 2376....

Question 10.5. Find the exact value of each of the following repeating decimals:

1. 3.4343434....

2. 12.213 213 213....

3. 148.78787....

Question 10.6. Suppose a patient takes a 200-milligram tablet of a particular drug every twelve hours. The patient's body retains only 20% of the drug after twelve hours. Determine the minimum drug levels in the patient's body after two days, a week, and in the long term.

Question 10.7. One gram of nutrient agar is added every hour to a petri dish in which bacteria are being cultivated. As they grow in numbers, the bacteria absorb 95% of all existing nutrients in the dish each hour. Compute the minimum amount of agar in the dish in milligrams, to five decimal places, after the experiment has been conducted for 2 hours, 1 day, and in the long term.

Question 10.8. In Figure 10.10 let us say the white square has sides 1 inch long. The largest shaded square has sides 1/2 inch long, the next one 1/4 inch, and so on, so that each consecutive square has sides half as long as the previous one. There are infinitely many squares in the sequence, going from the lower left corner of the white square to its upper right corner, all of them shaded. What is the total area of all shaded squares?

Figure 10.10: Squares on the diagonal.

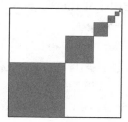

Question 10.9. Compute the fractal dimension of a solid equilateral triangle.

Question 10.10. Consider the following fractal. In the first step, on each side of a square, four smaller squares are added to the middle third of each side. The resulting shape has a perimeter of twenty segments. In the next step, a square is added to the middle third of each of these intervals, and so on to infinity. Figure 10.11 shows the first two steps of the construction.

1. Check that the perimeter of the obtained shape has infinite length.

2. If the area of the original square is 1, find the area of the shape obtained after infinitely many steps. Looking at the picture, can you guess the answer?

Question 10.11. Find the fractal dimension of the boundary of the shape described in Question 10.10.

Question 10.12. The curve in Figure 10.12 consists of a horizontal interval of length 2 inches, two vertical intervals of length 1 inch each, and two spirals consisting of infinitely many intervals, with the length of each next interval four-fifths of the length of the previous interval. The picture shows the first few steps of the spirals. Find the total length of the curve.

Figure 10.11: A square snowflake.

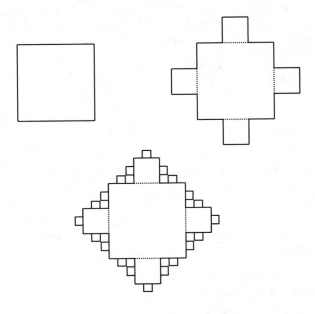

Figure 10.12: Square spirals.

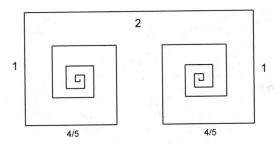

Question* 10.13. A well-known fractal is the *Cantor set*, defined as follows: start with the closed interval $[0, 1]$, then remove the middle third to obtain two intervals: $[0, 1/3]$ and $[2/3, 1]$. From each of these two intervals, we again delete the middle third to obtain four intervals:

$$[0, 1/9], \quad [2/9, 1/3], \quad [2/3, 7/9], \quad [8/9, 1].$$

Keep repeating this procedure. The remaining set of points after infinitely many steps is called the Cantor set. Check that the Cantor set has zero length. *Hint:* Find the length of the part removed from the interval at the nth step. Then check that the total length of the part removed from the interval is 1.

Figure 10.13: The first three steps in the Cantor set.

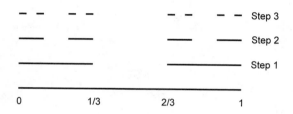

Question* 10.14. Consider the Cantor set constructed in Question 10.13.

1. Find a part of the set that is similar to the whole set, but smaller. How many such pieces does the set contain?

2. What is the magnification factor between the piece and the whole set?

3. Compute the fractal dimension of the Cantor set.

Question 10.15. When a helium-filled balloon gets punctured, it loses 1 cubic foot of helium in the first second. Each subsequent second, the balloon loses one-third of the volume lost in the previous second. Assuming that the balloon will lose all its helium after infinite time, estimate the initial volume of the balloon.

Question 10.16. Two trains that are 50 feet apart start braking while they are moving toward each other. In the first second, both trains cover 10 feet. In each subsequent second, the train approaching from the left covers three-fifths of the distance it covered in the previous second, while the train approaching from the right covers three-eighths of the distance it covered in the previous second.

1. Find the total distance covered by the train coming from the left.

2. Find the total distance covered by the train coming from the right.

3. Will the trains crash?

Chapter 11

Estimation: What is your first guess?

In this chapter, we will discuss a technique that allows us to make quick but relatively accurate estimates. This technique is called *first-order approximation*, or *linear approximation*. Historically, first-order approximations became popular because they were much easier to compute than precise formulas, and yet provided estimates of the results to a high degree of accuracy.

In today's world, where millions of people have access to calculators through smartphones and laptops that can perform exact calculations, first-order approximations are still important. Such approximations can be used to obtain a rough guess of an answer when precision is not required and time is crucial (e.g., in a business meeting).[1] Moreover, in science and engineering, using a precise formula often results in a complicated mathematical model. Finding an explicit algebraic solution for such a model might be hard or even impossible, while its first-order approximation leads to a much simpler model. It can be solved exactly, behaves better under small changes of parameters, and yet is practically applicable to a high degree of accuracy. Examples include a first-order approximation for the physical model of a pendulum, and for Einstein's relativistic mass-energy equation. The latter leads to the well-known formula for kinetic energy. We will discuss these examples at the end of the chapter.

---

[1] For more examples on how to make quick correct estimates, and why they are useful, see *Guesstimation*, by Laurence Weinstein and John A. Adams, Princeton University Press, 2008.

MATH

First-order approximation for exponents

You may have been amazed by someone claiming to compute powers and roots of numbers very quickly and without using a calculator or paper. For instance, what is 1.0001^{15}? We claim that it is quite close to 1.0015. Indeed, actual calculations show that the answer is $1.001501\ldots$, so our claim was correct to five decimal places. Or, what is the cube root of 1.0001, or, equivalently, $1.0001^{\frac{1}{3}}$? We claim that it is approximately $1.00003333\ldots$. This claim turns out to be accurate to eight decimal places.

These estimates are made using the principle of *first-order approximation for exponents*. It says that for a number $(1+x)$ which is very close to 1 (i.e., x is a very small positive or negative number), and all real t, the following approximation formula holds:

$$(1+x)^t \approx 1 + tx.$$

Let us justify the method when $t = 2$ and $t = 3$. Expanding the products and collecting the terms, we have

$$(1+x)^2 = 1 + 1 \cdot x + x \cdot 1 + x^2 = 1 + 2x + x^2,$$
$$(1+x)^3 = (1+x)^2(1+x) = (1+2x+x^2)(1+x)$$
$$= 1 + 2x + x^2 + x + 2x^2 + x^3 = 1 + 3x + 3x^2 + x^3.$$

Now suppose $x \approx 1/100$. Then the terms containing $x^2 \approx 1/10,000$ and $x^3 \approx 1/1,000,000$ would contribute only very small corrections to the result. Discarding these terms, we get

$$(1+x)^2 \approx 1 + 2x, \quad (1+x)^3 \approx 1 + 3x,$$

which are exactly the first-order approximations for $(1+x)^2$ and $(1+x)^3$. In fact, the name "first-order approximation" comes from this procedure: we expand the given power of $(1+x)$ as a polynomial and keep only the first-order terms, i.e., the constants and the terms containing x in the first power.

It is not difficult to show (see Question 11.11) that for any positive integer n, the first-order terms in the expansion of the product $(1+x)^n$ are $1 + nx$. (See the section on Pascal's triangle in Chapter 15 for a complete polynomial expansion formula for $(1+x)^n$.)

The case when t is not a positive integer is more complicated, but the result is the same: the terms that matter most in the expression of $(1+x)^t$ are $1 + tx$.

Example 11.1. Now we can explain how we made the two claims above: set $x = 0.0001$ and $t = 15$, or $1/3$. Then a simple calculation gives us each of the two claims:

$$(1 + 0.0001)^{15} \approx 1 + 15 \cdot 0.0001 = 1.0015,$$

$$(1 + 0.0001)^{1/3} \approx 1 + \frac{1}{3} \cdot 0.0001 \approx 1.0000333.$$

$\square$

Note that the above formula is merely an approximation and not a precise equality. The point is that the error term (i.e., the difference between the two sides) becomes smaller and smaller as x gets closer and closer to 0. Thus, the smaller the *perturbation x*, the closer our estimate $1 + tx$ is to the actual answer.

First-order approximation for products

Sometimes we have to deal with more than one independent small perturbation. Suppose we have two real values given in the form $a + \Delta a$, $b + \Delta b$, where Δa and Δb are pretty small compared with a and b, respectively. Then the first-order approximation of the product is given by the formula:

$$(a + \Delta a) \cdot (b + \Delta b) \approx a \cdot b + a \cdot \Delta b + b \cdot \Delta a.$$

Let us explain why this approximation gives a good estimate of the product. Suppose that $a, b, \Delta a, \Delta b$ are positive numbers, and $\Delta a/a \approx \Delta b/b \approx 1/100$. Then using the FOIL method (see Chapter 1), we have

$$(a + \Delta a) \cdot (b + \Delta b) = a \cdot b + a \cdot \Delta b + b \cdot \Delta a + \Delta a \cdot \Delta b.$$

Clearly the term $a \cdot b$ contributes the most to the product. Let us estimate the magnitude of the remaining terms with respect to $a \cdot b$. We have

$$\frac{a \cdot \Delta b}{a \cdot b} = \frac{\Delta b}{b} \approx \frac{1}{100}, \quad \frac{b \cdot \Delta a}{a \cdot b} = \frac{\Delta a}{a} \approx \frac{1}{100},$$

$$\frac{\Delta a \cdot \Delta b}{a \cdot b} \approx \frac{1}{100} \cdot \frac{1}{100} \approx \frac{1}{10,000}.$$

The last term, $\Delta a \cdot \Delta b$, is $10,000$ times less significant than $a \cdot b$, and 100 times less significant than any of the remaining terms. So by discarding it, we make the computation much simpler while getting an answer that is pretty close to the correct result.

Example 11.2. Estimate the product $1.001 \cdot 2.004$.

Solution: Here $1.001 = 1 + 0.001$, and the second summand is much smaller than the first. Similarly, $2.004 = 2 + 0.004$. Using the first-order approximation formula, we obtain

$$1.001 \cdot 2.004 \approx 1 \cdot 2 + 1 \cdot 0.004 + 2 \cdot 0.001 = 2 + 0.004 + 0.002 = 2.006.$$

Indeed, 2.006 is close to the actual value: $1.001 \cdot 2.004 = 2.006004$, so our approximation is correct up to five decimal places. □

The method works just as well if some of the numbers $a, b, \Delta a, \Delta b$ are negative. For instance, we can estimate the product $2.03 \cdot 1.98$ to be

$$(2 + 0.03) \cdot (2 - 0.02) \approx 2 \cdot 2 + 0.03 \cdot 2 - 2 \cdot 0.02 = 2 + 0.06 - 0.04 = 4.02.$$

The exact result is $2.03 \cdot 1.98 = 4.0194$. What matters for the method is that the values $\Delta a / a$ and $\Delta b / b$ should be close to zero.

Practice problem 11.3. Estimate each of the following quantities.

1. 1.004^8. *Answer:* 1.032.

2. $1.004 \cdot 2.015$. *Answer:* 2.023.

First-order approximation for the exponential and logarithmic functions

For small (compared to 1) values of x, the following approximation holds:

$$\exp(x) = e^x \approx 1 + x.$$

Indeed, if $x = 0$, we have $\exp(x) = \exp(0) = 1 + 0$. The fact that for a small x we have to add just x and not $3x$ or $10x$ is a consequence of an exceptionally fortunate choice of the number $e \approx 2.7182818...$ as the base of the exponential function. If we wanted to approximate a^x for a small x with a positive constant $a \neq 1$ different from e, the approximation formula would have been different and more complicated. The first-order approximation formula for e^x can serve as one of the ways to define e and is one of the reasons the number e holds a special status among all positive real numbers. Recall that the number e was defined in Chapter 5 as the limit of the values $(1 + \frac{1}{n})^n$ as n tends to infinity. It is possible to prove that the two definitions are equivalent, but we will not worry about it in this book. For the purpose of the applications considered further in this chapter, we will assume the first-order approximation formula for e^x and show in an example (see Example 11.11) that it can be used to produce reasonably accurate estimates.

Similarly, for the natural logarithm of $(1 + x)$, where x is small compared to 1, we have

$$\ln(1 + x) \approx x.$$

If $x = 0$, then $\ln(1) = 0$. Here again, the number e, the base of the natural logarithm, is responsible for the simple form of the approximation: the first-order term of $\ln(1 + x)$ is just x itself.

We will see many examples and applications of the above approximations in the remainder of this chapter. We also discuss some generalizations of these approximation formulas.

APPLICATIONS IN ENGINEERING

The use of first-order approximation is widespread in theoretical mathematics (e.g., in proving that something is true for "small enough" or "large enough" values of x) – but we will concentrate on applications in engineering, error analysis, and physics. We now provide several examples of such applications from different real-world settings. In some examples, we also mention the actual answer so that you can decide for yourselves which of the approximations are more accurate than others.

In real life, you may occasionally be required to perform first-order approximations with no recourse to a computer or calculator. The point of our providing the precise answer in the examples below is to reassure you via repeated demonstrations that, as long as you do not require extreme precision, a first-order approximation should provide a reasonably accurate answer.

Example 11.4. Find the approximate volume of a cubical box with edge length equal to:

1. 1.004 cm.

2. 1.0007 in.

3. 3.001 cm.

4. 2.003 in.

Solution: For the first two parts, the computations are similar, so we will elaborate only on the first one.

1. Use the first-order approximation formula with $x = 0.004$ and $t = 3$; thus,
   $$1.004^3 \approx 1 + 3 \cdot 0.004 = 1.012.$$

   Therefore the approximate volume of the box is 1.012 cubic centimeters.

2. Similarly, the volume of the box in the second question is approximately 1.0021 cubic inches. You can use a calculator to verify to what degree of precision this approximation is correct.

3. For the next two parts, the computations are a bit different. For the third problem, we cannot apply the first-order approximation for exponents, because it approximates only $(1 + x)^t$, and not $(3 + x)^t$. Thus, we have to reduce the calculation to the form $(1 + x)^t$ before using the formula. But this is easy using the properties of exponents:

   $$3.001^3 = \left(3 + \frac{1}{1,000}\right)^3 = 3^3 \left(1 + \frac{1}{3,000}\right)^3$$
   $$\approx 3^3 \left(1 + 3 \cdot \frac{1}{3,000}\right) = 3^3(1.001) = 27.027 \text{ cm}^3.$$

   Indeed, when we compute 3.001^3 using a calculator, the actual answer turns out to be $27.027009\ldots$.

4. Similar to the previous computation, we find
   $$2.003^3 = 2^3(1.0015^3) \approx 8(1.0045) = 8.036 \text{ in}^3.$$

   $\square$

Square roots and cube roots are useful in determining dimensions as well. For instance, taking square roots yields the length of a square from its area. Here is an example involving a cube root.

Example 11.5. *Computing the radius of a sphere from its volume.* A ball is dipped into a full container of water, and the amount of water spilled is 36.144π cubic centimeters. Given that the volume of a sphere with radius r is $\frac{4}{3}\pi r^3$, find the approximate radius of the ball.

Solution: We write: $\frac{4}{3}\pi r^3 = 36.144\pi$. Canceling π and dividing by 4, we obtain: $r^3/3 = 36.144/4 = 9.036$, and so $r^3 = 27.108$. Therefore we need to find the cube root of 27.108, which is a number close to 3 centimeters. Again, we have to bring the expression to the form $(1 + x)^t$ before we can use the first-order approximation formula. Thus,

$$27.108^{1/3} = 27^{1/3}\left(1 + \frac{1}{27}\cdot\frac{108}{1,000}\right)^{1/3} = 3\left(1 + \frac{4}{1,000}\right)^{1/3}$$
$$\approx 3\left(1 + \frac{1}{3}\cdot\frac{4}{1,000}\right) = 3 + \frac{4}{1,000}.$$

This implies that $r \approx 3.004$ cm. Just to compare, the actual value of the radius is: $\sqrt[3]{36.144\cdot(3/4)} = 3.0039947$ cm, which is accurate to four decimal places. □

The examples above made use of the formula for approximating exponents in the Math section above: $(1 + x)^t \approx 1 + tx$ when x is small compared to 1. Now we show how the product formula, $(a + \Delta a)(b + \Delta b) \approx ab + a\Delta b + b\Delta a$, can be used.

Example 11.6. *Computing the area of a rectangle.* The approximate dimensions of a rectangular plate of metal are 8.055×10.002 cm². Find its area.

Solution: Use $a = 8, \Delta a = 0.055, b = 10, \Delta b = 0.002$ and compute:

$$8.055\cdot 10.002 \approx 8\cdot 10 + 8\cdot 0.002 + 10\cdot 0.055 = 80 + 0.016 + 0.55 = 80.566 \text{ cm}^2.$$

The true answer can be verified as equal to 80.56611, so the approximation is accurate to three decimal places. □

The formula for multiplying two quantities with small perturbations extends to three quantities. Here is an example.

Example 11.7. *Computing the volume of a box.* The dimensions of a rectangular box are 5.01 ft $\times 4.02$ ft $\times 2.01$ ft. Compute its (approximate) volume.

Solution: In this case we use the approximation

$$(a_1 + \Delta a_1)(a_2 + \Delta a_2)(a_3 + \Delta a_3) \approx a_1 a_2 a_3 + a_2 a_3 \Delta a_1 + a_1 a_3 \Delta a_2 + a_1 a_2 \Delta a_3.$$

Substituting the given values, we obtain the volume to approximately equal

$$5.01 \cdot 4.02 \cdot 2.01 \approx 5 \cdot 4 \cdot 2 + 4 \cdot 2 \cdot 0.01 + 5 \cdot 2 \cdot 0.02 + 5 \cdot 4 \cdot 0.01$$
$$= 40 + 0.08 + 0.2 + 0.2 = 40.48 \text{ ft}^3.$$

The actual value of the volume is $40.4818\ldots$. $\square$

It is natural to ask if one can write an approximation for *any* number of terms multiplied by one another. Recall from the beginning of the chapter how we obtained the formula in the case of *two* terms:

$$(a_1 + \Delta a_1)(a_2 + \Delta a_2) = a_1 a_2 + a_1 \cdot \Delta a_2 + a_2 \cdot \Delta a_1 + \Delta a_1 \cdot \Delta a_2.$$

This is an exact statement; now we note that $\Delta a_1 \cdot \Delta a_2$ is very small compared to the remaining three terms, and hence we ignore it to obtain the approximation formula mentioned above.

Similarly if there are three factors, then we ignore all terms where two or more perturbations Δa are multiplied, as these terms are all quite small compared to the remaining terms. This is how we obtained the approximation formula in Example 11.7. We can rewrite it as follows:

$$(a_1 + \Delta a_1)(a_2 + \Delta a_2)(a_3 + \Delta a_3)$$
$$\approx a_1 a_2 a_3 + a_2 a_3 \Delta a_1 + a_1 a_3 \Delta a_2 + a_1 a_2 \Delta a_3$$
$$= a_1 a_2 a_3 \left(1 + \frac{\Delta a_1}{a_1} + \frac{\Delta a_2}{a_2} + \frac{\Delta a_3}{a_3} \right).$$

(Note that we are assuming here that the quantities a_1, a_2, a_3 are all nonzero.) In the same manner, it is not hard to show that the following first-order approximation holds for an arbitrary number of factors:

$$(a_1 + \Delta a_1)(a_2 + \Delta a_2) \cdots (a_n + \Delta a_n)$$
$$\approx a_1 a_2 \cdots a_n \left(1 + \frac{\Delta a_1}{a_1} + \frac{\Delta a_2}{a_2} + \cdots + \frac{\Delta a_n}{a_n} \right).$$

This approximation is valid whenever $a_1, \ldots, a_n$ are all nonzero, and moreover, $\Delta a_1, \ldots, \Delta a_n$ are small compared to $a_1, \ldots, a_n$, respectively. This last formula is quite general: in Example 11.6, we used the special case of this approximation where there are only two terms being multiplied, i.e., $n = 2$. This yielded the area of some rectangle. In Example 11.7, we multiplied three terms to get the volume, so we used the same formula with $n = 3$.

Error analysis

Our next application of first-order approximation is in error analysis. Often in industrial processes, a machine is programmed to manufacture a large number of copies of some object. There are specifications provided for the dimensions of the manufactured unit; however, it is not always the case that the copies produced conform exactly to these specifications. There may be manufacturing errors, and in turn, these minute errors in the dimensions of the object can get *compounded* or cause bigger errors in the area or volume of the object.

Example 11.8. A machine in a factory produces ball bearings of a pre-specified diameter. The manufacturing process allows for an error of 2% of the diameter – i.e., the actual diameter can vary between 98% and 102% of the specifications. Find the (approximate) maximum and minimum percentage deviation in the volume of the ball bearing from the pre-specified volume.

Solution: Suppose the pre-specified diameter is d centimeters. The actual diameter can vary between $0.98d$ and $1.02d$. Now the pre-specified radius would be $d/2$, so that the specified volume of the spherical ball bearing is given by $V = \frac{4}{3}\pi(d/2)^3 = \frac{\pi}{6}d^3$. Thus, we are to compute the percentage deviation between this amount and either of the two amounts:

$$\frac{\pi}{6}(0.98d)^3 = 0.98^3 \cdot \frac{\pi}{6}d^3 = 0.98^3\ V, \qquad \frac{\pi}{6}(1.02d)^3 = 1.02^3 \cdot \frac{\pi}{6}d^3 = 1.02^3\ V.$$

We compute the two quantities. By the first-order approximation formula,

$$0.98^3 = (1 + (-0.02))^3 \approx 1 + 3 \cdot (-0.02) \approx 1 - 0.06 = 0.94,$$

and similarly, check for yourself that $1.02^3 \approx 1.06$. This means that if the actual sizes of the ball bearings vary (approximately) between $0.94V$ and $1.06V$, there can be a 6% deviation in the volume of the ball, on either side. In other words, the volume can be up to 6% less or up to 6% more than the pre-specified amount V.   □

Note that the percentage of deviation is the same above and below the actual amount. This is because in taking powers or roots, the upper and lower error margins on the specified amount are equal in magnitude, but differ by a sign. Now, in applying the first-order approximation formula, we replace taking a power or root by multiplying the error margins by that power (or fractional power in the case of taking a root). Thus, the new margins are also equal in magnitude, but differ by a sign.

In our next example, two different manufactured components each have their relative errors (expressed in percentages of the specifications). We will see how one can combine both the power and the product approximation formulas in order to compute the *compound error* in terms of the individual errors.

Example 11.9. *Volume of a cylinder.* In a factory, open cylindrical barrels are manufactured by two different devices. One device creates the circular base of radius r, and the other creates a sheet of metal with height h, which is then bent and attached on top of the base. The pre-specified volume of the cylinder is therefore $\pi r^2 h$. Now suppose the device manufacturing the bases has a 2% manufacturing error in the radius, while the other device has a 3% error in the height. (We can assume that the width of the metal sheet can be made to correspond exactly to the circumference of the circular base by changing the width of the overlap.) Estimate the maximum percentage deviation in the volume of the cylinder from the pre-specified volume.

Solution: Similar to the previous example, we can compute either the maximum deviation above, or below, the actual amount $\pi r^2 h$. Let us compute it above so as to avoid any complications arising from minus signs. The maximum volume that can occur is

$$\pi(1.02r)^2(1.03h) = \pi r^2 h \cdot (1.02^2 \cdot 1.03).$$

We compute this last product in stages. The first step is to compute:

$$1.02^2 \approx 1 + 2 \cdot 0.02 = 1.04.$$

The second step is to then multiply this approximation by 1.03, to get:

$$1.02^2 \cdot 1.03 \approx 1.04 \cdot 1.03 = (1+0.04)(1+0.03) \approx 1 \cdot 1 + 1 \cdot 0.03 + 1 \cdot 0.04 = 1.07.$$

Thus, we obtain that $\pi(1.02r)^2(1.03h) \approx 1.07\pi r^2 h$. This indicates a maximum of 7% deviation from the pre-specified volume of the cylinder.   □

Remark. In the above example, we could have approximated the product $1.02^2 \cdot 1.03$ by treating it as a product of three perturbed terms, as in Example 11.7. In that case, we would have written the product as

$$1.02^2 \cdot 1.03 = (1 + 0.02)(1 + 0.02)(1 + 0.03)$$

and then used the first-order approximation formula for the product of three terms. If you carry out this computation carefully, you will obtain the same result, that the maximum deviation is 7% of the pre-specified volume.

APPLICATIONS IN PHYSICS

In the remaining part of this chapter, we discuss a few applications of first-order approximation in physics.

Constant speed motion

Here is how to estimate the average speed of a train (or a boat, or an airplane) easily and with pretty good precision.

Example 11.10. According to a timetable, a train takes 63 minutes to travel from city A to city B. If the distance between the cities is 70 miles, estimate the average speed of the train.

Solution: First note that if the train had traveled for 60 minutes, the speed would be exactly 70 miles per hour. Now, 63 minutes is a small perturbation of 60; 6 minutes is one-tenth of an hour, therefore 3 minutes is 5% of an hour, and so the time taken by the train is 1.05 hours. Then the speed equals

$$\frac{70}{1.05} = 70 \cdot 1.05^{-1} = 70 \cdot (1 + 0.05)^{-1}$$
$$\approx 70 \cdot (1 + (-1) \cdot 0.05) = 70 \cdot 0.95 = 66.5 \text{ mph}.$$

Exact calculation yields $70/1.05 = 66.67$ mph (to two decimal places). $\square$

Exponential decay

Exponential decay is a widespread phenomenon in physics. You have already seen the mathematics behind radioactive decay in Chapter 8. Here is another instance of exponential decay occurring in physics: the decrease of the voltage in electrical circuits. Consider a resistor and capacitor in a circuit, shown in the diagram below (RC circuit):

   A capacitor is a device used to store electric energy that can be discharged to create a current when a capacitor is connected to an external circuit, as in Figure 11.1. A resistor is a passive component that converts

Figure 11.1: Discharging capacitor

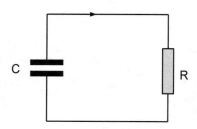

the energy carried by the electric current into heat. The voltage across the capacitor is a measure of how much electric energy it still contains. The voltage $V(t)$ across the capacitor decays exponentially with time t. More precisely, if V_0 is the initial voltage, then

$$V(t) = V_0 e^{-\frac{t}{RC}},$$

where R and C are the characteristics of the resistor and the capacitor, respectively. The resistance R is measured in ohms (Ω), and the capacitance C, in farads (f). For our considerations, it is sufficient to know that 1 farad times 1 ohm equals 1 second.

In order to estimate the decay of the voltage in an RC circuit with time, we will need the first-order approximation formula for the exponential function given at the beginning of this chapter.

Example 11.11. Suppose a capacitor in an RC circuit is charged with 500 volts. Suppose further that the characteristics of the capacitor and the resistor in the circuit are $C = r$ f and $R = 2/r$ Ω, respectively, for some fixed positive r. Find the voltage at $t = 0.05$ sec and $t = 0.1$ sec.

Solution: Because 1 farad multiplied by 1 ohm equals 1 second, we compute that $RC = 2$ sec, and therefore the voltage decays according to the (precise) formula $V(t) = 500e^{-t/2}$. Because we are interested in the small values of $t/2$, the first-order approximation formula can be applied. We compute for $t = 0.05 = 1/20$:

$$\begin{aligned} V(0.05) &= 500e^{-0.05/2} = 500\exp(-1/40) \\ &\approx 500 \cdot (1 + (-1/40)) = 500 \cdot 39/40 = 487.5 \text{ V.} \end{aligned}$$

(The exact value is 487.655 Volts.) Similarly, for $t = 0.1 = 1/10$ we compute:

$$V(0.1) = 500 \exp(-1/20) \approx 500 \cdot (1 + (-1/20)) = 500 \cdot 19/20 = 475 \text{ V}.$$

(The exact value is 475.615 V. In particular, the example shows that as t gets larger, the approximation becomes less precise.) □

We will describe the next two physics-based applications only briefly.

Pendulum

A pendulum is a weight attached by a light rod to a point, called the pivot, so that it can swing freely. If the weight is pulled away from the resting position, the gravitational force pulls it back down and causes it to oscillate. The *period* of the pendulum is the time T it takes for the weight to complete the cycle, in other words, to swing back and (almost) return to the initial position. In computing the period of a pendulum, one often sees the following formula, which relates T with the pendulum's length ℓ:

$$T = 2\pi \sqrt{\ell/g},$$

where g is the gravitational acceleration. Note that this formula does not even include the initial angle of the swing!

Figure 11.2: Pendulum of length ℓ. The formula works only for small swings.

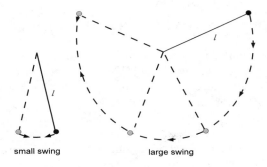

small swing   large swing

It is easy to see that the above formula cannot possibly approximate the period of a pendulum of a given length for every initial angle. For instance, if the initial position of the pendulum is almost vertically upward, then the time taken for a complete cycle can get pretty long. As the pendulum's starting position gets closer and closer to the unstable vertical equilibrium point, the period grows indefinitely.

Thus, the above formula for T is valid only for small swings of the pendulum. However, even in this case it is only an approximation and *not* an exact formula. In other words, $T \approx 2\pi\sqrt{\ell/g}$. The exact formula for T is more complicated and involves trigonometric functions. The approximation used in this case comes from considering the first-order approximation of a trigonometric function called sine, $\sin x \approx x$ (Figure 11.3). This is just another example of how first-order approximations help simplify mathematical models used in physics.

Figure 11.3: A linear approximation of $\sin(x)$.

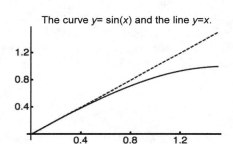

The curve $y= \sin(x)$ and the line $y=x$.

Mass-energy equivalence and kinetic energy

The last application of first-order approximation we will discuss relates a well-known equation from the special theory of relativity to the formula for kinetic energy used in classical mechanics. Recall that the kinetic energy of a particle with mass m and velocity v is $E_{\text{kinetic}} = \frac{1}{2}mv^2$. It turns out that from the special relativity viewpoint, this formula is only a first-order approximation, and we now explain why.

In the early 1900s, Albert Einstein, in one of his famous papers in *Annalen der Physik*, wrote down an equation relating mass and energy: $E = mc^2$. Here E is the energy of a particle, m is "rest mass" (i.e., the mass measured when the particle is at rest), and c is the speed of light. We present the last few steps in Einstein's derivation, and how it leads to a better understanding of kinetic energy.

Einstein claimed that the "total energy" of a particle is the sum of its "rest energy" (which is the intrinsic energy carried by the particle when stationary) and its kinetic energy. Further, Einstein showed that the total energy equals

$$E_{\text{total}} = E_{\text{rest}} + E_{\text{kinetic}} = \frac{mc^2}{\sqrt{1 - (v/c)^2}},$$

where v is the velocity of the particle and $c \approx 3 \cdot 10^8$ m/s is the speed of light. When the velocity is zero, one obtains: $E_{\text{rest}} = mc^2$, which yields the well-known mass-energy equivalence. What remains is the kinetic energy:

$$E_{\text{kinetic}} = E_{\text{total}} - E_{\text{rest}} = \frac{mc^2}{\sqrt{1 - (v/c)^2}} - mc^2.$$

Now, in the framework of classical mechanics, the speed of any particle is pretty small compared to the speed of light. For instance, the speed of a commercial jet, approximately 260 meters per second, is less than one millionth the speed of light. When v/c is small compared to 1, its square $(v/c)^2$ is even smaller. Therefore one can use a first-order approximation to compute:

$$\begin{aligned}
E_{\text{kinetic}} &= mc^2(1 - (v/c)^2)^{-1/2} - mc^2 \\
&\approx mc^2 \left(1 + \frac{-1}{2} \cdot \frac{-v^2}{c^2}\right) - mc^2 = mc^2 + \frac{1}{2}mv^2 - mc^2 \\
&= \frac{1}{2}mv^2.
\end{aligned}$$

Thus the familiar formula for the kinetic energy $\frac{1}{2}mv^2$ is nothing but the first-order approximation of the more general Einstein formula. This is a reasonable estimate if v/c is small compared to 1, i.e., the actual velocity of a particle is small compared to the velocity of light. In most practical situations considered in Newtonian mechanics, this is indeed the case. Einstein's formula provides a more precise result and represents the next step in our quest for an adequate model of nature.

Where the first-order approximation applies, and why*

We conclude with a brief discussion of the idea behind first-order approximations. Having read this far, you may ask: does every formula have a first-order approximation? Can every process that is modeled mathematically also have another, simpler formula that works in many cases?

The answer to these questions is yes, with rare exceptions. Basically, a first-order approximation exists for any dependence that is *smooth*, meaning that the process it models happens gradually and does not involve sudden drops and jumps.

In fact, to derive first-order approximations for most formulas used in this book, it is enough to use a few basic rules, starting with the four formulas given in the Math section at the beginning of this chapter. Namely, if x and y are small with respect to 1, then

$$(1 + x)^t \approx 1 + tx$$
$$(1 + x)(1 + y) \approx 1 + x + y$$
$$e^x = \exp(x) \approx 1 + x$$
$$\ln(1 + x) \approx x$$

Let us see how this and other first-order approximation formulas can arise. Suppose we have a geometric series $\sum_{k=0}^{\infty} ar^k$, where r is between -1 and 1, and a is a real number (see Chapter 10). Then the sum is given by the formula

$$\sum_{k=0}^{\infty} ar^k = \frac{a}{1 - r}.$$

However, if we know in addition that r is pretty small compared with 1, then we can use the first-order approximation for exponents and obtain

$$\sum_{k=0}^{\infty} ar^k = \frac{a}{1 - r} = a \cdot (1 - r)^{-1} \approx a \cdot (1 + r).$$

These are exactly the first two terms of the series! Thus the first-order approximation method tells us that if r is very small compared to 1, then the sum of an infinite geometric series is almost the same as the sum of its first two terms.

Figure 11.4: Functions and their linear approximations.

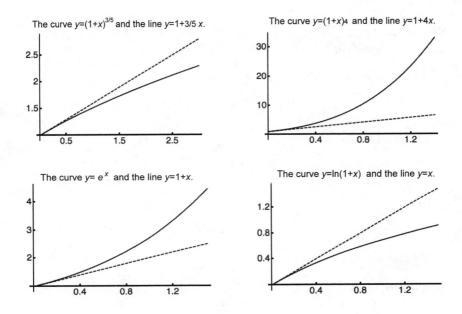

The curve $y=(1+x)^{3/5}$ and the line $y=1+3/5\ x$.

The curve $y=(1+x)4$ and the line $y=1+4x$.

The curve $y=e^x$ and the line $y=1+x$.

The curve $y=\ln(1+x)$ and the line $y=x$.

What might be less obvious is that the example with the geometric series actually reveals the general mechanism of the first-order approximation. Namely, a *smooth* process can *always* be represented as an (infinite) sum of consecutive powers of a variable with some numerical coefficients. This is a difficult mathematical statement, and defining the series associated with a given process might be highly nontrivial. But assuming we have the series, the next step is to estimate the variable. If we know that the variable in question is small compared to 1, we can cut the series down to the first two summands. The resulting formula provides a quick and practical way of getting an approximate answer. This is the essence of the first-order approximation method.

EXERCISES

Question 11.1. Compute approximately the following expressions using first-order approximation:

1. The cube and cube root of 1.015.

2. The cube of 2.999. *Note:* $2.999 = 3 + (-0.001)$.

3. The number $e^{0.012}$.

Question 11.2. Compute approximately the following expressions using first-order approximation:

1. $4.001^{3.5}$.

2. $\ln(\sqrt{e} + 0.04)$.

3. The fourth power of 2.007.

Question 11.3. The volume of a (right circular) cylinder is $\pi r^2 h$, where r and h denote the radius and height of the cylinder, respectively.

1. Suppose a cylinder is 3.995 inches high, with a radius of 4.001 inches. Using first-order approximation, estimate the volume as a multiple of π. (Meaning, the answer should be close to 64π, and you are supposed to compute a number close to 64. Don't multiply it by π using a calculator; keep it as a decimal number times π.)

2. Carry out the same calculations if the height is 4.002 inches and the radius is 1.001 inches. (This time, your answer should be close to 4π cubic inches.)

Question 11.4.

1. Suppose a measuring device makes (at most) a 2% error while measuring either the radius or the height of a cylinder. Approximately what percentage error would it make in computing the volume? What if we had 3% error for all measurements instead of 2%?

2. Estimate the volume of a box with dimensions 1.001 inches, 2.011 inches, and 0.999 inches.

Question 11.5. An assembly line manufactures a rectangular table, with separate machines creating the square tabletop and the four legs. Suppose the device creating the tabletops can have up to a 4% error in the length and in the width, while the device creating the four legs can have up to a 1% error in the height. Compute (approximately) the maximum percentage error in the volume under the rectangular table.

Question 11.6. In this question we will see that simple interest is the first-order approximation of compound interest.

1. Suppose I deposit a thousand dollars in a bank that compounds quarterly at an annual rate of 1%. What is the amount in the bank after two years?

2. Suppose instead that the bank added simple interest at the same annual rate. What is the amount in the bank after two years? By how many dollars does it differ from the previous answer? *Note:* the amount deposited was a thousand dollars.

3. Now consider the general question. Suppose a dollar amount A deposited in a bank gets compounded n times a year at an annual rate of $r\%$ per year.

   Write down an expression for the total amount after t years, and then verify that its first-order approximation is indeed the amount after t years under a simple interest scheme – under the hypothesis that r/n is small compared to 100.

   Now subtract the initial deposit A from both sides; hence the compound interest is approximated by the simple interest.

4. Verify that the first-order approximation does not depend on n, the number of compoundings per year.

Question 11.7. Show that simple interest is also the first-order approximation of continuous compounding. See Chapter 5 for the formula for continuous compounding.

Question 11.8. Suppose an object travels in a straight line at a velocity of 51 miles per hour for 57 minutes. Estimate the distance traveled using first-order approximation. What if the object was traveling for 63 minutes?

Question 11.9. The relative rate of decay of radium-226 is $k = -4.28 \cdot 10^{-4}/\text{yr}$. A man unearths a lead can containing radium-226 a hundred years after it was placed in the ground; $t = 100$ yr ago, it contained 1 gram of radium-226. Estimate (without using a calculator) how much radium-226 remains in the can, recalling that the model for exponential decay is given by: $A(t) = A(0) \exp(kt)$.

Question 11.10. Suppose a resistor and capacitor are arranged in series in an RC circuit (see Example 11.11), with resistance $R = 10\ \Omega$ and capacitance $C = 0.01$ f, respectively. If the capacitor is charged with 100 volts, find the approximate voltage at times $t = 0.2$ sec and $t = 0.5$ sec after the circuit is completed.

Question* 11.11. Show that in the polynomial expansion of the expression $(1 + x)^n$ for a variable x and any positive integer n, the first-order part is $1 + nx$. *Hint*: Use the examples with $n = 2$ and $n = 3$ in the beginning of the chapter. Then suppose you know the claim is true for integers up to n. Can you prove it for $n + 1$? (This is the method of *mathematical induction*.) See the end of Chapter 15 for another way to obtain the coefficients in the expansion of $(1 + x)^n$.

Question* 11.12.   1. Use the first-order approximation for exponents to estimate the sum of the series

$$\sum_{k=0}^{4} (0.998)^k = 1 + 0.998 + 0.998^2 + 0.998^3 + 0.998^4.$$

2. Can you derive the formula for a first-order estimate for the sum of a geometric series

$$\sum_{k=0}^{n} x^k,$$

assuming that x is pretty close to 1?

Chapter 12

Modular arithmetic I: Around the clock and the calendar

In this chapter we will introduce a new way of looking at integer numbers: we will consider them up to a multiple of a given number. We call the branch of mathematics that studies the behavior of such numbers *modular arithmetic*. Without knowing it, we often use modular arithmetic when we consider time-related questions arising in everyday life, for example, what time it is 40 hours from now, or what day of the week you were born on. The same methods apply to any cyclic process.

MATH

Modular arithmetic deals only with integer (whole) numbers and studies their divisibility properties. For two integers a and n, we say that a is *divisible* by n (or a is a multiple of n) if $a = nk$ for some integer k. For example, 14 is divisible by 2 and by 7, but not divisible by 3 or 4.

Here is the definition that gives modular arithmetic its name: let a and x be integers and n an integer greater than 1. We say that a is *congruent* to x *modulo* n if $(a - x)$ is divisible by n. The notation is

$$a \equiv x \pmod{n}.$$

In other words, $a \equiv x \pmod{n}$ if and only if a and x differ by a multiple

213

of n.

Example 12.1. For instance, $25 - 4 = 21$ is a multiple of 7, so $25 \equiv 4$ (mod 7). Also, $30 - (-5) = 35$ is a multiple of 7, so $30 \equiv -5$ (mod 7). At the same time, $30 - 2 = 28$ is a multiple of 7 as well; therefore $30 \equiv 2$ (mod 7). □

Properties of congruences

Many properties of modular congruences are "the same" as those of the usual equalities. You can add, subtract, and multiply congruences *modulo the same integer*. Suppose

$$a \equiv x \pmod{n} \qquad \text{and} \qquad b \equiv y \pmod{n},$$

where a, b, x, y are integers and n an integer greater than 1. Note that n is the same in both congruences. Then

1. $(a + b) \equiv (x + y)$ (mod n),

2. $(a - b) \equiv (x - y)$ (mod n),

3. $ab \equiv xy$ (mod n),

4. $a^s \equiv x^s$ (mod n), where s is a positive integer.

Let us see why (1) is true. By the definition of a congruence, we have

$$a - x = nk, \qquad b - y = nm$$

for some integers k and m. Adding the two equalities, we get

$$(a + b) - (x + y) = nk + nm = n(k + m).$$

Because the right-hand side is a multiple of n, we get $(a + b) \equiv (x + y)$ (mod n). You can check property (2) similarly.

To check (3), let us rewrite the conditions $a - x = nk$ and $b - y = nm$ as follows:

$$a = nk + x, \qquad b = nm + y.$$

Then, taking the product ab and using the distributive law, we have

$$ab = (nk+x)(nm+y) = n^2km+nmx+nky+xy = n(nkm+mx+ky)+xy.$$

Therefore, $ab - xy = n(nkm + mx + ky)$ is a multiple of n, which means that $ab \equiv xy \pmod{n}$.

Property (4) is a direct consequence of (3): taking the product of congruences $a \equiv x \pmod{n}$ s times, we obtain $a^s \equiv x^s \pmod{n}$ for any positive integer s. □

Example 12.2. Find the smallest positive integer which is congruent to $1,000$ modulo 7.

Solution: Because $1,000 = 10^3$, we can use property (4) to simplify the computation. We have $10 = 7 + 3$, therefore $10 \equiv 3 \pmod{7}$. Then by property (4),

$$10^3 \equiv 3^3 \pmod{7} \implies 10^3 \equiv 27 \equiv 21 + 6 \equiv 6 \pmod{7}.$$

Therefore $1,000 \equiv 6 \pmod{7}$. We can check this using a calculator: $1,000 - 6 = 994 = 7 \cdot 142$. □

Remark: You cannot divide congruences. For example, we have $24 \equiv 3 \pmod{7}$ and $12 \equiv 5 \pmod{7}$, but $\frac{24}{12} = 2 \not\equiv \frac{3}{5} \pmod{7}$. In particular, $\frac{3}{5}$ is not an integer and it cannot appear in a congruence.

Solving congruences

The basic question we are trying to solve in modular arithmetic is to find an integer x such that for given positive integers a and n,

$$a \equiv x \pmod{n}.$$

This means that $(a - x)$ is divisible by n. Of course, the answer to this question is not unique: for example,

$$46 \equiv 4 \pmod{7}, \quad 46 \equiv -3 \pmod{7}, \quad 46 \equiv 18 \pmod{7}, \quad \text{and so on.}$$

In fact there are infinitely many solutions. What we really want to find is the smallest nonnegative x such that the congruence holds. With this additional condition, the answer is unique:

$$46 \equiv 4 \pmod{7}.$$

In general, by "solving a congruence" we will mean finding the unique integer x such that for given integers a and $n > 1$,

$$a \equiv x \pmod{n} \quad \text{and} \quad 0 \leq x \leq n - 1.$$

This number x is the *remainder* of the integer division of a by n.

Practice problem 12.3. Solve the following congruences:

1. $53 \pmod 6$. *Answer:* $53 \equiv 5 \pmod 6$.

2. $1,014 \pmod 4$. *Answer:* $1,014 \equiv 1,012 + 2 \equiv 2 \pmod 4$.

3. $45 \cdot 43 \cdot 51 \pmod 7$. *Answer:* By property (3), $45 \cdot 43 \cdot 51 \equiv 3 \cdot 1 \cdot 2 \equiv 6 \pmod 7$.

APPLICATIONS

Congruences tell us that there is more than one way of adding integers. You can add positive integers the usual way. Then the result will get bigger and bigger each time.

Figure 12.1: Multiples of 5 on the infinite number line.

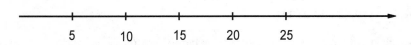

Or you can add them *modulo a given number*; then the result will stay within a finite interval and you can imagine that your addition is going "around a circle." For example, let us try adding 5 modulo 12:

$$5 \equiv 5 \pmod{12}, \quad 5 + 5 \equiv 10 \pmod{12}, \quad 5 + 5 + 5 \equiv 3 \pmod{12},$$

$$5 + 5 + 5 + 5 \equiv 8 \pmod{12}, \quad 5 + 5 + 5 + 5 + 5 \equiv 1 \pmod{12}.$$

Why would anyone ever consider such a bizarre operation? Below we will consider applications of congruences to counting time, describing cyclic motion, and deriving divisibility tests for integers. We will consider more advanced applications in cryptography and computer security in Chapter 13. But first we would like to discuss an interesting parallel between the concepts of modular arithmetic and the ideas of universal vs. relative truth in philosophy and literature.

Figure 12.2: Multiples of 5 modulo 12.

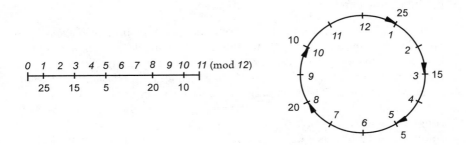

Easy as $2 + 2 = 4$

Contrary to popular belief, 4 is not the only correct answer to what is $2 + 2$. If you count modulo 3, you will have $2 + 2 \equiv 1 \pmod 3$. Moreover, $2 \cdot 2 \equiv 1 \pmod 3$ as well. If you count modulo 4, then $2 + 2 \equiv 2 \cdot 2 \equiv 0 \pmod 4$.

Modular arithmetic provides a perfect framework for understanding how different answers to a given question can be true at the same time. It shows that there are systems where 4 is equivalent to 1 or to 0; therefore these answers to the question "What is $2 + 2$?" have the right to coexist. At the same time, modular arithmetic reminds us that the variety of coexisting truths must be limited. In George Orwell's *Nineteen Eighty-Four*, the Inner Party requires everyone to acknowledge that $2 + 2 = 5$.[1] Let us try to find a counting system where this might make sense. Modulo what number is 5 congruent to 4? If $5 \equiv 4 \pmod a$, then $1 \equiv 0 \pmod a$. Then 1 should be divisible by a, which can happen only if $a = 1$. But then *all* integers are equivalent because the difference between any two integers is divisible by 1 (strictly speaking, congruences are considered only modulo numbers greater than 1, precisely because congruences modulo 1 are not very discerning). If you assume $2 + 2 = 5$, you will have to assume as well that all integer numbers are one and the same, and any number can be the right answer to any arithmetic operation. In this system, $124 - 76 = 11$ and $3 \cdot 5 = 7$. Too much diversity leads to the loss of all meaning. Then any statement can be true, exactly as it happened in the dystopian dicta-

---

[1] George Orwell, *Nineteen Eighty-Four*, Martin Secker and Warburg, 1949.

torship described in *Nineteen Eighty-Four*.

Practice problem 12.4. You can design your own trick in a similar way: modulo what number is $2 \cdot 3 \equiv 2$? *Answer: 4.*

Here is another example from a well-known book. In *The Hitchhiker's Guide to the Galaxy*, Arthur Dent pulls out random letters from a Scrabble bag, to get the sentence "What do you get if you multiply six by nine?"[2] As you may know, the "answer" is 42, to which another character says that there is something "fundamentally wrong with the universe." We hope not. Let us instead try to answer the question: modulo which number is $6 \cdot 9$ congruent to 42? We have $6 \cdot 9 = 54 \equiv 42 \pmod{a}$. Then $54 - 42 = 12 \equiv 0 \pmod{a}$, so 12 has to be divisible by a. There are many positive integers that satisfy this condition, in particular, $2, 3, 4, 6, 12$. To obtain the richest and most discerning counting system, we will choose the biggest number $a = 12$. Then we have $6 \cdot 9 \equiv 42 \pmod{12}$, and the universe is saved. However, this comes at a price: any two numbers that differ by a multiple of 12 have to be declared the same.

Practice problem 12.5. In subsequent novels, the "answer" 42 is declared to be off by 1, and the actual "answer" is found to be 43. Modulo which number is $6 \cdot 9$ congruent to 43? *Answer: 11.*

Clocks and hours

Now we turn to more conventional and direct applications of modular arithmetic. You might have noticed that the picture we used to illustrate modular addition in Figure 12.2 looks exactly like a clock dial. This is because hours are counted modulo 12 (or modulo 24). For example, what time is it, 16 hours after 10 am? We compute: $10 + 16 = 26$, but this answer does not make much sense because the time is measured in a 24-hour cycle. To find the time of day, we have to consider the number of hours modulo 24, or modulo 12 if you specify am or pm. This is equivalent to subtracting the biggest multiple of 24 that can fit into the given number of hours: $26 - 24 = 2$ (am). The fact that 16 hours after 10 am is 2 am is represented by the following congruence:

$$(10 + 16) \equiv 2 \pmod{24}.$$

Example 12.6. What time is it 100 hours from noon?

---

[2]Douglas Adams, *The Hitchhiker's Guide to the Galaxy*, Pan Books, 1979.

Solution: We need to find 100 (mod 24). Because $100 = 4 \cdot 24 + 4$, we have

$$100 \equiv 4 \quad (\text{mod } 24).$$

Disregarding the 4 complete days, we just say: 4 hours after noon, or 4 pm.   □

When we have to deal with larger numbers, properties of congruences become useful.

Example 12.7. What time is it 10,000 hours from noon? Instead of computing the remainder of division of $10,000$ by 24, we can use properties of congruences. We already know that $100 \equiv 4$ (mod 24), so we have

$$10,000 \equiv 100^2 \equiv 4^2 \equiv 16 \quad (\text{mod } 24).$$

Though 4^2 actually equals 16, we write $4^2 \equiv 16$ because we are working modulo 24. So the answer is: 16 hours past noon, or 4 am.   □

In this last example, we didn't just leave the answer as 16 hours past noon. We immediately converted it into a more familiar form, 4 am. The operation we performed was taking 16 modulo 12:

$$16 \equiv 4 \quad (\text{mod } 12).$$

Because clocks traditionally have a 12-hour cycle, we are used to reporting time modulo 12; in other words, the number of hours goes from 1 to 12. This is suitable for everyday life, because usually we have no trouble telling day from night. However, in situations where nothing can be taken for granted, it is safer to count hours modulo 24 – this way of counting time is called *military time*. For example, when we say 1345 hours (military time), we mean 1:45 pm. In congruences, this corresponds to $13 \equiv 1$ (mod 12). In the age of commercial jets people routinely cross many time zones in a day, and confusing am and pm can become a real problem. This is why airplane schedules use military time.

Practice problem 12.8. (a) What time is it 43 hours from midnight? *Answer:* 7 pm.
(b) What time is it $4,300$ hours from midnight? *Hint:* Use part (a) and Example 12.6 to solve the congruence modulo 24. *Answer:* $43 \cdot 100 \equiv -5 \cdot 4 \equiv -20 \equiv 4$ (mod 24). The time is 4 am.

Months and seasons

Months and seasons satisfy cyclic patterns as well. The questions we can ask in this case are similar to those we asked about the time of day, with the month cycle having length 12, and the season cycle length 4.

Example 12.9. What month is it 99 months after April?

Solution: We need to compute 99 modulo 12:

$$99 \equiv 12 \cdot 8 + 3 \equiv 3 \pmod{12}.$$

Ninety-nine months after April is July.   □

Practice problem 12.10. What season is it 35 seasons after spring?
Answer: $35 \equiv 3 \pmod 4$. It is winter.

Perpetual calendar

Another situation in which we routinely use modular arithmetic is when we try to figure out the day of the week of a given date. For example, what day of the week was June 17, 2013? Suppose you remember the day of the week some time before or after that date – say, you remember that May 25, 2013, was a Saturday. Then you can count the number of days between the two dates, and take it modulo 7. Since there are 31 days in May, we find

$$(31 - 25) + 17 \equiv 6 + 17 \equiv 23 \equiv 2 \pmod 7.$$

Therefore, counting 2 days from a Saturday, we find that June 17, 2013, was a Monday.

Example 12.11 (*Doomsday rule*)**.** The Doomsday rule is a well-known algorithm for calculating the day of the week of any date. It was designed by the British mathemtician John Conway. Here is the basic observation underlying this rule: *Check that in any year, the dates 4/4, 6/6, 8/8, 10/10, and 12/12 all fall on the same day of the week.*

Solution: To check the statement, we need to find the number of days between the given dates, and take it modulo 7. Note that April, June, September, and November all have 30 days, while May, July, August, and October have 31 days. Therefore the number of days between the consecutive pairs of the given dates is the same: it is always $30+31+2 = 63$. Now,

$$63 \equiv 0 \pmod 7,$$

because $63 = 7 \cdot 9$, so there is a whole number of weeks between each two consecutive dates. Therefore, they all fall on the same day of the week in any given year. For instance, in 2000, April 4 was a Tuesday, and so were June 6, August 8, October 10, and December 12. □

Practice problem 12.12. Check that in any given year, July 4 also falls on the same day of the week as the dates listed in Example 12.11.

You can also check that the last day of February and May 2 fall on the same day as all of the dates listed in Example 12.11 (see Question 12.3). This is good to know, because now we can use these dates as *markers* to find the day of the week of any given date in the same year.

Example 12.13. Suppose that June 5 fell on a Sunday one year. What day of the week was October 28 in the same year?

Solution: Because June 5 was a Sunday, we know that June 6 was a Monday, and so was October 10 (see Example 12.11). Then for October 28, we find

$$28 - 10 \equiv 18 \equiv 4 \pmod 7.$$

Therefore, October 28 was the fourth day after a Monday, which is a Friday. □

Now let us consider how the day of the week of a given date changes over multiple years.

Example 12.14. Suppose this year March 12 is a Wednesday. What day of the week will it be the next year?

Solution: The answer depends on the particular year. If the *next* year is not a leap year, then the number of days between March 12 this year and the next is 365. In this case, we compute

$$365 \equiv 350 + 14 + 1 \equiv 7 \cdot 50 + 7 \cdot 2 + 1 \equiv 1 \pmod 7.$$

Therefore, March 12 will be the next day after Wednesday, which is a Thursday.

If the next year is a leap year, then there are 366 days between March 12 this year and the next. Then we have

$$366 \equiv 350 + 14 + 2 \equiv 2 \pmod 7.$$

In this case, March 12 the next year will be a Friday.

In general, the day of the week of any given date moves one day ahead if the year in between is common, and two days ahead if the year in between is a leap year.   □

We can reformulate the last conclusion as follows: the day of the week of a given date moves one day forward each year, and one additional day forward each leap year.

Let us recall the rule that defines a leap year. It has three parts:

> All years divisible by 4 are leap years,
> Except those divisible by 100,
> But including those divisible by 400.

For example, years 1968 and 2004 were leap years; and so was the year 2000, because it is divisible by 400. But the year 1900 was not a leap year, because 1900 is divisible by 100 but not by 400.

Now we can use the rule derived in Example 12.14 to find the day of the week of any date in the past, or of a particular event that happened some time ago, or will happen in the future.

Example 12.15. Margaret was born on September 3, 1986, and in 2013 her birthday fell on a Tuesday. On what day of the week was Margaret born?

Solution: The leap years between 1986 and 2013 are: 1988, 1992, 1996, 2000, 2004, 2008, 2012. Therefore, there are 7 leap years and $2013 - 1986 = 27$ total years. We have to subtract 1 for each year, and 1 more for each leap year from the day of the week of Margaret's birthday in 2013. We compute

$$-7 - 27 \equiv -34 \equiv -6 \equiv 1 \pmod 7.$$

Moving one day forward from a Tuesday, we obtain a Wednesday. Margaret was born on a Wednesday.   □

To find the day of the week of an event a long time ago, it is convenient to combine the "one day per year, one additional day per leap year" rule with the Doomsday rule.

Example 12.16. President Lincoln was assassinated on April 14, 1865. What day of the week was that?

Solution: Suppose we remember that New Year's Day fell on a Wednesday in 2014. Then we can find the day of the week of 12/12/2013:

$$-31 - 1 + 12 \equiv -20 \equiv -6 \equiv 1 \pmod 7;$$

therefore 12/12/2013 was a Thursday. By the Doomsday rule, 4/4/2013 was a Thursday as well. Then we compute for 4/14/2013:

$$14 - 4 \equiv 10 \equiv 3 \pmod 7;$$

therefore 4/14/2013 was a Sunday. Now we need to count back to the same date in 1865. There are $2013 - 1865 = 148$ full years in between. The leap years closest to the given dates are 1868 and 2012. Dividing $2012 - 1868 = 144$ by 4, we obtain 36. Therefore, there were 37 years divisible by 4 between 1865 and 2013. Of these, only 1900 was not a leap year, which leaves 36 leap years. Then from a Sunday, we need to count back one day for each year, and one additional day for each leap year:

$$-148 - 36 \equiv -184 \equiv -7 \cdot 26 - 2 \equiv -2 \pmod 7.$$

Counting 2 days back from a Sunday, we obtain a Friday. President Lincoln was assassinated on a Friday. □

Now let us try to apply the same methods to bigger time spans.

Example 12.17. On which days of the week did the second and the third millennia start?

Solution: Suppose again that we know January 1, 2014, was a Wednesday; count the days back from then. If you have another favorite date for which you remember the day of the week, you can count back from that date instead.

Let us begin with the third millennium, which started on January 1, 2001. There are $2014 - 2001 = 13$ years from then to January 1, 2014. Of these, only three years, 2004, 2008, and 2012, were leap years. Just as before, we compute

$$-13 - 3 \equiv -16 \equiv -2 \pmod 7.$$

Counting two days back from a Wednesday, we get a Monday. January 1, 2001, was a Monday.

Strictly speaking, we cannot use the same argument to find the first day of the second millennium. Back then, people in different parts of the world used many different calendars, and none of them was exactly the same as the Gregorian calendar we use today. But if we extend the current calendar into the past, then the second millennium would have started on January 1, 1001. This extended calendar is called the *proleptic Gregorian calendar*. Let us find the day of the week of January 1, 1001, according to this calendar.

There are $2014 - 1001 = 1,013$ full years between then and January 1, 2014. Because $4 \cdot 24 = 96$, we get 24 leap years in each full hundred between years 1001 and 2001, plus the leap years 2004, 2008, 2012. Now we also have to add the leap years divisible by 400: these are 1200, 1600 and 2000. Totally we get $1,013$ years, of which

$$24 \cdot 10 + 3 + 3 = 246$$

were leap years. Then counting one day back for each year and another day back for each leap year, we get

$$-1,013 - 246 \equiv -1,259 \equiv -7 \cdot 179 - 6 \equiv -6 \equiv 1 \pmod{7}.$$

Moving one day forward from a Wednesday, we get a Thursday. So according to the proleptic Gregorian calendar, the second millennium started on a Thursday.   □

Now let us try to get a little more historical. Pope Gregory XIII introduced the Gregorian calendar, used today, in 1582. The Italian doctor and astronomer Aloysius Lilius designed this system and presented it to the calendar reform commission in 1575. For a millennium and a half before that, the commonly adopted calendar in Europe was the Julian calendar, introduced by Julius Caesar in 46 BC. The Julian calendar assumed the length of a year to equal 365.25 days, and considered each fourth year a leap year. The Gregorian calendar corrected the length of a year to equal $365 + \frac{97}{400} = 365.2425$ days, which is about 11 seconds less, and corresponds better to the time the Earth takes to complete its orbit around the Sun. This resulted in demoting the years divisible by 100, but not by 400, from leap years to common years. In addition, the current date was corrected by skipping 10 days: October 4, 1582, was followed immediately by October 15, 1582, while the cycle of weekdays was not affected. On this day, the reform was implemented in most of Italy, Spain, Portugal, and the Polish-Lithuanian Commonwealth. It took several centuries for the Gregorian calendar to get adopted throughout Europe (in particular, in the British Empire, and what is now the eastern US, the change took place in 1752, when September 2, 1752, was immediately followed by September 14). The upshot of the reform was that the months stayed reasonably in sync with the seasons: spring in April, summer in July, and so on. In particular, Pope Gregory's concern was to keep the Easter celebration, which was tied with the spring equinox, from drifting back in the calendar. Today the Gregorian calendar is accepted as a civil calendar almost everywhere in the world.

Example 12.18. What day of the week was the last day of the Julian calendar in Italy?

Solution: We will find the day of the week of October 15, 1582, and then count one day back. We already found that 12/12/2013 was a Thursday (see Example 12.16). By the Doomsday rule, 10/10/2013 was a Thursday as well. Then 10/15/2013 was five days forward, yielding a Tuesday. The total number of years between 1582 and 2013 is $2013 - 1582 = 431$. The years divisible by 4 closest to the given dates are 1584 and 2012. Dividing $2012 - 1584 = 428$ by 4, we get 107, meaning there were 108 years divisible by 4 in between. Of these, we have to exclude 1700, 1800, 1900 (but not 1600 and 2000). Therefore, there were 105 leap years. Counting back one for each year, and another one for each leap year, we find

$$-431 - 105 \equiv -536 \equiv -7 \cdot 76 - 4 \equiv -4 \equiv 3 \pmod{7}.$$

Counting three days forward from a Tuesday, we obtain a Friday. So, the last day of the Julian calendar in Italy, October 4, 1852, was a Thursday. □

With this in mind, let us try to find a more meaningful answer to the question about the beginning of the second millennium.

Example 12.19. What day of the week did the second millennium start on according to the Julian calendar?

Solution: We need to find the day of the week of January 1, 1001, counted according to the Julian calendar. We know that October 4, 1582, was a Thursday. The lengths of the months in the Julian calendar were the same as in the current Gregorian system. Therefore, we can use the Doomsday rule. Because 10/10/1582 would have been a Wednesday according to the Julian calendar, it was also Wednesday on the last day of February, namely, 2/28/1582. Then counting back to January 1, 1582, we find

$$-28 - 30 \equiv -58 \equiv -7 \cdot 8 - 2 \equiv -2 \pmod{7}.$$

January 1, 1582, was two days back from a Wednesday, which is a Monday. Now, there are 581 years between 1001 and 1582. Of these, 145 years are divisible by 4. Because each fourth year was a leap year in the Julian calendar, we have 145 leap years in between. Counting back one day for each year and one additional day for each leap year, we have

$$-581 - 145 \equiv -726 \equiv -7 \cdot 103 - 5 \equiv -5 \equiv 2 \pmod{7}.$$

Two days after a Monday is a Wednesday. Therefore, according to the Julian calendar, the second millennium started on a Wednesday.

Alternatively, we can obtain the same answer using the fact that January 1, 1001 was a Thursday according to the proleptic Gregorian calendar (see Example 12.17). In the sixteenth century, the two calendars differed by 10 days (the adjustment implemented at the time of the reform). In the fifteenth century, the difference was 9 days, because the Julian calendar added one day for the leap year 1500, and the proleptic Gregorian calendar didn't. Further in the past, we have to subtract one day from the difference when crossing each whole 100, except for the years divisible by 400. The whole hundreds not divisible by 400 between years 1001 and 1582 are 1100, 1300, 1400, and 1500. Then in the eleventh century, the difference between the proleptic Gregorian and Julian calendars was $10 - 4 = 6$ days. Counting 6 days from a Thursday, we obtain a Wednesday.   □

Now when the Internet knows everything, you can simply type any date into an online application and immediately obtain the corresponding day of the week. However, for dates far back in the past, you need to pay attention to which calendar is being used; the default is the proleptic Gregorian calendar. Even when you use a Web application to find the answer to your question, it is satisfying to know how this answer was obtained. Making your computer's responses to your inquiries a little less mysterious is one of the goals of this book.

Periodicity is essential for our perception of time, and it is reflected in the structure of calendars. The imperfect correlation between the periods of rotation of the Earth about its own axis (a day), the phase cycle of the Moon (a month), and rotation of the Earth about the Sun (a year) makes combining all these features in a calendar a challenging task. All calendars count days; most of them try to take into account the solar and lunar cycles. In addition to the cycles, calendars usually include a linear measure of time (years), which is counted from a particular point and can be continued indefinitely in the future.

In astronomy, where precision of time intervals is extremely important, scientists often use a linear day count. Any date is determined by the number of days since November 24, 4714 BC, 12 hours Greenwich Mean Time (GMT) according to the Gregorian proleptic calendar. This day was chosen because it was a simultaneous beginning of certain solar and lunar cycles that form the Julian calendar.[3] At that instant, the Julian

---

[3]For more details on the Julian day count, and an overview of ancient calendars, see James Evans, *The History and Practice of Ancient Astronomy*, Oxford University Press, 1998.

day count is set to 0, and any subsequent day's number is counted linearly from it. For example, January 1, 2001, is Julian day $2,451,910.5$. A half day appears because a Julian day starts at noon GMT, while our conventional day starts at midnight. An advantage of a linear day count for scientists is that time intervals between dates many years apart can be determined quickly and precisely by simple subtraction. No month lengths or leap years need to be taken into account. The *Julian day count* was proposed by Josephus Justus Scaliger, a sixteenth-century Dutch historian. Despite its simplicity and convenience, the system is used only in science. Software engineers often use *unix time*, a linear count of seconds starting from midnight on January 1, 1970. But in everyday life, our messy cyclic organization of time with the seven-day week, twelve unequal months, and occasional leap years turns out to be the most natural.

Even though most people probably consider conventional addition to be more natural than modular addition, a cyclic calendar is perceived as more natural than a linear one. When it comes to measuring time, we prefer to count days, months, and hours as if they are modular numbers. This shows that both the linear and modular ways of counting are useful and meaningful in the real world.

Round-the-world travel and other circular motion

Time wraps around in many different cycles determined by the cyclic motion of celestial bodies. So does space, in some sense. Our home, Earth, is round, and if you go too far in the same direction, you are bound to get back to where you started.

Example 12.20. Suppose you fly $100,000$ miles due west, starting from a point on the equator. Where will you finish?
Solution: The circumference of Earth is approximately $24,901$ miles. If you go that far, you will finish up at your starting point. Therefore, to answer the question, we should consider the congruence

$$100,000 \equiv x \pmod{24,901}.$$

The number $24,901$ is almost a quarter of $100,000$. More precisely, $24,901 \cdot 4 = 99,604$. Then we have

$$100,000 = 4 \cdot 24,901 + 396 \equiv 396 \pmod{24,901}.$$

You will finish 396 miles to the west of your starting point.   □

Here is an example of a puzzle that uses the idea of round-the-world travel.

Example 12.21. Suppose you go 50 miles south, then 50 miles east, then 50 miles north, and find yourself at the point where you started. Where is this point?

Solution: One answer that immediately comes to mind is the North Pole, because all the meridians intersect there. Going down from the north pole along any meridian, then east along a latitude, and then back north along another meridian brings you right back where you started. This is the only point that works in the Northern Hemisphere.

Figure 12.3: Solutions in the Northern and Southern Hemispheres.

There are many more solutions in the Southern Hemisphere. First we need to find a latitude near the South Pole such that the length of the circumference along this latitude is exactly 50 miles. It will be pretty close to the pole, at about 89.9°S. Now take any point 50 miles north from this latitude. Starting at this point and going 50 miles south gets you down to the chosen latitude. Then going 50 miles east along the latitude brings you back to the same point. Then after going 50 miles north you are back at the starting point.

In fact there are many more solutions. The length of the circumference along the latitude does not have to be 50 miles, but any integer fraction of it. If the circle is $\frac{50}{n}$ miles long, going 50 miles east along it will make you complete exactly n rounds. For example, a 25-, 12.5-, 10- or even 1-mi-long circle along a latitude will work as well. There is no limit to how short the circle can be. This realization adds a modular flavor to the discussion: roughly speaking, we need to find all numbers x such that $50 \equiv 0 \pmod{x}$, if we extend the definition of "modulo" to include fractions as well.   □

On a bigger scale, the same ideas apply to orbital travel in the Solar System, where the orbits of the planets wrap around the Sun in ellipses.

The real-world phenomena that we are able to perceive are finite rather than infinite, and many have cyclic periodicity. Even computers, being physical machines, have to have their integer calculations done modulo the *largest integer value*, which is set to equal to 2^{32} (or $2^{31} - 1$ for signed integers). Of course there are ways to overcome this restriction when needed, but this requires redefining an integer as a more complicated object. Because we are creatures living in the real world, our brains should naturally prefer finite modular arithmetic to the infinite arithmetic of whole numbers. The problem is that in reality different processes happen modulo different numbers. It is impossible to choose one that would cover all natural phenomena. This is one of the reasons the idea of an infinite array of integers with the conventional addition and multiplication is widely popular, while modular arithmetic remains a subject of special interest. Another reason, we hope, is humankind's intrinsic love for abstraction and generalization, best reflected in the invention of mathematics.

Divisibility tests

Most of this chapter was devoted to applications of modular arithmetic to real-world processes with cyclic periodicity. Now we turn to a different kind of application that is useful in computations. Namely, we will consider how methods of modular arithmetic allow us to develop and check divisibility tests of integers by a given number.

You might remember from elementary school a well-known test of divisibility by 3 and by 9: a number is divisible by 3 (respectively, by 9) if and only if the sum of digits of the number is divisible by 3 (respectively, by 9).

Example 12.22. Is $6,237,813$ divisible by 3 and by 9?
Solution: $6 + 2 + 3 + 7 + 8 + 1 + 3 = 30$ is divisible by 3, but not by 9. Therefore the number itself is divisible by 3, but not by 9.   □

Let us see why this criterion holds in the case of a 5-digit number. Let $n = a_4 a_3 a_2 a_1 a_0$ be the standard decimal notation of a 5-digit number that can be written in terms of its digits as follows: $n = a_4 \cdot 10^4 + a_3 \cdot 10^3 + a_2 \cdot 10^2 + a_1 \cdot 10 + a_0$. The criterion says that this number is divisible by 3 if and only if the number $s = a_4 + a_3 + a_2 + a_1 + a_0$ is divisible by 3. In terms of congruences, this means that

$$n - s \equiv 0 \pmod{3}.$$

Indeed, if $s \equiv 0$ (mod 3), then the congruence would mean that $n \equiv 0$ (mod 3) as well. On the other hand, if s is not congruent to zero modulo 3, then neither is n, and we get the "if and only if" statement that is required for the test. Let us consider the number $n - s$:

$$n - s$$
$$= a_4 \cdot 10^4 + a_3 \cdot 10^3 + a_2 \cdot 10^2 + a_1 \cdot 10 + a_0 - (a_4 + a_3 + a_2 + a_1 + a_0)$$
$$= 9{,}999a_4 + 999a_3 + 99a_2 + 9a_1.$$

This difference is always divisible by 3 (and by 9), and therefore the congruence is true. By the same computation, we have obtained the congruence $n - s \equiv 0$ (mod 9), and therefore the same test holds for divisibility by 9.

Even though we considered a 5-digit number n, it is clear that the same argument will work for a number of any length. In general, any number minus the sum of all its digits equals a sum of the digits times numbers of the form $99\ldots9$, and such a sum is always divisible by 3 and by 9.

Practice problem 12.23. Which of the following numbers are divisible by 3 and by 9? (a) $777{,}777$, (b) $8{,}543$, (c) $200{,}901$, (d) $654{,}327$. *Answer:* (a), (c), (d) are divisible by 3; (d) is also divisible by 9.

Another well-known divisibility test is for 11.

Example 12.24. Check the criterion: a number is divisible by 11 if and only if the alternating sum of its digits is divisible by 11.
Solution: Take a number of any length, for example $n = a_4a_3a_2a_1a_0$ and set $s = a_0 - a_1 + a_2 - a_3 + a_4$. Let us consider the difference $n = a_4 \cdot 10^4 + a_3 \cdot 10^3 + a_2 \cdot 10^2 + a_1 \cdot 10 + a_0$ modulo 11. We know that $10 \equiv -1$ (mod 11). By property (4) of congruences, this implies that $10^{\text{odd}} \equiv -1$ (mod 11) and $10^{\text{even}} \equiv 1$ (mod 11). Using properties (1) and (3), we can write

$$n = a_4 \cdot 10^4 + a_3 \cdot 10^3 + a_2 \cdot 10^2 + a_1 \cdot 10 + a_0$$
$$\equiv a_4 \cdot 1 + a_3 \cdot (-1) + a_2 \cdot 1 + a_1 \cdot (-1) + a_0 \equiv s \pmod{11}.$$

Therefore, $n \equiv s$ (mod 11). It is clear that the same computation works for any number of digits. In general, n is divisible by 11 if and only if s is. $\qquad\square$

For some numbers, divisibility tests can be quite complicated. For example, let us consider the following test of divisibility by 7: multiply the digits of the number, starting from the first digit, by $1, 3, 2, 6, 4, 5,$

repeating the cycle as long as necessary. Add the products. Then the number is divisible by 7 if and only if this sum is divisible by 7.

This is an example of a "backward" divisibility test, because we read the digits of the number from right to left and multiply each digit by a suitable number and then add up these products. There exist also "forward" divisibility tests, where you start with the last digit of the number.[4]

Example 12.25. Consider the number $4,074,800,352$. Is it divisible by 7?
Solution: According to the criterion, we compute:

$$s = 2 \cdot 1 + 5 \cdot 3 + 3 \cdot 2 + 0 + 0 + 8 \cdot 5 + 4 \cdot 1 + 7 \cdot 3 + 0 + 4 \cdot 6$$
$$= 2 + 15 + 6 + 40 + 4 + 21 + 24 = 112.$$

If we are still not sure, we can repeat the test with the obtained number, which is much smaller than the initial one:

$$s_1 = 2 \cdot 1 + 1 \cdot 3 + 1 \cdot 2 = 2 + 3 + 2 = 7.$$

Therefore, 112 is divisible by 7, and so is $4,074,800,352$. □

Let us try to explain this criterion using a strategy similar to the test for 11. It is clear that $10 \equiv 3 \pmod 7$. Using property (3) of congruences, we obtain for the next few powers of 10:

$$10^2 \equiv 3 \cdot 3 \equiv 9 \equiv 2 \pmod 7,$$
$$10^3 \equiv 2 \cdot 3 \equiv 6 \pmod 7,$$
$$10^4 \equiv 6 \cdot 3 \equiv 18 \equiv 4 \pmod 7,$$
$$10^5 \equiv 4 \cdot 3 \equiv 12 \equiv 5 \pmod 7,$$
$$10^6 \equiv 5 \cdot 3 \equiv 15 \equiv 1 \pmod 7.$$

Suppose we have a 7-digit number $n = a_6 a_5 a_4 a_3 a_2 a_1 a_0$. Set $s = 1 \cdot a_0 + 3 \cdot a_1 + 2 \cdot a_2 + 6 \cdot a_3 + 4 \cdot a_4 + 5 \cdot a_5 + 1 \cdot a_6$. Then using properties (1) and (3), we have

$$n \equiv a_6 \cdot 10^6 + a_5 \cdot 10^5 + a_4 \cdot 10^4 + a_3 \cdot 10^3 + a_2 \cdot 10^2 + a_1 \cdot 10 + a_0 \equiv$$

---

[4]In my junior year in high school (1996), I found general "backward" and "forward" divisibility tests that work *for every number* – this was my first independent research project. To keep the discussion from getting too abstract, we will not formulate it here, referring interested readers to my article
http://math.furman.edu/~mwoodard/fuejum/content/1997/1997paper1.pdf.
Individual cases of these tests keep popping up in various books of mathematical puzzles. (*A.K.*)

$$\equiv a_6 \cdot 1 + a_5 \cdot 5 + a_4 \cdot 4 + a_3 \cdot 6 + a_2 \cdot 2 + a_1 \cdot 3 + a_0 \cdot 1 \equiv s \pmod{7}.$$

Because each of the summands of n is congruent to each of the summands of s modulo 7 separately, it is clear that the test works for any number with fewer than 7 digits. It takes just a little more work to show that the test works in general (see Question 12.13).

Strategy for solving congruences with large numbers*

Often in applications it is necessary to solve a congruence, or in other words, to find the remainder of the division of one number by another.

If the numbers in question are small, the answer is easy to find. For example,

$$712 \equiv 700 + 7 + 5 \equiv 5 \pmod{7}.$$

But when the numbers are very large, the question becomes more complicated. Below we will propose some methods of solving congruences.

We can first try the *floating point division* approach. If the number a is not too large, you can use your calculator to divide a by n and obtain the number $y = \frac{a}{n}$. It is important that your calculator gives the answer with precision at least up to the unit's digit. If $\frac{a}{n}$ is an integer, then

$$a \equiv 0 \pmod{n}.$$

In general, the answer is not an integer. Take its integer part $[y]$, discarding the digits that appear after the decimal point. Multiplying $[y]$ by n, we get the largest integer divisible by n that is still smaller than a. Then the difference $a - n \cdot [y]$ is the remainder of the integer division of a by n:

$$a - n \cdot [y] = x,$$

and

$$a \equiv x \pmod{n}.$$

Example 12.26. If $a = 7^{10}$ and $n = 18$, we have

$$y = \frac{7^{10}}{18} = \frac{282,475,249}{18} = 15,693,069.388888\ldots.$$

Then the integer part of y is

$$[y] = 15,693,069,$$

and
$$a - n \cdot [y] = 282,475,249 - 15,693,069 \cdot 18 = 7.$$
Therefore,
$$7^{10} \equiv 7 \pmod{18}.$$

$\square$

Floating point division can fail due to overflow in case of very large numbers. For example, if $a = 558,844,796,231 \cdot 544,895$, and $n = 17$, then the calculator returns $a = 3.04511735242291\,E17$. Even after dividing by 17, the calculator does not return the number with a precision sufficient to calculate its integer part. A way to overcome this difficulty is by using the multiplicative property of congruences:

if $a \equiv x \pmod{n}$ and $b \equiv y \pmod{n}$ then $a \cdot b \equiv x \cdot y \pmod{n}$.

This means that if a is given as a product of two or more smaller numbers, we can find the congruences of these smaller factors modulo n, and then multiply them to obtain the congruence of a modulo n.

Example 12.27. Let $a = 558,844,796,231 \cdot 544,895$ and suppose we need to find $a \pmod{17}$. To each of the factors separately we can apply the method of floating point division, and we find

$$558,844,796,231 \equiv 12 \pmod{17}, \qquad 544,895 \equiv 11 \pmod{17}.$$

Therefore,

$$558,844,796,231 \cdot 544,896 \equiv 12 \cdot 11 \equiv 132 \pmod{17}.$$

Now the problem is reduced to finding 132 mod 17, which can be easily solved by division:
$$132 = 7 \cdot 17 + 13.$$
Finally, we get

$$558,844,796,231 \cdot 544,896 \equiv 132 \equiv 13 \pmod{17}.$$

Alternatively, we can use negative congruences. In this case, it makes the computations simpler:

$$558,844,796,231 \equiv 12 \equiv -5 \pmod{17}, \quad 544,895 \equiv 11 \equiv -6 \pmod{17}.$$

Therefore,

$$558,844,796,231 \cdot 544,896 \equiv (-5) \cdot (-6) \equiv 30 \equiv 13 \pmod{17}.$$

$\square$

This method also works if the number a is given as a product of more than two factors. In Chapter 13 we will discuss more advanced methods of solving congruences that can be used when the number a is given as a power of a smaller number.

EXERCISES

Question 12.1. Solve the following congruences:

1. $3 \cdot 3 \cdot 3 \pmod 8$, and $\pmod 9$.

2. $5 \cdot 5 \cdot 5 \pmod{100}$.

3. $5 \cdot 5 \cdot 5 \pmod 7$.

4. $5^8 \pmod 7$.

Question 12.2. Solve the following congruences:

1. $25 \cdot 77 \cdot 16 \pmod 9$.

2. $45^3 \pmod{11}$.

3. $16 \cdot 17 \cdot 18 \pmod 5$.

Question 12.3. Check that the last day of February, May 2 and July 4 fall on the same day of the week as April 4 in any given year. Conclude that 3/0 (last day of February), 4/4, 5/2, 6/6, 7/4, 8/8, 10/10, and 12/12 all fall on the same day of the week in any year (see Example 12.11).

Question 12.4.

1. On which day of the week were you born?

2. Find out on which day of the week the fourth millennium will begin (assuming we continue with the Gregorian calendar then).

Question 12.5.

1. What day of the week was Christmas 1990?

2. On what day of the week did *Apollo 11* land on the Moon? (July 20, 1969)

3. On what day was the Declaration of Independence adopted? (July 4, 1776)

Question 12.6. Find the days of the week on which each of your parents and your siblings was born.

Question 12.7. On what day of the week did the crew of the Columbus expedition first see the shore of America (October 12, 1492)

1. according to the proleptic Gregorian calendar?

2. according to the Julian calendar?

Question 12.8.

1. What season is it 100 seasons from a summer?

2. What season was it 57 seasons back from a summer?

Question 12.9.

1. What time is it 5,000 hours from 7 pm?

2. What month is it 10,000 months from January?

Question 12.10. Moving along its orbit, Mars covers approximately 2,142,000 kilometers per day. It completes the entire orbit in 687 days. How far from its present position will Mars be along its orbit 1 billion days from now?

Question 12.11. A hamster is running counterclockwise along a circular track with a circumference of 8 feet at a constant speed of 25 feet per minute (see Figure 12.4). If he started at the east side of the track, in what direction on the track will he be located

1. after 33 minutes?

2. after 51 minutes?

Question 12.12.

1. Let $n = a_1 a_0$ be a 2-digit number. What condition do the digits a_1 and a_0 satisfy if and only if n is divisible by 4?

2. Derive a divisibility test by 4 for numbers of any length.

Question 12.13. Show that the test of divisibility by 7 works for numbers with any number of digits. *Hint:* Use the fact that $10^6 \equiv 1 \pmod 7$, therefore $10^7 \equiv 10^6 \cdot 10 \equiv 3 \pmod 7$, and so on. Deduce that the powers of 10 have periodically repeating congruences modulo 7.

Figure 12.4: Hamster on an 8-foot-long circular track.

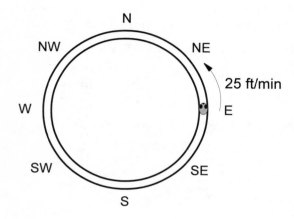

Question 12.14.

1. Use floating point division to find x between 0 and 36 such that $8^{12} \equiv x \pmod{37}$.

2. Use the multiplicative property of congruences to find x such that

$$57,777,777 \cdot 63,333,333 \cdot 2,888,888 \equiv x \pmod 6, \qquad 0 \le x \le 5.$$

Question 12.15.

1. Use floating point division to find x between 0 and 24 such that $8^{12} \equiv x \pmod{25}$.

2. Use the multiplicative property of congruences to find x such that

$$98,989,898 \cdot 74,747,474 \cdot 12,121,212 \equiv x \pmod 6, \qquad 0 \le x \le 5.$$

Chapter 13

Modular arithmetic II: How to keep (and break) secrets

In this chapter, we will study some applications of modular arithmetic in cryptography, the art of keeping and breaking secrets. In particular, we will explain how the Diffie-Hellman algorithm allows two parties to jointly establish a secret password over an insecure communication channel.

MATH

Recall that a congruence is the following relation between integer numbers a and x and an integer number n greater than 1:

$$a \equiv x \pmod{n}.$$

It means that the difference $a - x$ is a multiple of n. Congruences satisfy many properties, listed in Chapter 12. In this chapter, we will mostly use the following two:

1. The multiplicative property, which includes

$$a \equiv x \pmod{n}, \quad b \equiv y \pmod{n} \quad \Longrightarrow \quad ab \equiv xy \pmod{n},$$

and

$$a^s \equiv x^s \pmod{n} \quad \text{for any positive integer} \quad s.$$

237

2. Another important fact is the existence and uniqueness of a solution for a congruence. For the given a and n, there is a unique number x that satisfies the congruence

$$a \equiv x \pmod{n}, \qquad 0 \le x \le n - 1,$$

and x can be computed algorithmically. However, if x and n are given, there are infinitely many numbers a that satisfy the congruence.

These two properties lie at the heart of the use of modular arithmetic in cryptography and computer security.

Example 13.1. (a) What is 10^{18} modulo 9? (b) What is 8^{18} modulo 9?

Solution: (a) Because $10 = 9 + 1$, we have $10 \equiv 1 \mod 9$. Then by the multiplicative property, $10^{18} \equiv 1^{18} \equiv 1 \mod 9$. This works for any natural power of 10: $100 \equiv 1 \pmod 9, 1,000 \equiv 1 \pmod 9$, and so on. If we subtract 1 from any natural power of 10, we get a number whose decimal form is a string of 9's. For instance, $1,000 - 1 = 999$. Clearly, any such number is divisible by 9.

(b) Here we have no shortcut to find the answer: powers of 8 look quite complicated when written in decimal form. However, property (3) gives us an immediate answer. Because $8 = 9 - 1$, we have $8 \equiv -1 \pmod 9$. Then by property (3), $8^{18} \equiv (-1)^{18} \equiv 1 \pmod 9$. □

When we discuss divisibility properties, it is natural to pay particular attention to the prime numbers. A positive integer $p > 1$ is *prime* if it has exactly two positive divisors: the number p itself, and 1. The prime numbers can be considered as elementary "building blocks" for all other positive integers, because any number can be written as a product of primes. For example, 3 and 11 are prime numbers, and 9 and 12 are not because they are divisible by 3. But both 9 and 12 can be written as a product of prime numbers:

$$9 = 3 \cdot 3, \qquad 12 = 2 \cdot 2 \cdot 3.$$

If we are looking to solve a congruence modulo a prime number p, our task is simplified by the following statement: if p is a prime and a any integer number, then

$$a^p \equiv a \pmod p.$$

This statement is called *Fermat's Little Theorem* to distinguish it from *Fermat's Last Theorem*,[1] a well-known statement proposed by the French mathematician Pierre de Fermat (1601-1665). We will not prove Fermat's Little Theorem, referring interested readers to Question 13.10, which contains an outline of a proof. Here are two examples:

$$a = 2, \ p = 11 \quad \implies \quad 2^{11} = 2,048 = 186 \cdot 11 + 2 \equiv 2 \pmod{11}.$$
$$a = 7, \ p = 3 \quad \implies \quad 7^3 = 343 = 114 \cdot 3 + 1 \equiv 1 \equiv 7 \pmod 3.$$

If a is not divisible by p, then the statement is equivalent to

$$a^{p-1} \equiv 1 \pmod p.$$

Multiplying this congruence by a (where a is not congruent to 0 modulo p) yields the original statement. Here is how Fermat's Little Theorem (FLT) can help solve congruences modulo a prime.

Example 13.2. Find 42^{73} (mod 11).

Solution: By FLT, because 42 is not divisible by 11, we have

$$42^{11-1} = 42^{10} \equiv 1 \pmod{11}.$$

Then raising this congruence to the 7th power, we have

$$(42^{10})^7 = 42^{70} \equiv 1 \pmod{11}.$$

It remains to find 42^3 (mod 11). Proceeding as in Example 13.1, we have $42 \equiv -2$ (mod 11). Then $42^3 \equiv (-2)^3 \equiv -8 \equiv 3$ (mod 11). Therefore,

$$42^{73} \equiv 42^{70} \cdot 42^3 \equiv 1 \cdot 3 \equiv 3 \pmod{11}.$$

□

Practice problem 13.3. Recall that by solving a congruence a (mod n) we mean finding the smallest nonnegative integer x such that $a \equiv x$ (mod n). Use FLT to solve the following congruences.

1. 9^6 (mod 7). *Answer:* 1.

2. 9^{37} (mod 7). *Answer:* 2.

---

[1] Fermat's Last Theorem states that there are no positive integers a, b, c such that $a^n + b^n = c^n$, where $n \geq 3$. Fermat left no proof of this statement, and for more than 300 years many mathematicians and enthusiasts tried to find it, until finally, in 1994, the British mathematician Sir Andrew Wiles published a proof.

3. 15^{16} (mod 17). *Answer:* 1.

Solving congruences for large exponents is especially useful in cryptography and computer security.

APPLICATIONS

Cryptography: the art of keeping and sharing secrets

Suppose you have to transmit a secret message under threat of a third party intercepting it. To avoid revealing the secret, you need to disguise, or *encipher*, the message. The original message is a *plaintext* (the spaces and punctuation in the original message are usually omitted), a disguised message a *ciphertext*, the process of disguising *enciphering*, and the reverse process *deciphering*. A *cryptosystem* consists of two parts: a *key* to turn a plaintext message into a ciphertext, and an *inverse key* to turn a ciphertext into a plaintext message. The goal of a cryptosystem is to transfer a message in a way that only the designated destination party can read it. The goal of the *code breakers* is the opposite: to break the code or, equivalently, to decipher the message without knowing the key. Cryptographers study methods of enciphering and deciphering texts.

Cryptosystems are compared for their relative strength, or security – the ability to withstand code breaker attacks. Known ciphers range from simple encodings, some of which we will discuss below, to very complicated and virtually unbreakable ones. In Chapter 8 we encountered a formidable example of an enciphered (we hope!) text, the Voynich manuscript, written in an unknown language using an unknown alphabet that remains as yet undeciphered. Another famous old manuscript dating from the eighteenth century, the Copiale cipher, was deciphered in 2011 using modern methods of cryptography.[2] More recent examples of complicated ciphers include the Enigma cipher, used by the Nazis during World War II and cracked by the Allied code breakers. Today, the art of building and breaking codes is no longer confined to the world of mysterious texts and international intelligence. Anyone shopping online relies on the strength of the Advanced Encryption Standard cryptosystem used in Internet security to transfer

---

[2] "How revolutionary tools cracked a 1700s code" by John Markoff, *New York Times*, October 24, 2011.

their credit card information. In the rest of this chapter, we will outline some of the ideas and methods that are used in modern cryptography.

Historical ciphers

The simplest example of a cipher is the *translation cipher*, where each letter of a plaintext is replaced by the letter standing n positions farther ahead in the alphabet. If we are at the end of the alphabet, it is assumed that the letter following Z is A again, and so on, listing the alphabet in a *cyclic order*. For example, the *Caesar cipher* is a translation cipher with $n = 3$:

$$
\begin{array}{ccccccccc}
A & B & C & D & \ldots & X & Y & Z \\
\downarrow & \downarrow & \downarrow & \downarrow & \ldots & \downarrow & \downarrow & \downarrow \\
D & E & F & G & \ldots & A & B & C
\end{array}
$$

The key of this cipher is "shift by three to the right in the cyclic alphabetical order," and the inverse key "shift by three to the left in the cyclic alphabetical order." A key can be given by a letter-mapping scheme as shown above.

Practice problem 13.4. Decipher the message PDWKLVDODQJXDJH written in the Caesar cipher. *Hint:* If your answer makes sense, it should be correct.

Practice problem 13.5. How many different nontrivial translation ciphers can be made using the English alphabet? *Answer:* 25.

A generalization of a translation cipher is a *permutation cipher*, where each letter in the alphabet is mapped to a "partner" letter that replaces it in the enciphered message. Each letter in the alphabet can be assigned one and only one partner. (For example, one cannot have A → X and B → X.) The easiest way to describe a permutation cipher is by giving a permutation (a new order) of the letters in the alphabet. Then A corresponds to the first letter in the new sequence, B to the second, and so on. The key of the cipher is given by the list of all 26 correspondences, or by a permutation of 26 letters. The inverse key is the same correspondence read in the opposite way. For example, the key

$$
\begin{array}{cccccccccccccccccc}
A & B & C & D & E & F & G & H & I & J & K & L & M & N & O & P & Q & R \\
\downarrow & \downarrow & \downarrow & \downarrow & \downarrow & \downarrow & \downarrow & \downarrow & \downarrow & \downarrow & \downarrow & \downarrow & \downarrow & \downarrow & \downarrow & \downarrow & \downarrow & \downarrow \\
Q & W & E & R & T & Y & U & I & O & P & L & K & J & H & G & F & D & S
\end{array}
$$

$$
\begin{array}{cccccccc}
\text{S} & \text{T} & \text{U} & \text{V} & \text{W} & \text{X} & \text{Y} & \text{Z} \\
\downarrow & \downarrow & \downarrow & \downarrow & \downarrow & \downarrow & \downarrow & \downarrow \\
\text{A} & \text{Z} & \text{X} & \text{C} & \text{V} & \text{B} & \text{N} & \text{M}
\end{array}
$$

is used to encipher the plaintext message CLOUDYSKY into EKGXR-NALN. Note that all translation ciphers are special cases of the permutation ciphers, when the permutation is a cyclic shift of the alphabet.

Example 13.6. How many different permutation ciphers can be made using the English alphabet?

Solution: The number of different permutation ciphers equals the number of different permutations of 26 letters, or $26 \cdot 25 \cdot 24 \cdot \ldots \cdot 2 \cdot 1 = 26!$.[3] $\square$

The number 26! is very large, but it includes, for example, the trivial permutation, where all letters are mapped to themselves, and many almost-trivial permutations, where almost every letter is mapped to itself and only a few are permuted. Even after discarding these permutations, the security of a cryptosystem based on a permutation cipher is low. Such ciphers can be easily broken by a *frequency analysis*: matching ciphertext letters with plaintext letters according to their relative statistical frequency of appearance in a text. This frequency depends only on the language of the text, and it is known for each letter in all major languages. For example, count the letters "e" and "a" in the current paragraph. The result is: 122 "e"'s and 69 "a"'s. The letter "e" is statistically the most frequent in an English text.[4] Replacing the most frequent letter in the ciphertext with "e", and trying to match other letters according to their frequency while watching for parts of a meaningful text to appear often leads to deciphering the text.

To avoid this weakness, *polyalphabetic substitution ciphers* are used. Suppose that you have several lists of 26 symbols of any nature (a slavic alphabet, pairs of English letters, etc.) and you use any of the lists randomly when replacing each letter of the plaintext message. Then any given English letter corresponds to several symbols and the frequency analysis is much harder to perform. For example, if M → TU, A → LV, T → KS,

---

[3] This is the number of permutations of 26 elements, $_{26}P_{26}$. We will discuss the methods of counting permutations in Chapter 15.

[4] It is also the most frequent letter in French. All the more impressive is the achievement of the French writer Georges Perec, who wrote a three-hundred-page novel, *La Disparition (The Disappearance)*, which uses no "e" whatsoever. Moreover, a few years later he wrote the counter-novel, *Les Revenentes*, in which "e" was the only vowel used. Both books are translated into English, following the same rules of letter usage: *A Void*, translated by Gilbert Adair (The Harvill Press), and *The Exeter Text*, translated by Ian Monk and published in *Three by Perec* (David R. Godine).

H → LQ are some of the substitutions in the first list and M → IS, A → VS, T → PA, H → VN in the second list, then MATH can be enciphered as any of the following strings:

$$\text{TULVKSLQ}, \quad \text{TUVSKSVN}, \quad \text{ISLVPAVN}, \quad \ldots,$$

totalling $2^4 = 16$ different strings of ciphertext. Provided the necessary keys, each ciphertext above can be unambiguously deciphered to obtain the plaintext message MATH. In this case, the key consists of several ordered lists of 26 symbols. This improves the strength of the system, but makes the transfer of the key a more cumbersome and potentially vulnerable procedure.

Modular arithmetic in cryptography

Let us reconsider the translation cipher from the viewpoint of modular arithmetic. It is easy to see that the key of any translation cipher is actually a modular addition of a given number. Enumerate the letters of the alphabet by integers from 0 to 25. Then the Caesar cipher replaces the nth letter of the alphabet with the letter number $(n + 3)$ (mod 26). We have to add "(mod 26)" to account for the cyclic order that we are using: after the alphabet is exhausted, we start reading it again from the beginning. Then, for example, the letter number 24, which is Y, is mapped to the letter number $(24 + 3) \equiv 27 \equiv 1$ (mod 26), and we obtain the letter B. The key of the Caesar cipher is given by "+3 (mod 26)," and the inverse key by "−3 (mod 26)."

In many polyalphabetic ciphers with multiple keys, the choice of the key to use on each letter is determined by the position of the letter within the plaintext message. Here is how it works.

Example 13.7. Encipher the plaintext FRIDAYMAYSIX using the key that shifts the first letter of the message 1 position forward in the alphabet, the second letter 2 positions forward, the third 3 positions, and so on to the end of the message. Write the key of the cipher as a modular operation.

Solution: We can use the alphabet circle (Figure 13.1) to find the shifted letters. Then F is enciphered as G, R as T, and so on. We obtain the ciphertext

$$\text{GTLHFETIHCTJ}$$

The key can be written as $+n$ (mod 26), where n is the number of the letter within the text.   □

Figure 13.1: The Caesar cipher.

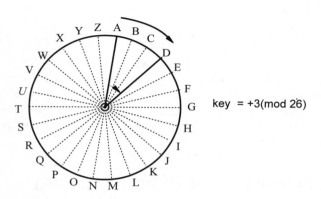

key = +3(mod 26)

Note that the cipher we just considered resists a frequency analysis attack: the letter T stands first for R, then for M, and finally for I. However, if a message is sufficiently long, adding a bigger number each time can become tedious. For longer messages, a *periodic cipher* can be used. To use this cipher, you have to choose a sequence of, say, m keys and apply each key to the consecutive letters of the text, repeating the sequence until the message is exhausted.

A simple example of a periodic substitution cipher is the *Vigenère cipher*: in this case each of the keys is a translation by a fixed number of positions in the cyclic alphabetic order. Consider, for example, the Vigenère cipher with period 5 and keys $(+15, +1, +2, +3, +4)$ (mod 26). This means that the enciphering replaces the first letter in the plaintext by the one 15 positions to its right in the alphabet, the second by the one following it in the alphabet, the third by the letter 2 positions to the right, the fourth by 3, and the fifth by 4. Then again, the sixth letter is replaced by the one 15 positions to the right, the seventh by the letter following it in the alphabet, the eighth by the letter shifted by 2 positions, and so on.

Example 13.8. Using the Vigenère cipher with the key $(+15, +1, +2, +3, +4)$ (mod 26), encipher the plaintext

DESTROYANDFORGET

Solution: First we encode the plaintext as a string of numbers from 0 to 25 according to the alphabetical order and space them in groups of 5 (because the period of the key is 5):

$$3, 4, 18, 19, 17, \quad 14, 24, 0, 13, 3, \quad 5, 14, 17, 6, 4, \quad 19.$$

Now write the periodic key under the string of numbers and perform addition modulo 26:

3	4	18	19	17		14	24	0	13	3
+15	+1	+2	+3	+4		+15	+1	+2	+3	+4
18	5	20	22	21		3	25	2	16	7

	5	14	17	6	4		19
	+15	+1	+2	+3	+4		+15
	20	15	19	9	8		8

Converting the obtained string back into letters, we get:

$$\text{SFUWVDZCQHUPTJII}$$

□

Note that the first E in the plaintext is enciphered as F, and the second as I. Moreover, I stands for both E and T in the ciphertext. This shows that although the enciphering and deciphering (if you know the key) with the Vigenère cipher is easy to perform, it provides a good protection from a direct frequency analysis attack. It also has an advantage of being easily converted into a machine algorithm. In fact, the Enigma cipher, though more complicated than the Vigenère cipher, was an example of a machine-generated polyalphabetic cipher.

The Enigma machine

This name can be used for any of the variety of mechanical-electrical rotor-based cipher machines used in the first half of the twentieth century. The main principle was invented almost simultaneously by Dutch, American, Swedish, and German cryptographers.

Let us first imagine a machine that consists of a keyboard to enter plaintext letters, a lamp board to read off the enciphered letters, and a rotor wheel that implements the enciphering. The rotor wheel is a thick metal wheel with 26 contacts spread along the rim on one side, randomly

connected by electric wires to 26 contacts on the other side. If we assume
that each contact corresponds to a letter of the alphabet, then one side
of the wheel will look approximately like the alphabet cycle in Figure
13.1. If we connect the keys of the keyboard to the contacts on one side
of the wheel and the lamps of the lamp board to the other side, then
pressing a letter key will send an electric signal through the wheel and
will cause one of the letters on the lamp board to light up. This will be
the enciphered letter. So far, our imaginary machine performs a simple
permutation cipher whose key is given by the wiring of the wheel.

Figure 13.2: A keyboard, a rotor wheel, and a lamp board.

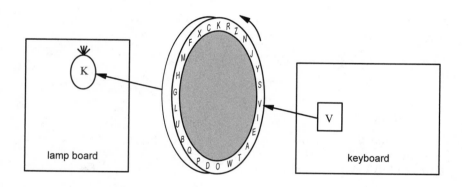

Now imagine that each time a key is pressed, the wheel turns one step
counterclockwise. This means that the same letter pressed twice will be
enciphered differently. For example, if the wheel enciphers V as K the
first time, the next time it will encipher V as if it is the next letter on the
wheel, in our case I, and so the result will be different. This feature of the
rotor wheel is called *stepping*, and it allows the machine to implement a
periodic polyalphabetic substitution cipher with period 26.

A simplified model of an Enigma machine consists of a keyboard, three
or four rotor wheels, a reflector, and a lamp board. The reflector is a static
wheel that permutes the obtained signals and sends them back to the rotor
wheels. It is designed so that no letter can be enciphered to itself, and it
allows a message to decipher automatically, if the machine with the same
design and initial settings was used for enciphering.

Pressing a key on the keyboard sends an electric signal to the first wheel, which scrambles the letters and sends the signal to the second wheel, which sends it to the third wheel, and then to the reflector. From the reflector, the signal goes back through the third, the second, and the first wheel, and finally to the lamp board, which indicates the resulting letter. All rotor wheels can step. The first wheel steps each time a key is pressed. When the first wheel makes a full circle, it causes the second wheel to step. When the second wheel completes a circle, the third wheel steps. Thus there are 26^3 different permutations applied consecutively to the letters of the message.

Figure 13.3: Simplified Enigma model.

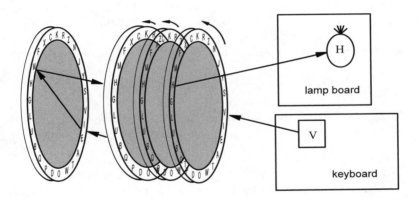

To decipher the message, a machine of the same design has to be set up in the exact same way. The settings include the order of the wheels and the initial positions of each of the three wheels. To complicate the matter, machines used for military purposes during WWII were equipped with additional features: it was possible to move the contacts of the wheel independently of its alphabet, and it was possible to swap any number of pairs of letters. The resulting key consisted of all the initial settings and the design of a particular machine. Breaking the Enigma cipher required advanced methods of modular arithmetic and group theory, and was a remarkable achievement of Polish and Allied cryptographers.

Modular arithmetic remains the main mathematical tool used in cryptography and, with the advent of computers, also in computer security.

Diffie-Hellman public key exchange

As an example of the application of modular arithmetic to computer security, let us consider the *Diffie-Hellman key exchange.* For two parties to exchange secret messages, it is necessary to agree on a cryptosystem and to have a common key. But how to secretly exchange a key? Say two parties want to exchange secret messages using a cipher whose key can be represented as a single number. This assumption does not impose much restriction because virtually anything that can be expressed in a language can also be represented as a single number.[5]

Suppose Alice and Bob[6] need to establish a common secret key that no one else would know, and they have only an insecure communication channel. Below we give an outline of the solution published by the American cryptographers Whitfield Diffie and Martin Hellman in 1976. Before formulating the mathematical algorithm, let us explain the main idea by an analogy with mixing colors. Suppose Alice and Bob need to establish a common secret color or shade of a color. They can proceed as follows:

1. They openly exchange a common base color, say, a particular shade of green.

2. Alice secretly chooses a color, say, a particular shade of yellow. Bob secretly chooses another color, say, a particular shade of red.

3. Alice mixes some amount of her secret color with the base color, obtaining a light green mixture. Similarly, Bob mixes some amount of his secret color with the base color, obtaining a brownish mixture.

4. They exchange jars of the obtained mixtures over an insecure communication channel.

5. Now Alice adds the same amount of her secret color as in the previous step to the mixture she obtained from Bob. Bob adds the same amount of his secret color to the mixture he obtained from Alice.

Because both mixtures contain the same amounts of the base color, of Alice's secret color, and of Bob's secret color, they are exactly the same. This is the common secret color of Alice and Bob.

Let us see why the third party would not be able to reproduce the common secret color. Anyone observing the communication would have

---

[5]We leave it to our readers to figure out why and how.

[6]These names are commonly used to denote two parties in cryptography and computer security, and we follow the fashion.

Figure 13.4: How to produce a common secret color.

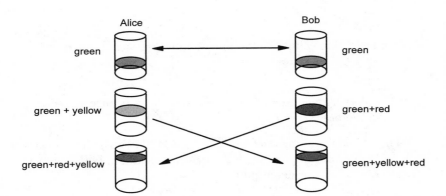

access to the base color and to the mixtures of the base color with the respective secret colors chosen by Alice and Bob. But if we assume that color separation is difficult, the secret colors could not be separated from the mixtures, and the three available ingredients cannot be combined to make the secret common color. Indeed, the secret common color contains base green, Alice's yellow, and Bob's red. The third party can use a mixture of any of three solutions: base green and Alice's yellow, or base green and Bob's red, or just the base green. Mixing these ingredients together never produces a mixture with the right proportions: any mixture containing both Alice's and Bob's secret colors will always contain too much of the base color, resulting in a different hue.

Modular arithmetic allows us to implement the same idea using numbers instead of colors. The mathematical Diffie-Hellman algorithm, or *protocol*, goes as follows:

1. Alice and Bob openly exchange two positive integer numbers: a prime number p and a number g not divisible by p. (The numbers g and p play the role of the base color.) To improve security, there is a more restrictive condition for the choice of g, which we will mention later.

2. Now Alice secretly chooses an integer number a. Bob secretly chooses an integer number b. (These numbers play the role of Alice's and Bob's respective secret colors.)

3. Alice solves the congruence g^a (mod p). In other words, she computes the unique integer x such that

$$g^a \equiv x \mod p, \qquad 0 \le x \le p-1.$$

Similarly, Bob computes the unique integer y such that

$$g^b \equiv y \pmod{p}, \qquad 0 \le y \le p-1.$$

(This corresponds to adding Alice's and Bob's respective secret colors to the base color.)

4. They openly exchange the results: now Alice knows y and Bob knows x. Because the communication channel is insecure, both numbers x and y can be considered publicly known.

5. Now Alice computes the unique integer number K_1 such that

$$y^a \equiv K_1 \pmod{p}, \qquad 0 \le K_1 \le p-1,$$

and Bob computes the unique integer K_2 such that

$$x^b \equiv K_2 \pmod{p}, \qquad 0 \le K_2 \le p-1.$$

These numbers are equal: $K_1 = K_2 = K$. This is the *common secret key K*. (This corresponds to adding their secret colors to the mixture and obtaining the common secret color.)

First, let us see why the numbers K_1 and K_2 are equal. Using the properties of exponents, we have

$$K_1 \equiv y^a \equiv (g^b)^a \equiv g^{a \cdot b} \equiv g^{b \cdot a} \equiv (g^a)^b \equiv x^b \equiv K_2 \pmod{p}.$$

With the additional condition that K_1 and K_2 should be between 0 and $p-1$, such a number is unique. Therefore, $K_1 = K_2 = K$ is indeed a common key.

Now, why is this protocol secure? The third party could have acquired the following information: g, p, x, y. To build the secret key, they would have to know $g^{a \cdot b}$ (mod p), but both a and b are unknown. They know $x \equiv g^a$ (mod p), and $y \equiv g^b$ (mod p), but there are infinitely many numbers congruent to a given number modulo p. Which of them are powers of g? As of today, there is no efficient algorithm to find an exponent b such that $g^b \equiv y$ (mod p), if g, p and y are given. A straightforward way to find b

Figure 13.5: The Diffie-Hellman protocol.

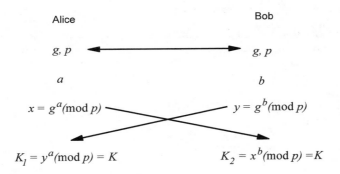

would be to compute all consecutive powers of g: $g, g^2, g^3, g^4, \ldots$ until you hit the one that is congruent to y modulo p. However, the numbers used in the Diffie-Hellman algorithm are typically very large. Moreover, the number g can be chosen to satisfy an additional condition that its powers g^n for different n are congruent to every possible number between 1 and $p-1$ modulo p. Modular arithmetic makes the inverse operation to taking an exponent very difficult to perform. One can compute y such that $y \equiv g^b$ (mod p), but one cannot (except by trial and error) find b such that $y \equiv g^b$ (mod p). The second operation is called the *discrete logarithm problem*, by analogy with $\log_g$, the inverse operation to exponentiation. In real-world situations, trial and error methods are made inefficient by taking p, a and b very large, so that p can be several hundred digits long, and a and b about a hundred digits long. A straightforward approach of trying all powers of g modulo p might not lead to a solution in many years or even decades.

Modular arithmetic plays the central role in the Diffie-Hellman protocol. Indeed, if Alice and Bob openly exchanged $g, x = g^a$ and $y = g^b$, they still would be able to compute a common number $K = x^b = y^a = g^{ab}$. But in this case the communication would not be secure and the key would not be secret. The third party, obtaining g, x and y, would easily be able to take the logarithm $\log_g(x) = a$ and $\log_g(y) = b$, and then compute the key $K = g^{ab}$. It is the operation of taking a congruence modulo p that makes the communication secure.

Here is a toy example with small p, a and b, to demonstrate how the

Diffie-Hellman algorithm works.

Example 13.9. Suppose $p = 13$, $g = 4$, $a = 80$, $b = 29$. Compute x, y and the common key K according to the Diffie-Hellman algorithm.

Solution: Because $p = 13$ is a prime, we can apply Fermat's Little Theorem to find a power of $g = 4$ that is congruent to 1 modulo $p = 13$. We have

$$4^{13-1} \equiv 1 \pmod{13}.$$

Then any power that is a multiple or 12 is congruent to 1 modulo 13. For x, we have

$$x \equiv g^a \equiv 4^{80} \equiv 4^{72} \cdot 4^8 \equiv 1 \cdot 4^8 \equiv 4^8 \quad \text{mod } 13.$$

Now, $4^2 \equiv 16 \equiv 3 \pmod{13}$, and $4^6 \equiv 3^3 \equiv 27 \equiv 1 \pmod{13}$. Using again the multiplicative property of congruences, we have

$$4^8 \equiv 4^6 \cdot 4^2 \equiv 27 \cdot 3 \equiv 1 \cdot 3 \equiv 3 \pmod{13}.$$

Therefore, $x = 3$. For y, we compute

$$y \equiv g^b \equiv 4^{29} \equiv 4^{24} \cdot 4^5 \equiv 1 \cdot 4^5 \equiv 4^5 \pmod{13}.$$

As before, $4^2 \equiv 3 \pmod{13}$, so that $4^5 \equiv 4^4 \cdot 4 \equiv 3^2 \cdot 4 \equiv 36 \equiv 10$ (mod 13). Therefore, $y = 10$.

Now to find the common key, we need to compute $x^b \equiv 3^{29} \pmod{13}$. Again by FLT, we have $3^{12} \equiv 1 \pmod{13}$. Then we compute

$$3^{29} \equiv 3^{24} \cdot 3^5 \equiv 1 \cdot 3^5 \equiv 3^5 \pmod{13}.$$

Because $3^3 \equiv 27 \equiv 1 \pmod{13}$, we have $3^5 \equiv 3^3 \cdot 3^2 \equiv 1 \cdot 9 \equiv 9 \pmod{13}$. The common secret key is $K = 9$.

Let us check that we would obtain the same number K if we solved the congruence $10^{80} \pmod{13}$ instead. By FLT, we have

$$10^{80} \equiv 10^{72} \cdot 10^8 \equiv 1 \cdot 10^8 \equiv 10^8 \pmod{13}.$$

Now, $10^2 \equiv 100 \equiv 8 \cdot 13 - 4 \equiv -4 \pmod{13}$. Then

$$10^8 \equiv (10^2)^4 \equiv (-4)^4 \equiv (-1)^4 \cdot 16^2 \equiv 1 \cdot 3^2 \equiv 9 \pmod{13}.$$

So we have obtained the same secret key $K = 9$. $\square$

In real-world computer security, the numbers used are much bigger, and powerful computers perform the analogs of the above computation. This effort pays off because it ensures secure communications.

The properties of modular arithmetic operations are used in many other cryptography and computer security applications. Note that the Diffie-Hellman key exchange by itself is not designed to allow Alice to send a secret message to Bob. It only allows for both of them to arrive at the same secret number, that can be later used to send messages. Often in everyday life we need to send a secret message to another party without first agreeing on a common secret key, for example when we send our credit card number to an online retailer. This can be accomplished using a modification of Diffie-Hellman. Many other cryptosystems that allow quick, efficient and secure data transmission, for instance the RSA cryptosystem[7], are based on more advanced methods of modular arithmetic.

Many applications of modular arithmetic, including the Diffie-Hellman algorithm, require solving congruences for large numbers that are given as exponents. We will conclude this chapter with an overview of methods and examples of solving such problems.

Solving congruences for large exponents*

When we are faced with a problem to solve a congruence a^s (mod n), where a, s, and n are possibly large positive integers, the first question to ask is whether n is a prime number. If it is, the most efficient way to solve the congruence is by applying FLT. Here is an example.

Example 13.10. Find 6^{234} (mod 11).

Solution: The number 6^{234} is way too big to use the direct division method. The number 11 is a prime: it is not divisible by any positive integer except 1 and itself. Then by FLT, we have

$$6^{11-1} = 6^{10} \equiv 1 \pmod{11}.$$

Now we need to represent the power 234 as a multiple of 10 plus a remainder. This is easy: we have $234 = 23 \cdot 10 + 4$. Then

$$6^{234} \equiv 6^{23 \cdot 10} \cdot 6^4 \equiv (6^{10})^{23} \cdot 6^4 \equiv 1^{23} \cdot 6^4 \equiv 6^4 \pmod{11}.$$

---

[7]Named after its designers, American computer scientists Ron Rivest and Leonard Adleman and Israeli cryptographer Adi Shamir.

We need to find 6^4 (mod 11). Using the multiplicative properties of congruences, we compute:

$$6^4 \equiv (6^2)^2 \equiv 36^2 \equiv 3^2 \equiv 9 \pmod{11}.$$

Therefore, we have

$$6^{234} \equiv 6^4 \equiv 9 \pmod{11}.$$

$\square$

What if the modulus n in the congruence is not prime? The next best case is if we are trying to solve a^s (mod n), where n is not prime but a and n are *relatively prime*, meaning that they have no common divisors except 1.

In this case, the eighteenth-century Swiss mathematician Leonhard Euler proved a generalization of Fermat's statement: if a and n have no common divisors, then

$$a^{\varphi(n)} \equiv 1 \pmod{n},$$

where $\varphi(n)$ is the number of positive integers less than n that are relatively prime to n. For example, $\varphi(8) = 4$, because the positive numbers that have no common divisors with 8 are all odd positive integers less than 8, that is, $1, 3, 5, 7$.

In this chapter we will call this statement *Euler's theorem*.

Practice problem 13.11. Find $\varphi(n)$ for $n = 10$ and $n = 18$. *Answer:* $\varphi(10) = 4$, $\varphi(18) = 6$.

Example 13.12. Find 15^{184} (mod 26).

Solution: The numbers 15 and $26 = 2 \cdot 13$ are relatively prime, so we can apply Euler's theorem. To compute $\varphi(26)$, we observe that 26 is divisible only by 2 and 13. Therefore, we need to exclude all positive integers less than 26 that are even (13 are left) and that are divisible by 13 (12 are left). Then $\varphi(26) = 12$, and we have

$$15^{12} \equiv 1 \pmod{26}.$$

Now $184 = 12 \cdot 15 + 4$, and hence

$$15^{184} \equiv (15^{12})^{15} \cdot 15^4 \equiv 1 \cdot 15^4 \equiv 15^4 \pmod{26}.$$

We could solve the remaining congruence using a calculator. However, we can also do it without a calculator. Because $15^2 = 225 \equiv 26 \cdot 8 + 17 \equiv 17$ (mod 26), we have

$$15^4 \equiv 17^2 \equiv 289 \equiv 260 + 26 + 3 \equiv 3 \quad (\text{mod } 26).$$

Finally we have,

$$15^{184} \equiv 3 \quad (\text{mod } 26).$$

$\square$

Sometimes it is easier to use a more straightforward approach. If we can guess which power of a is congruent to 1 modulo n, then we can solve the congruence.

Example 13.13. Find 3^{514} (mod 26).

Solution: We know by Euler's theorem that there is a power of 3 that is congruent to 1 modulo 26, because 3 and 26 are relatively prime. Instead of computing the number $\varphi(26)$, we can observe that

$$3^3 \equiv 27 \equiv 1 \quad (\text{mod } 26).$$

Now $514 = 3 \cdot 171 + 1$; therefore

$$3^{514} = 27^{171} \cdot 3 \equiv 1^{171} \cdot 3 \equiv 3 \quad (\text{mod } 26).$$

$\square$

Note that Euler's theorem would have worked as well in the last example, but we would have to compute the number $\varphi(26)$ first. The solution in Example 13.13 is even simpler. Euler's theorem provides a power of a that is congruent to 1 modulo n, but this power is not necessarily the smallest.

Finally, suppose that you are looking for the smallest positive number x congruent to a^s (mod n), where a and n have nontrivial common divisors. Because n is a finite number, there are only $n - 1$ residues less than n that a power of a can be congruent to modulo n. Therefore, if we consider enough consecutive powers of a:

$$a, a^2, a^3, \ldots \quad (\text{mod } n),$$

sooner or later we will hit a residue that we have already seen before. But if $a^t \equiv x$ (mod n) and $a^s \equiv x$ (mod n), then the next powers are also congruent to the same residue:

$$a^{s+1} \equiv x \cdot a \quad (\text{mod } n), \qquad a^{t+1} \equiv x \cdot a \quad (\text{mod } n).$$

This means that the sequence of residues is necessarily periodic beyond a certain point. This observation allows us to solve congruences for any a and n by finding a periodic pattern of residues modulo n in the powers of a. We just have to compute the first few powers of a, take them modulo n, and with some luck we will be able to to find a pattern.

For example, the above method is useful when you want to find the last digit of a number given as an exponent with an even base.

Example 13.14. Find the last digit of $12^{4,676}$.

Solution: Finding the last digit is equivalent to solving the congruence

$$12^{4,676} \equiv x \pmod{10}.$$

To find the last digit of an exponent, only the last digit of the base matters. This should be clear from the rules of multiplication and congruences: in a product, the last digit of the result is completely determined by the last digits of the factors. So we are looking for x such that

$$2^{4,676} \equiv x \pmod{10}.$$

Trying the first few powers of 2, we find

$$2 \equiv 2 \pmod{10}, \quad 2^2 \equiv 4 \pmod{10}, \quad 2^3 \equiv 8 \pmod{10},$$

$$2^4 \equiv 6 \pmod{10}, \quad 2^5 \equiv 2 \pmod{10}.$$

The first and the fifth powers of 2 are congruent to 2 modulo 10. This means that the congruences of the powers of 2 modulo 10 are periodic with period 4: they are equal to $2, 4, 8, 6, 2, 4, 8, 6, \ldots$ and so on to infinity. The power $4,676$ is a multiple of 4: $4,676 = 4 \cdot 1,169$, and every fourth power of 2 is congruent to 6 modulo 10. Therefore we have

$$12^{4,676} \equiv 6 \pmod{10}.$$

Similarly, we can also deduce that

$$12^{4,677} \equiv 2 \pmod{10}, \quad 12^{4,678} \equiv 4 \pmod{10},$$

$$12^{4,679} \equiv 8 \pmod{10}.$$

$\square$

Each time you use your credit card online or send an email message, your computer performs the same kind of operations that we just encountered on a much larger scale to ensure that the transaction is secure. An ingenious application of modular arithmetic operating on a bunch of very large numbers is at work to give you peace of mind.

EXERCISES

Question 13.1. A translation cipher converts T to B. Decipher the message

BWLIGIBVQVM

written in this cipher.

Question 13.2. A message was enciphered using a Vigenère cipher with the key
$(+5, -5, +12)$ mod 26. The ciphertext is

YCQLJXIWGLWKJYSFMMQGMSKAJ

Decipher the message. The answer contains an example of a substitution cipher.

Question 13.3.   1. Check the pattern for powers of 5 modulo 7. Is there a repetition?

2. What is 5^{27} modulo 7? How about 5^{601} modulo 7?

Question 13.4.   1. Check the pattern for powers of 5 modulo 8. Is there a repetition?

2. What is 5^{143} (mod 8)?

Question 13.5.   1. Find x between 0 and 12 such that
$9^{36} \equiv x$ (mod 13).

2. Find x between 0 and 10 such that $5^{132} \equiv x$ (mod 11).
Hint: Use FLT.

Question 13.6.   1. Find x between 0 and 6 such that
$19^{37} \equiv x$ (mod 7).

2. Find x between 0 and 7 such that $5^{618} \equiv x$ (mod 8).

Question 13.7.   1. Find the last digit of 323^{516}.

2. Find the last digit of 78^{117}.

Question 13.8.   1. What is the last digit of 134^{467}?

2. What is the last digit of 67^{46}?

Question 13.9.   1. Find x between 0 and 10 such that
$3^{111} \equiv x \pmod{11}$.

2. Find x between 0 and 8 such that $7^{111} \equiv x \pmod 9$.

Question* 13.10. This question outlines a proof of Fermat's Little Theorem. Suppose $a \geq 2$ is an integer and p is a prime number.

1. Consider all possible words of length p in an alphabet of a letters. Here by a "word" we mean any string of letters of length p (even if it is not a word in any language). Check that there are a^p such words.

2. Show that there are a words of length p that use exactly one letter of the alphabet.

3. Suppose we divide all words of a given length into groups according to the following rule: two words are in the same group if and only if one can be obtained from the other by cyclic permutation of letters. For example, the words

    XYYZX,   XXYYZ,   ZXXYY,   YZXXY,   YYZXX

    are members of one group, and there are no other members in the same group. If you imagine the letters of a word arranged in a circle, then two words are in the same group if and only if one can be obtained from another by rotation. Show that if a word cannot be divided into two or more *equal* shorter words, then the group it belongs to has as many members as there are letters in the word. (An example of such a group is given above).

4. Now consider a word of length p which contains more than one letter of the alphabet. Show that if p is a prime, such a word cannot be divided into equal shorter words. Using (3), conclude that any such word belongs to a group that has exactly p members.

5. Combine (1), (2), and (4) to show that $a^p - a$ is divisible by p.

Question 13.11. Suppose that $p = 17$, $g = 3$, $a = 50$, and $b = 67$. Find x, y, and the common secret key K according to the Diffie-Hellman algorithm.

Question 13.12. Suppose that $p = 11$, $g = 2$, $a = 33$, and $b = 74$. Find x, y, and the common secret key K according to the Diffie-Hellman algorithm.

Chapter 14

Probability: Dice, coins, cards, and winning streaks

Games of chance are common in the real world – including many card games, roulette, slot machines, and lotteries. However, the applications of probability are far more widespread; probabilistic models are crucial to finance and trading in the stock market or commodity markets. Probability theory is also used in the environmental sciences and in insurance modeling, as well as in predicting election results via exit polls.

In this chapter, we will discuss the basic laws of probability and learn how to calculate probabilities in simple cases. Our models include rolling dice, tossing coins, and drawing cards from a deck.

MATH

In a very basic sense, probability theory is the art of counting. When we say that we have a one-in-two chance of obtaining a head by tossing a coin, we are counting two things:

- The total number of *possible outcomes* that an event – in this case, tossing a coin – can have. There are two outcomes – a head and a tail – and we are assuming that the coin is *fair*, i.e., it is just as likely to come up heads as tails.

- The number of outcomes that are *favorable* – in this case, just the one outcome: a head.

Now the chance, or *probability*, of obtaining a head is simply the ratio: 1/2.

Example 14.1. Suppose we want to calculate the probability of getting exactly two heads in three coin tosses. Again, we need to count the number of all possible outcomes and the number of favorable outcomes. The total number of (possible) outcomes can be computed by *multiplying* the number of outputs for each coin, because no coin toss affects the output of any other coin toss.[1] There are two outputs – head (H) and tail (T) – and three tosses. So the total number of outcomes is:

$$2 \cdot 2 \cdot 2 = 8 : \ \{HHH, HHT, HTH, HTT, THH, THT, TTH, TTT\}.$$

Out of these, how many outcomes are favorable? Clearly, there are three of them: the outcomes where there are exactly two heads:

$$\{HHT, HTH, THH\}.$$

Therefore the probability of getting exactly two heads in three tosses of a fair coin is precisely $3/8 = 0.375$.   □

It will be convenient to use the language of *sets* to discuss   probabilities. Let Ω denote the set (or collection) of all possible outcomes from an experiment. We will assume that such an experiment can only have finitely many possible outcomes – i.e., that Ω is finite.

An *event* is just a subset: $E \subset \Omega$. It consists of the *favorable* outcomes. We will denote the number of outcomes in the set E by $\#E$.

Now the probability of E is defined to be the proportion of outcomes that lie in E:

$$P(E) = \frac{\text{number of favorable outcomes}}{\text{total number of outcomes}} = \frac{\#E}{\#\Omega}.$$

For instance, in the above example,

$$\Omega = \{HHH, HHT, HTH, HTT, THH, THT, TTH, TTT\}$$

---

[1] In this book, we will distinguish between the *output* from a coin toss and the *outcome* of two coin tosses. The latter is a pair of outputs, one from each coin toss. We make the same distinctions for rolling dice or drawing cards.

is the set of all possible outcomes from the three coin tosses, and

$$E = \{\text{HHT}, \text{HTH}, \text{THH}\}$$

is the set of favorable outcomes (exactly two heads). Hence,

$$P(E) = \frac{\#E}{\#\Omega} = \frac{3}{8}.$$

In general, the method of calculating probabilities is the same: compute the number of possible outcomes and the number of favorable outcomes. Divide the number of favorable outcomes by the number of possible outcomes. Of course, as the problems get increasingly complicated, sometimes you have to find clever ways of counting without actually writing down the entire list of possible and favorable outcomes. Let us illustrate with an example.

Example 14.2. Suppose you roll two dice. Find the probability that:

1. The sum of the two outputs is 4.

2. The sum of the two outputs is at least 4.

3. Exactly one of the two outputs is even.

Solution: We know that the output from each die is a number from 1 to 6. Each outcome is a pair of outputs (x_1, x_2), from rolling the first and second dice, respectively. Now compute the total number of outcomes, i.e., the size of Ω. Each die has six outputs and there are two dice, so $\#\Omega = 6 \cdot 6 = 36$. We do not need to write down all of these outcomes; we just need their number: 36. However, you may write out Ω if you wish: it goes from $(1, 1)$ to $(1, 2), (2, 1)$ and so on, all the way to $(6, 6)$.

1. How can two positive integer numbers add up to 4? The only possibilities are: $(1, 3), (2, 2), (3, 1)$. In other words, E is precisely this set:

$$E = \{(1, 3), (2, 2), (3, 1)\}.$$

Hence $P(E) = \#E/\#\Omega = 3/36 = 1/12$.

2. What are the possible sums of the two dice outputs? Check that the sum can go from $1 + 1 = 2$ to $6 + 6 = 12$. Here's how to say this precisely: if x_1 and x_2 are the outputs from rolling the dice (as above), then

$$1 \leq x_1 \leq 6, \qquad 1 \leq x_2 \leq 6.$$

Adding up both the inequalities, $2 \leq x_1 + x_2 \leq 12$.

So to compute (the size of) E, one would have to write down all the pairs of outputs whose sum is $4, 5, 6, \ldots, 12$. That is a lot of writing! Is there a simpler way, perhaps?

Indeed there is. Instead of writing down which outcomes are favorable, write down the ones that are *not* so. This is a much smaller set – namely, the set of pairs of outputs whose sum is 2 or 3 (i.e., *not* 4 or more):

$$E^c = \text{(sum is 2 or 3)} = \{(1,1), (1,2), (2,1)\}.$$

Here E^c stands for the *complement* of E – the outcomes that are in Ω but not in E, i.e., unfavorable. So E^c has size 3 and Ω has size 36, meaning that E has size $36 - 3 = 33$. Finally, $P(E) = \#E/\#\Omega = 33/36 = 11/12$.

3. Once again, we could have written down all thirty-six pairs of outputs, and then checked by inspection which pairs have exactly one even output and one odd one. However, once again, there is a simpler way to count. Every favorable outcome (x_1, x_2) is of exactly one of the following two types:

   - x_1 is even and x_2 is odd. Denote the set of such (favorable) outcomes by E_1.

   - x_2 is even and x_1 is odd. Denote the set of such (favorable) outcomes by E_2.

   Let us "cleverly" count how many outcomes there are in E_1. If x_1 is even, then it can be only one of $\{2, 4, 6\}$. Similarly, x_2 is odd, so it is 1, 3, or 5. Because neither die affects the output of the other, how many outcomes lie in E_1? The answer is: $3 \cdot 3 = 9$.

   Similarly, calculate that E_2 also has nine outcomes in it. We have just calculated these numbers without having to write down all nine outcomes in each case.

   Thus, how many favorable outcomes are there? Because x_1 cannot be even and odd at the same time, no outcome is common to E_1 and E_2. In probabilities, we call such events *disjoint*: two events are disjoint if there is no outcome common to both. Then to find the number of outcomes in either E_1 or E_2, we have to add the numbers of outcomes in each of them. The answer is: $\#E = \#E_1 + \#E_2 = 18$. Finally,

$$P(E) = \#E/\#\Omega = 18/36 = 1/2.$$

$\square$

Laws of probability

Example 14.2 illustrates a couple of very important techniques of "cleverly counting" the number of favorable outcomes. In many textbooks, these techniques are known as the *laws of probability*. We will discuss these techniques in general and offer very brief – and simple – explanations for each of them.

First let us introduce the notation. We say that the set of outcomes of an event E_1 *contains* the set of outcomes of another event E_2, and write $E_2 \subset E_1$, if every outcome in E_2 is also an outcome of E_1. For instance, the event of getting at least two heads in three coin tosses contains the event of getting three heads in three coin tosses. For any events E_1 and E_2, the *union* $E_1 \cup E_2$ of E_1 and E_2 is the set of outcomes that are either in E_1 or in E_2 (or in both). For example, if E_1 is getting HT in two coin tosses, and E_2 is getting TH, then $E_1 \cup E_2$ is the event of getting one head and one tail in two coin tosses in any order.

Law 1. If E is any event (of favorable outcomes), then $0 \le P(E) \le 1$.

Reason: Every set of outcomes E is contained in Ω (the set of all possible outcomes) and contains the empty set $\emptyset$ (the set of *no* outcomes). So, the number of elements in E is at most $\#\Omega$, and at least 0. Now dividing by $\#\Omega$, we get the same inequality for probabilities:

$$0 = \frac{0}{\#\Omega} \le P(E) = \frac{\#E}{\#\Omega} \le \frac{\#\Omega}{\#\Omega} = 1.$$

Law 2. If events E and E' are such that E' contains E, then $P(E) \le P(E')$. (For instance, the chances of getting a diamond when you draw a card from a deck are less than the chances of getting a red card.)
Reason: If $E \subset E'$, then E' has more outcomes than E: $\#E \le \#E'$. Now divide both sides by $\#\Omega$. Then we get: $P(E) \le P(E')$.

Law 3. Given two sets E_1 and E_2, their union $E_1 \cup E_2$ is the set of all outcomes that are either in E_1 or in E_2, or common to both events. If E_1

and E_2 are *disjoint* events – i.e., there is no outcome common to both – then their union $E \cup E'$ has probability:

$$P(E_1 \cup E_2) = P(E_1) + P(E_2).$$

For instance, in Example 14.2(3) E_1 and E_2 were the events that the first die had an even output but not the second, and that the second die had an even output but not the first, respectively. Their "union" is the event that exactly one die has an even output.

Reason: If E_1 and E_2 have no outcomes in common, then their sizes clearly add up:

$$\#(E_1 \cup E_2) = \#E_1 + \#E_2.$$

Now divide both sides by $\#\Omega$. Then we get: $P(E_1 \cup E_2) = P(E_1) + P(E_2)$.

Law 4. If E is any event, and E^c is its *complementary* event – i.e., the set of outcomes in Ω that are not in E, then

$$P(E^c) + P(E) = 1.$$

(In other words, $P(E^c) = 1 - P(E)$.) We applied this rule in Example 14.2(2). The advantage is that we can solve for either $P(E)$ or $P(E^c)$ in terms of the other.

Reason: It is clear that E and E^c are disjoint, because no outcome can be in both of them. Moreover, $E \cup E^c = \Omega$, which has probability $\#\Omega/\#\Omega = 1$. So by Law 3 (or dividing both sides of the equation $\#E + \#E^c = \#\Omega$, by $\#\Omega$),

$$1 = P(\Omega) = P(E) + P(E^c).$$

APPLICATIONS

The next few examples illustrate the use of the definition and laws of probability to cleverly count the number of possible and favorable outcomes.

Example 14.3. Suppose a class consists of forty-five female students and thirty-three male students. A student is randomly chosen from among this class. What is the probability that the student is female? If two students are randomly chosen, what is the probability that they are both male?

Solution: The answer to the first question is clear: E is the set of female students, so $\#E = 45$ and $\#\Omega = 45 + 33 = 78$. Hence, $P(E) = 45/78 = 15/26$.

As for the second question, the first student can be chosen in seventy-eight ways, and for each choice there are seventy-seven ways to choose the second. Thus, $\#\Omega = 78 \cdot 77$. Similarly, in order for both students chosen to be male, the set of favorable outcomes can be calculated to have size $\#E = 33 \cdot 32$. Thus,

$$P(E) = \frac{\#E}{\#\Omega} = \frac{33 \cdot 32}{78 \cdot 77} = \frac{1 \cdot 16}{13 \cdot 7} = \frac{16}{91}.$$

$\square$

Practice problem 14.4.   1. What are the odds that a ball chosen randomly from a bag containing five green balls, three red balls, and seven blue balls, is (a) green? (b) not blue? *Answer:* (a) $\frac{1}{3}$, (b) $\frac{8}{15}$.

2. Suppose two students are chosen at random from a group consisting of seventeen sophomores and thirteen freshmen. What is the probability that they are not both in the same year? *Answer:* $\frac{17 \cdot 13 + 13 \cdot 17}{30 \cdot 29} = \frac{221}{435}$.

More on disjoint events

One of the most common and efficient ways to simplify a computation of a probability is to break the event into a union of two or more disjoint events and compute their probabilities separately. If E_1 and E_2 are disjoint, then their union $E_1 \cup E_2$ (meaning all outcomes that lie either in E_1 or in E_2) has the probability: $P(E_1 \cup E_2) = P(E_1) + P(E_2)$ (Law 3). In practice it is important to check that the events E_1 and E_2 are disjoint – otherwise the formula cannot be applied! In set theory language, this means that the *intersection* $E_1 \cap E_2$ should be an empty set; i.e., no outcome can lie in both E_1 and E_2. The examples below illustrate the use of disjoint events, along with other laws of probability.

Example 14.5. Let us roll two dice. What are the chances that the sum of the outputs is at most 3?

Solution: Note that the outputs both have to be 1 or more, so the sum of the outputs cannot be 1. It must be 2 or 3. How can it be 2? This happens via only one possible outcome: (1,1). How can it be 3? The dice rolled yield (1,2) or (2,1). Thus, our event is now $\{(1,1),(1,2),(2,1)\}$, and the total number of outcomes is $6 \cdot 6 = 36$ (because each die has six outcomes, two dice rolled independently yield thirty-six outcomes). The probability of getting 2 or 3 is $P(2 \text{ or } 3) = 3/36 = 1/12$.

The same result can be computed as the probability of a union of two disjoint events. The event of getting at most 3 is the disjoint union of two events: getting *exactly* 2 and *exactly* 3 as the sum of the two dice. Clearly, the events are disjoint. The outcomes that yield 2 are simply (1,1), so the probability of getting 2 is 1/36. The outcomes that yield 3 are (1,2) and (2,1), so $P(3) = 2/36 = 1/18$. Adding them up,

$$P(2 \text{ or } 3) = P(2) + P(3) = \frac{1}{36} + \frac{2}{36} = \frac{1}{12}.$$

□

Practice problem 14.6. Suppose you roll two dice. Find the probability that:

1. The sum of the two outputs is 6. *Answer:* 5/36.

2. The sum of the two outputs is at most 9. *Answer:* 5/6.

3. Both outputs are odd. *Answer:* 1/4.

Example 14.7. Suppose we toss six fair coins. How many total outcomes are possible? Also compute the probability that:

1. At least five of the coins are heads.

2. Exactly four coins are heads.

3. The first and last coins are tails.

Solution: First, the total number of outcomes is $2 \cdot 2 \cdots \cdots 2$ (6 times) $= 64$ (because there are six coins, each of whose outputs does not affect the others). Thus for all three parts, $\#\Omega = 64$ (and Ω itself contains the sixty-four possible results, from HHHHHH to TTTTTT).

1. For at least five coins to be heads, either exactly five or exactly six coins must be heads. There is only one outcome with six heads: HHHHHH. How about five heads? A "clever" way to count five heads is to realize that there is exactly one tail. This tail can occur in any of the six tosses, so there must be exactly six such outcomes. Thus, we get seven favorable outcomes in all. Hence, $P(E) = \#E/\#\Omega = 7/64$.

   Alternatively, the same calculation could have been done using Law 3 of probability. The events of obtaining exactly five heads and exactly six heads are, of course, disjoint. Hence:

$$P(\text{at least five heads})$$
$$= P(\text{exactly five heads}) + P(\text{exactly six heads})$$
$$= \frac{6}{64} + \frac{1}{64} = \frac{7}{64}.$$

2. As above, it may be simpler to look for exactly two tails, rather than four heads. In the next chapter, we will introduce a formula that directly calculates in how many ways one can choose two tails in six coin tosses. This formula uses *permutations* and *combinations*, the tools that we will introduce and study in Chapter 15. For now, there is no choice but to try and write out all the possibilities:

TTHHHH, THTHHH, THHTHH, THHHTH, THHHHT,
HTTHHH, HTHTHH, HTHHTH, HTHHHT,
HHTTHH, HHTHTH, HHTHHT,
HHHTTH, HHHTHT,
HHHHTT.

In writing out such lists, it is helpful to be systematic: first write all ways in which to obtain four heads and two tails, where the first tail occurs in the *first* coin toss. (This is the entire first row.) Then move on to all outcomes where the first tail occurs in the *second* coin toss (there are four such outcomes, all written in the second row), and so on.

Counting the outcomes above, $\#E = 15$, so $P(E) = 15/64$.

3. If the first and the last coins are tails, we can again write out all the options – or we can "count cleverly": there are no restrictions of any kind on the middle four coin toss outputs, and no individual coin toss output affects any other output. So we simply multiply the number of possible outputs for each toss:

$$\#E = 1 \cdot 2 \cdot 2 \cdot 2 \cdot 2 \cdot 1 = 16.$$

Therefore, $P(E) = \#E/\#\Omega = 16/64 = 1/4.$ $\square$

Example 14.8. Suppose I draw two cards randomly from a deck of fifty-two cards (without replacement). Compute the probability that:

1. Both cards are spades.

2. The first card is a diamond.

3. The second card is a diamond.

Compare the answers in (2) and (3).

Solution: First compute the total number of outcomes. There are fifty-two outputs for the first card, and for each chosen card there are fifty-one possible outputs for the second card. Thus for all parts, $\#\Omega = 52 \cdot 51$ (let us not multiply this out just yet.)

1. For both cards to be spades, the first card can be one of thirteen. After drawing any one of these cards, there are twelve spades remaining to be chosen as the second card. Therefore,

$$P(\text{both cards are spades}) = P(E) = \frac{\#E}{\#\Omega} = \frac{13 \cdot 12}{52 \cdot 51} = \frac{1}{4} \cdot \frac{4}{17} = \frac{1}{17}.$$

2. Even without any calculations, the probability that the first card is a diamond should not depend at all on the second card, so it should just be $13/52 = 1/4$.

   If we want to be more thorough, for favorable outcomes the first card has to be one of thirteen cards. Choosing any of these cards leaves fifty-one cards for the second drawing. Thus, $\#E = 13 \cdot 51$, and:

$$P(\text{first card is a diamond}) = \frac{13 \cdot 51}{52 \cdot 51} = \frac{13}{52} = \frac{1}{4}.$$

3. In order to calculate the number of outcomes where the second card is a diamond, it seems that we need to analyze whether or not the first one is also a diamond. There are two disjoint events:

   • The first card is a diamond (hence there are thirteen choices), and the second card is also a diamond (hence there are twelve choices). There are $13 \cdot 12$ outcomes of this kind.

   • The first card is not a diamond (hence there are $52 - 13 = 39$ choices), but the second card is a diamond (hence there are thirteen choices for each of the above thirty-nine choices). There are $39 \cdot 13$ outcomes of this kind.

So the total number of favorable outcomes is:

$$(13 \cdot 12) + (39 \cdot 13) = 13 \cdot (12 + 39) = 13 \cdot 51.$$

Therefore, the probability that the second card is a diamond, is:

$$\frac{13 \cdot 51}{52 \cdot 51} = \frac{13}{52} = \frac{1}{4}.$$

It turns out that the answers to parts (2) and (3) are equal – i.e., the probability of the first card being a diamond equals that of the second card being a diamond! Why?

Here is one way to think about it. Draw a card from the deck and place it *face down* in front of you. Now draw a second card and place it face down next to the first. Repeat this with a third card, a fourth card... all the way to the fifty-second card, all in a row on your (long) desk. (Do not turn over any of these cards.)

Thus, we have a line of fifty-two cards – which is simply our original deck – all face down. We want to know what the chances are that if we turn over *only* the second card, it will be a diamond. But the chances of any one card being a diamond, given no information about any other card, are precisely 1/4. (Because this is equivalent to drawing just one card out of the deck.)

In fact, this explanation shows that the probability that the twenty-ninth card drawn from the deck is a diamond is also 1/4. We figured this out without any calculations whatsoever!   □

Systems with and without memory

It might seem that getting all heads in six, eight, or ten consecutive coin tosses is a very unlikely event. Indeed, their probabilities are pretty small: 1/64, 1/256, and 1/1,024, respectively. What might be less intuitively clear is that the probabilities of getting sequences of *alternating* heads and tails, starting from a head, in the same numbers of tosses are just as small. In general, the question about the odds of getting sequences of one and the same output in consecutive experiments is more subtle than it seems, and the answer depends on the setting. Consider two experiments:

1. Repeatedly tossing a fair coin.

2. Pulling out fruits one by one at random from a bag containing fifty apples and fifty oranges (without replacement).

In some sense, the two experiments are similar. In both cases there are equal chances of getting one of the two possible outcomes: heads or tails in the first case, an apple or an orange in the second. However, the probabilities of getting longer sequences of the same outputs in these two situations are different. Let us consider the two cases in more detail.

The probability of pulling an apple (A) from the bag containing an equal number of apples and oranges is equal to the probability of getting heads (H) on a single toss of a coin:

$$P(\text{A}) = P(\text{H}) = \frac{1}{2}.$$

But if we continue pulling fruit and tossing coins, the probabilities become different:

$$P(\text{AA}) = \frac{50 \cdot 49}{100 \cdot 99} \simeq 0.24747 \quad < \quad P(\text{HH}) = \frac{1}{2} \cdot \frac{1}{2} = \frac{1}{4} = 0.25.$$

The longer the sequence, the bigger the difference in the probabilities:

$$P(\text{AAAAA}) = \frac{50 \cdot 49 \cdot 48 \cdot 47 \cdot 46}{100 \cdot 99 \cdot 98 \cdot 97 \cdot 96} \simeq 0.02814$$

$$< P(\text{HHHHH}) = \frac{1}{2^5} = \frac{1}{32} = 0.03125.$$

The reason for this lies in the fact that the consecutive tosses of a fair coin are independent, but the consecutive pulls of fruit from the bag (containing a finite number of apples and oranges) are not. Pulling one apple out of the bag reduces the probability of pulling another one out. In coin tosses, flipping a head does not reduce the probability of flipping another head because the coin has *no memory* of the previous tosses. This is the difference between systems *with* and *without memory*: the bag of fruit "knows" how many apples are missing, but the coin remembers nothing of its past tosses. Another way to think about it is that the supply of apples in the bag is finite, but the supply of heads in coin tosses is infinite. The same considerations can be applied in other situations. For example, a die has no memory of the previous rolls, but a deck of cards remembers which cards have been pulled out (without replacement).

Another difference between systems with and without memory is in the probabilities of getting sequences of the same versus mixed outputs. If a coin is fair, the probability of getting any particular sequence of heads and tails in five consecutive tosses is the same:

$$P(\text{HTHTH}) = P(\text{HHHHH}) = \frac{1}{32}.$$

In the case of apples (A) and oranges (O), sequences that look more "random" actually have a higher probability of occurring:

$$P(\text{AOAOA}) = \frac{50 \cdot 50 \cdot 49 \cdot 49 \cdot 48}{100 \cdot 99 \cdot 98 \cdot 97 \cdot 96} \simeq 0.03189$$
$$> P(\text{AAAAA}) = \frac{50 \cdot 49 \cdot 48 \cdot 47 \cdot 46}{100 \cdot 99 \cdot 98 \cdot 97 \cdot 96} \simeq 0.02814.$$

The longer the sequence, the more pronounced the effect. This might be one of the reasons for the persistence of the *gambler's fallacy* – a belief that if a long streak of a particular output has happened in a random process, then the opposite output is "due." This assertion is false when applied to systems without memory: no matter how many heads have been flipped, the next flip of a coin yields heads or tails with probability $\frac{1}{2}$. Another way to see it is to compute the probabilities of the sequences of the kind HHHHHH and HHHHHT. They are the same:

$$P(\text{HHHHHH}) = P(\text{HHHHHT}) = \frac{1}{2^6} = \frac{1}{64}.$$

However, the probability of getting an orange after a long sequence of apples is much higher than the probability of getting yet another apple. For example, after pulling five apples in a row, there are forty-five apples and fifty oranges left in the bag. The probability of pulling a sixth apple is

$$P(\text{sixth A}) = \frac{45}{95} \simeq 0.47368,$$

while the probability of pulling an orange is

$$P(\text{O after five A}) = \frac{50}{95} \simeq 0.52632.$$

Note that these are complementary events, so their probabilities add up to 1.

Another reason for the bias of the gambler's fallacy might lie in an incorrect interpretation of the *law of large numbers*. According to this law, proved by J. Bernoulli in 1713, the *proportion* of successes in repeated (independent) trials of a random process tends to the probability of a success in this process, as the number of trials tends to infinity. In particular, the *proportion* of heads in a large number of tosses actually approaches $\frac{1}{2}$ as the number of tosses tends to infinity. However, the *difference* between the numbers of heads and tails does not need to decrease as the number of tosses grows, which makes it impossible to predict the result of the next toss based on the previous tosses.

The main point of this discussion is that the probability of getting a long streak with the same output depends on the setting in which this output occurs. When rolling dice, flipping coins, or playing roulette, do not expect that a frequent occurrence of a certain kind of output makes this output less likely to appear in the future; but in card games, such as blackjack, counting cards that are already out can help you estimate the odds of drawing any given card.

EXERCISES

Question 14.1. You toss seven fair coins. Compute the probability that:

1. at least two coins are tails.

2. at least one of the coins yields a head.

3. the third and fifth coins yield tails, while the second coin yields a head.

Question 14.2. You draw two cards randomly from a deck (without replacing the first card). Compute the probability that:

1. the cards are both aces.

2. the cards are an ace and a king, in some order.

3. the cards are not both clubs.

Question 14.3. A bag contains three balls of green yarn, four of red yarn, and five of blue yarn. A cat snatches one ball of yarn from the bag. What is the probability that it is blue? What is the probability that it is not red?

Question 14.4. Find the probabilities of the following events:

1. drawing one card from a deck and obtaining a number between 5 and 10.

2. drawing one card from a deck and obtaining a heart or a picture card (J, Q, K).

Question 14.5. Find the probabilities of the following events:

1. tossing a fair coin five times and obtaining exactly two heads.

2. rolling three dice and obtaining the sum of 3 or 4.

Question 14.6. Suppose two people did not show up for an event where ten freshmen, eight sophomores, and twelve juniors were invited. Assuming that all invitees had equal chance of not showing up, compute the probability that

1. both of them are sophomores.

2. neither of them is a sophomore.

3. they are a freshman and a junior, in any order.

Question 14.7. Your friend tosses four fair coins. Compute the probability that:

1. exactly two of the outputs are heads.

2. at least two of the outputs are heads.

3. an even number of tails occurs.

4. an odd number of heads occurs. *Hint:* Can you use part (3) to simplify your calculations?

Question 14.8. A box contains seven milk, five dark, and eight white chocolates. Drawing two chocolates at random, what is the probability that

1. both are dark?

2. the first is white?

3. neither is white?

4. there is one dark and one milk, in any order?

5. the second is milk?

Question 14.9. Suppose I roll three dice. What is the probability that the sum of the outputs is at most 5? At least 5?

Question 14.10. An illusionist draws two cards from a deck without replacement. Find the probability that:

1. the first is a diamond and the second is an ace.

  2. they are exactly a jack and a queen, in any order.

Question 14.11. An experiment consists of rolling a fair die and tossing a fair coin. The outcome is computed as follows: if a coin comes up heads, 1 is added to the output of the die; otherwise nothing is added. Find the probability that the outcome

  1. equals 6.

  2. is less than 3.

  3. is odd.

Question 14.12. David Copperfield draws three cards from a deck, without replacement. Find the probability that:

  1. all of the cards are diamonds.

  2. all of the cards are aces.

  3. the cards are a heart, a diamond, and a spade, in that order.

  4. the cards are all of different suits.

  5. they are two diamonds and one spade, in any order.

  6. they are two aces of spades and one diamond ring!

Chapter 15

Permutations and combinations: Counting your choices

In this chapter we will develop efficient methods of counting the number of choices we have when we put together certain ordered and unordered collections of items. The same methods can be applied to counting possible and favorable outcomes of an event, and thus to computing probabilities. The main goal of the chapter is to discuss such applications. In addition, similar constructions appear in many other areas of mathematics and applications. We will mention some of them at the end of the chapter.

MATH

Permutations

A *permutation* of n items is an arrangement of these items in a particular order. For example, a permutation of three colors, blue, white, and red, is (WRB). Another permutation of the same items is (RWB). We will be interested in the *number* of possible permutations of n items. It is denoted by $_nP_n$. In the case of three items (e.g, B, W, and R), there are six possible permutations:

$$(BWR), (BRW), (RWB), (RBW), (WBR), (WRB),$$

so we have $_3P_3 = 6$. In general, when counting the number of permutations of n items, there are n choices for the first item, $(n-1)$ for the second, and so on, until the last item, for which there is only one choice. To compute this number, we have to multiply the individual numbers of choices:

$$_nP_n = n \cdot (n - 1) \cdot (n - 2) \cdots 2 \cdot 1 = n!$$

where $n!$ is the notation for the product of consecutive whole numbers from 1 to n, which is read "n factorial."

A *permutation* of k out of n items is an arrangement of any k out of these n items in a particular order. How many different permutations, or arrangements, are there of three items selected from a set of five items? For example, you can think of the ordered arrangements of three colors selected out of five available colors. Then we have five choices for the first item, four for the second, and three for the last item, so the number of possible choices of ordered triples of items is

$$_5P_3 = 5 \cdot 4 \cdot 3 = 60.$$

We can express the same number using the factorial notation. Indeed, the number of permutations in this case is the product of all natural numbers from 3 to 5, which is the product of the natural numbers up to 5 divided by the product of the natural numbers up to 2:

$$_5P_3 = 5 \cdot 4 \cdot 3 = \frac{5 \cdot 4 \cdot 3 \cdot 2 \cdot 1}{2 \cdot 1} = \frac{5!}{2!} = 60.$$

Following the same reasoning, we find that in general, the number of permutations of k items out of n items is

$$_nP_k = n \cdot (n - 1) \cdot (n - 2) \cdots (n - k + 1) = \frac{n!}{(n - k)!}.$$

For convenience, it is agreed that $0! = 1$. Then for $k = n$ we get

$$_nP_n = \frac{n!}{0!} = n!,$$

as before. By convention, the number of permutations of 0 out of n items that does not a priori make sense is assigned the value 1: $_nP_0 = \frac{n!}{n!} = 1$.

Now suppose we need to count the number of permutations of the letters in the word "bee." Presumably, there should be $_3P_3 = 3! = 6$ permutations, but in fact we have only three:

$$(bee), (ebe), (eeb).$$

This happens because two of the letters are identical – call them e_1 and e_2 – and so the number of permutations of 3 items has to be divided by 2 to account for the identical permutations of the kind (e_1be_2) and (e_2be_1). In general, the number of permutations of n items of which r are identical is the quotient of the number of permutations of n items over the number of permutations of r identical items. It is equal to

$$\frac{{}_nP_n}{{}_rP_r} = \frac{n!}{r!}.$$

If in addition there are p other identical items, then the number of distinct permutations will be

$$\frac{n!}{r!p!},$$

and so on.

Combinations

A *combination* of k out of n items is a choice of an *unordered* collection of k out of n items. It is denoted by ${}_nC_k$. Let us count the number of combinations of three out of three items. There are six possible permutations, but because the order of the items does not matter, they constitute just one combination. Therefore, there is only one combination of three out of three items. The same is true for any number of items to choose from: because the collections are considered regardless of the order in which the items are chosen, the number of combinations of n out of n items is always 1:

$${}_nC_n = 1 \qquad \text{for all natural } n = 1, 2, 3, \dots.$$

Now let us count the number of combinations of three out of nine items, denoted by ${}_9C_3$. For example, you want to take three of your favorite nine books with you on a trip. What is the number of possible choices? First, let us count the number of permutations of three out of nine:

$${}_9P_3 = \frac{9!}{6!} = 9 \cdot 8 \cdot 7 = 504.$$

This number counts a collection of any three given books A, B, and C six times:
(ABC), (ACB), (BAC), (BCA), (CAB), (CBA). For our purposes, they are indistinguishable. Therefore, to get the number of distinct collections of

three books out of nine books, we have to divide $_9P_3$ by 6, which is the number of permutations of three items:

$$_9C_3 = \frac{_9P_3}{_3P_3} = \frac{504}{6} = 84.$$

In general, to count the number of combinations $_nC_k$ of k out of n items, we have to divide the number of permutations of k out of n items $_nP_k$ by the number of permutations of k items $_kP_k$:

$$_nC_k = \frac{_nP_k}{_kP_k} = \frac{n!}{(n-k)! \cdot k!}.$$

Looking at the formula for $_nC_k$, we immediately see that the number of choices of k out of n items equals the number of choices of $(n-k)$ out of n items:

$$_nC_k = \frac{n!}{(n-k)! \cdot k!} = \frac{n!}{k! \cdot (n-k)!} = \frac{n!}{(n-(n-k))! \cdot (n-k)!} = {}_nC_{(n-k)}.$$

This property has a clear explanation in terms of the choices of unordered collections: each time you choose an unordered collection of k out of n items, the remaining $(n-k)$ items also form an unordered collection, and different collections of k items correspond to different collections of $(n-k)$ items. So in both cases, the number of choices is the same.

Because by convention $0! = 1$, we have $_nC_n = \frac{n!}{0! \cdot n!} = {}_nC_0 = 1$. Indeed, there is only one way to pick n unordered items out of n – namely, pick all of them.

EXAMPLES

Example 15.1. Cinderella has six dresses and the ball goes on for three nights. How many ways are there for Cinderella to dress? Assume that she wants to wear a different dress each night. The order in which she wears the dresses is important.

Solution: The number we are looking for is the number of permutations of three out of six items. It is given by

$$_6P_3 = \frac{6!}{3!} = 6 \cdot 5 \cdot 4 = 120.$$

$\square$

Practice problem 15.2. How many distinct permutations can be made using the letters of the word MASSACHUSETTS? *Answer:* $\frac{13!}{4!2!2!} = 64,864,800$, because the repeated letters are four S's, two T's, and two A's.

Example 15.3. How many ways are there to choose (a) four courses, (b) thirteen courses out of sixteen courses offered this semester? (The order does not matter.)

Solution: For question (a), the number we are looking for is the number of combinations of 4 out of 16:

$$_{16}C_4 = \frac{16!}{12!4!} = \frac{16 \cdot 15 \cdot 14 \cdot 13}{4 \cdot 3 \cdot 2 \cdot 1} = 4 \cdot 5 \cdot 7 \cdot 13 = 1,820.$$

For question (b), we would need to compute

$$_{16}C_{13} = \frac{16!}{3!13!} = \frac{16 \cdot 15 \cdot \dots \cdot 4}{13 \cdot 12 \cdot \dots \cdot 1}.$$

This looks like a lengthy computation. However, if we recall that $_{16}C_{13} = {}_{16}C_3$ or, equivalently, if we swap the order of the factorials in the denominator, we can simplify the computation and obtain

$$_{16}C_{13} = {}_{16}C_3 = \frac{16!}{13!3!} = \frac{16 \cdot 15 \cdot 14}{3 \cdot 2 \cdot 1} = 8 \cdot 5 \cdot 14 = 560.$$

In general, it is a good idea to cancel the largest factorial in the denominator with a part of the factorial in the numerator first, and then compute the remaining products. □

Practice problem 15.4. Compute $_5C_4$, $_5C_1$, $_5C_3$, $_5P_2$, $_5P_3$. *Answer:* $_5C_4 = {}_5C_1 = 5$, $_5C_3 = 10$, $_5P_2 = 20$, $_5P_3 = 60$. Can you think of real-world examples where counting the number of choices results in these numbers? Remember that permutations count the choices when the order of the items matters, and combinations when the order does not matter.

The next two examples illustrate the difference between permutations and combinations.

Example 15.5. How many three-scoop bowls of ice-cream with three different flavors can be made out of fifteen possible flavors?

Solution: Clearly, in this case the order does not matter: a bowl of strawberry, vanilla, and chocolate is the same as a bowl of chocolate, strawberry, and vanilla. The answer is given by the number of combinations of three items out of fifteen,

$$_{15}C_3 = \frac{15!}{12!3!} = \frac{15 \cdot 14 \cdot 13}{3 \cdot 2 \cdot 1} = 5 \cdot 7 \cdot 13 = 455.$$

□

Example 15.6. How many three-scoop bowls of ice-cream with two different flavors can be made out of possible fifteen flavors?

Solution: Here we have to pick two flavors out of fifteen, but the order of the two matters: we will ask for two scoops of the first chosen flavor, and one scoop of the second flavor. A bowl with two scoops of vanilla and one scoop of strawberry is different from a bowl of one scoop of vanilla and two scoops of strawberry. Therefore, the number of possible choices is equal to the number of permutations of two out of fifteen:

$$_{15}P_2 = \frac{15!}{13!} = 15 \cdot 14 = 210.$$

□

Practice problem 15.7.   1. How many different pizzas can be chosen, each having four different toppings out of seven available choices? *Answer:* $_7C_4 = 35$.

2. Little Clara eats one tub of yogurt every morning. She has yogurts of eight different flavors in the fridge, one tub of each. How many ways are there for her to eat five yogurts on five consecutive days? (The order of eating different flavors matters to Clara!) *Answer:* $_8P_5 = 6,720$.

APPLICATIONS TO COUNTING CHOICES

Some situations require a more creative approach to permutations and combinations. Here is an example.

Example 15.8. Suppose you are the head of R&D at a pastry factory and you work on designing new flavors for your pastries. The company has bought a five-nozzle machine that can drop five different flavors and mix them together. Each nozzle is designed to release one drop of a flavor. You have fifteen flavors available and you want to run evaluation tests on all possible flavor combinations that can be achieved with five nozzles. For example, one of the mixtures can contain three drops of vanilla, one drop of apple flavor and one drop of cinnamon. Another mixture can contain five drops of five different flavors. It is assumed that you always have to use each nozzle, so each mixture contains five drops of flavors, some of which might be repeated. You need to estimate the cost of testing all possible flavor combinations. The cost is proportional to the number of flavor combinations that can be obtained using this machine.

Solution: The number of flavors used in one mixture can be anything from one to five. We will consider the question case by case, according to the number of flavors used.

(1) How many different 5-drop mixtures can be made with just one flavor out of 15? This is easy: just pick one flavor and take five drops of it. The number of choices of one out of fifteen is given by either $_{15}P_1$ or $_{15}C_1$, which are equal (why?):

$$_{15}P_1 = {}_{15}C_1 = 15.$$

(2) Now suppose the mixture has two flavors out of fifteen. For example, if we choose strawberry and vanilla, the resulting mixtures might be:

1 drop of strawberry	4 drops of vanilla
2 drops of strawberry	3 drops of vanilla
3 drops of strawberry	2 drops of vanilla
4 drops of strawberry	1 drop of vanilla

So for each choice of two unordered flavors, we will have four different flavor combinations. The total number of flavor combinations is

$$_{15}C_2 \cdot 4 = \frac{15!}{13! \cdot 2!} \cdot 4 = 420.$$

(The same number can be obtained using permutations: count the number of *ordered* choices of two flavors and multiply it by 2 to account for the two ways to divide 5 into two unequal parts, $5 = 4 + 1$, and $5 = 3 + 2$. Then we obtain: $2 \cdot {}_{15}P_2 = 2 \cdot \frac{15!}{13!} = 420$.)

(3) Next, suppose the mixture has three different flavors. The number of drops of each flavor can be chosen in the following six ways: $(3, 1, 1)$, $(1, 3, 1)$, $(1, 1, 3)$, $(2, 2, 1)$, $(2, 1, 2)$, and $(1, 2, 2)$. The total number of flavor combinations is then the number of unordered choices of three items out of fifteen, multiplied by 6:

$$_{15}C_3 \cdot 6 = \frac{15!}{12! \cdot 3!} \cdot 6 = 15 \cdot 14 \cdot 13 = 2,730.$$

(Think how you can obtain the same number using permutations rather than combinations.)

(4) Now suppose the mixture has four flavors out of fifteen. This can happen only if one of the four flavors is used twice. So, we need to choose one distinguishable flavor, which will be used in two nozzles, and three indistinguishable flavors. The first operation can be done in $_{15}P_1 = 15$ ways. Of the remaining fourteen flavors, we need to choose three whose order does not matter:

$$_{14}C_3 = \frac{14 \cdot 13 \cdot 12}{3 \cdot 2 \cdot 1} = 14 \cdot 13 \cdot 2 = 364.$$

Therefore, the total number of mixtures with four flavors is

$$15 \cdot 364 = 5,460.$$

(5) Now suppose the mixture has to have five different flavors in it. The order does not matter, and the number of possibilities is given by the number of combinations of five items out of fifteen:

$$_{15}C_5 = \frac{15!}{10!5!} = \frac{15 \cdot 14 \cdot 13 \cdot 12 \cdot 11}{5 \cdot 4 \cdot 3 \cdot 2 \cdot 1} = 7 \cdot 13 \cdot 3 \cdot 11 = 3,003.$$

Finally, we have to sum up the numbers obtained in cases (1) through (5):

$$15 + 420 + 2,730 + 5,460 + 3,003 = 11,628.$$

This is the number of possible flavor combinations that can be obtained using the five-nozzle machine with fifteen flavors available. $\qquad\square$

APPLICATIONS TO PROBABILITY

Now we will apply the newly acquired methods of counting to the computation of probabilities. Recall that Ω denotes the set of all outcomes of an experiment. For example, if you randomly draw one card from a deck, then Ω is the set of all outcomes: from the 2 of clubs to the ace of spades. An event is a set of outcomes; in other words, a subset of Ω.

To compute the probability of an event E, we need to divide the number of outcomes that lie in E by the total number of possible outcomes:

$$P(E) = \#E/\#\Omega.$$

Let us see how permutations and combinations can be used to compute probabilities. In some cases, using combinations is the only possible way to compute probabilities, short of explicitly counting dozens or even hundreds of possible outcomes.

Example 15.9. In Chapter 14, Example 14.1, we computed the probability of getting exactly two heads in three tosses of a fair coin. We found that the desired outcomes were HHT, HTH, THH. There are $2^3 = 8$ possible outcomes. Therefore, the probability is $3/8$.

There is another way to compute this probability: the number of favorable outcomes is equal to the number of choices of two out of three items – in this case, two out of three positions where H will be placed in the sequence of the outputs. This number is given by the number of combinations, because the order in which the chosen positions are listed does not matter: $_3C_2 = \frac{3!}{2!1!} = 3$. Dividing by the number of all possible outcomes, we get $3/8$.

The use of combinations allows us to compute with ease the probability of getting a particular number of heads in a large number of tosses. For example, what is the probability of getting exactly five heads in twelve coin tosses? The total number of possible outcomes is easy to find: it is 2^{12}, for each toss provides two possibilities. It is more complicated to find the number of favorable outcomes. We need to count the number of strings of the kind

HTTHTHTTTHHT,

where H occurs five times, and T seven times. Listing all such strings is possible in principle, but quite tedious. Instead we can try to use combinations. Indeed, the number of such strings is the number of choices of five out of twelve positions in the string that will be filled with H's, and the remaining positions will be filled with T's. In other words, the number of such strings is the number of choices of five out of twelve items, or the number of combinations of 5 out of 12. We have:

$$_{12}C_5 = \frac{12!}{7! \cdot 5!} = \frac{12 \cdot 11 \cdot 10 \cdot 9 \cdot 8}{5 \cdot 4 \cdot 3 \cdot 2 \cdot 1} = 11 \cdot 9 \cdot 8 = 792.$$

Then the probability of having exactly 5 heads in 12 coin tosses is

$$P(5\,\mathrm{H}\ \mathrm{in}\ 12) = \frac{792}{2^{12}} = \frac{792}{4,096} \simeq 0.193.$$

We can try to consider an even longer sequence of tosses. For example, what is the probability of getting twenty heads in forty tosses? It is not $\frac{1}{2}$! To find the number of favorable outcomes, compute

$$_{40}C_{20} = \frac{40!}{20!20!} = 137,846,528,820.$$

The total number of outcomes is 2^{40}. Therefore, the probability of getting exactly twenty heads in forty coin tosses is

$$P(20\,\text{H in }40) = \frac{{}_{40}C_{20}}{2^{40}} \simeq 0.125.$$

The reason this probability is so small is that we asked for a very precise result: the number of heads is equal to the number of tails. But it is just about as likely that they will differ by two: nineteen heads and twenty-one tails, or the opposite. You can check that the probability of getting anywhere between eighteen and twenty-two heads in forty tosses is greater than $\frac{1}{2}$. The formula you need is

$$P(18\text{ to }22\,\text{H in }40) = \frac{2 \cdot {}_{40}C_{18} + 2 \cdot {}_{40}C_{19} + {}_{40}C_{20}}{2^{40}} \simeq 0.57,$$

where we took into account that ${}_{40}C_{22} = {}_{40}C_{18}$ and ${}_{40}C_{21} = {}_{40}C_{19}$. □

Many of the questions we considered in Chapter 14 can be solved using combinations or permutations: instead of counting the number of outcomes directly, as we did before, now we can interpret them as numbers of combinations or permutations. For example, this method can be used to count the number of choices in electing members of a committee out of several groups of people, or in drawing cards from a deck. When the numbers are relatively small, the straightforward way we used in Chapter 14 works equally well. But for larger numbers, the use of combinations and permutations is preferable. In any case, it is a faster and easier way to find the answer.

Example 15.10. What is the probability of drawing three spades in a row from a standard deck of fifty-two cards?

Solution: The most direct way to solve this problem is as in Chapter 14: the number of ways to draw three of the thirteen spades in a deck is $13 \times 12 \times 11$, out of a total number of $52 \times 51 \times 50$ ways to draw any three cards from a deck. Thus the probability is

$$\frac{13 \cdot 12 \cdot 11}{52 \cdot 51 \cdot 50} = \frac{4 \cdot 11}{4 \cdot 17 \cdot 50} = \frac{11}{850}.$$

Here is another way to solve the same problem, this time using combinations. The number of ways to draw 3 cards out of 52 (order does not matter) is ${}_{52}C_3$. The number of ways to draw 3 cards out of 13 spades in the deck (order does not matter) is $_{13}C_3$. Therefore, the probability of drawing three spades in a row is

$$P = \frac{_{13}C_3}{_{52}C_3} = \frac{\frac{13 \cdot 12 \cdot 11}{3!}}{\frac{52 \cdot 51 \cdot 50}{3!}} = \frac{11}{17 \cdot 50} = \frac{11}{850}.$$

□

Example 15.11. A club has five men and seven women members. Of these, three members are selected at random to attend a conference. What is the probability of selecting three women?

Solution: The number of ways to select three people out of twelve people is given by the number of combinations of 3 out of 12 (order does not matter):

$$_{12}C_3 = \frac{12!}{9!3!} = \frac{12 \cdot 11 \cdot 10}{3 \cdot 2} = 220.$$

The number of ways to select three women out of seven women is $_7C_3$:

$$_7C_3 = \frac{7!}{4!3!} = \frac{7 \cdot 6 \cdot 5}{3 \cdot 2} = 35.$$

Therefore, the probability is

$$P(3 \text{ women}) = \frac{_7C_3}{_{12}C_3} = \frac{35}{220} = \frac{7}{44}.$$

□

Example 15.12. A box contains twenty juice bottles, of which six are mango and fourteen orange. If three bottles are selected at random, find the probability that

1. all are mango.

2. none are mango.

3. there are at least two bottles of mango.

Solution: In all cases, the numbers of total and favorable outcomes are naturally given by numbers of combinations (the order of the bottles does not matter). The total number of possible outcomes is the number of choices of three out of twenty bottles:

$$_{20}C_3 = \frac{20!}{17! \cdot 3!}.$$

(1) To have all bottles contain mango juice (M), we need to choose three items out of six. The number of favorable outcomes is

$$_6C_3 = \frac{6!}{3! \cdot 3!}.$$

The probability of having three bottles with mango juice is

$$P(3\,\mathrm{M}) = \frac{\frac{6!}{3! \cdot 3!}}{\frac{20!}{17! \cdot 3!}} = \frac{6! \cdot 17!}{3! \cdot 20!} = \frac{6 \cdot 5 \cdot 4}{20 \cdot 19 \cdot 18} = \frac{1}{19 \cdot 3} = \frac{1}{57}.$$

(2) Having zero bottles with mango juice is the same as having three bottles of orange. We should choose them all from the 14 remaining bottles. The number of favorable outcomes is

$$_{14}C_3 = \frac{14!}{11! \cdot 3!}.$$

The probability is

$$P(0\,\mathrm{M}) = \frac{\frac{14!}{11! \cdot 3!}}{\frac{20!}{17! \cdot 3!}} = \frac{14! \cdot 17!}{11! \cdot 20!} = \frac{14 \cdot 13 \cdot 12}{20 \cdot 19 \cdot 18} = \frac{7 \cdot 13}{5 \cdot 19 \cdot 3} = \frac{91}{285}.$$

(3) Having at least two bottles of mango is a union of two events: having exactly two bottles of mango, and having three bottles of mango. The events are clearly disjoint, therefore the probability of their union is the sum of their probabilities. We already know that the probability of having three bottles of mango is $P(3\,\mathrm{M}) = \frac{1}{57}$. We need to find the probability of the other event. The favorable outcome consists of pulling two bottles of mango, and one of orange (O), in any order. The number of such events is the product of the number of choices of two items out of six times the number of choices of one item out of fourteen:

$$_6C_2 \cdot {}_{14}C_1 = \frac{6!}{4! \cdot 2!} \cdot \frac{14!}{13! \cdot 1!} = \frac{6!}{4! \cdot 2!} \cdot 14.$$

The probability of this event is

$$P(2\,\mathrm{M},\ 1\,\mathrm{O}) = \frac{\frac{6!}{4! \cdot 2!} \cdot 14}{\frac{20!}{17! \cdot 3!}} = \frac{6 \cdot 5 \cdot 14 \cdot 3 \cdot 2}{2 \cdot 20 \cdot 19 \cdot 18} = \frac{7}{38}.$$

Adding the probabilities of the disjoint events, we get

$$P(\geq 2\,\mathrm{M}) = P(3\,\mathrm{M}) + P(2\,\mathrm{M},\ 1\,\mathrm{O}) = \frac{1}{57} + \frac{7}{38} = \frac{2 + 21}{114} = \frac{23}{114}.$$

$\square$

The birthday paradox

Suppose you are hosting a big party. How many people do you have to invite to make sure that two of them share a birthday? To be absolutely certain, you will have to have 367 participants, because there are at most 366 different dates for a birthday to fall on.

Now, how many guests do you have to invite to have 50% probability that two of them share a birthday? The answer is surprisingly low – this is why the question was given the name "paradox" – it is enough to have twenty-three guests! Let us try to understand why.

Consider the following question: in a group of n unrelated people with $n \leq 365$, what is the probability of two of them sharing a birthday? We assume that a birthday can fall with equal probability on any of 365 days of the year. (For simplicity, we drop February 29 from our considerations – and apologize to the readers whose birthday falls on this day. Including it would lead to a more complicated model, but the answer is roughly the same.) Let us consider the probability of the complementary event: all n people have birthdays on different days. The number of favorable outcomes for this complementary event equals the number of ordered collections of n birthdays chosen out of 365 possible dates. Therefore, we are looking for the number of permutations of n out of 365:

$$_{365}P_n = \frac{365!}{(365-n)!} = 365 \cdot 364 \cdot 363 \cdot \ldots \cdot (365 - n + 1).$$

The total number of possible outcomes is 365^n. Thus, the probability of the complementary event is

$$P(n \text{ different birthdays}) = \frac{365 \cdot 364 \cdot 363 \cdot \ldots \cdot (365 - n + 1)}{365^n}.$$

This probability rapidly decreases as n grows: the probability of $n + 1$ different birthdays is the product of the probability of n different birthdays times the fraction $\frac{365-n}{365}$, which gets smaller and smaller with larger and larger n.

We are interested in the probability of two people out of n sharing a birthday. This probability is

$$P(n) \equiv P(2 \text{ out of } n \text{ share a birthday }) = 1 - P(n \text{ different birthdays}).$$

The table below lists $P(n)$ for different values of n.

n	2	7	15	20	22	23	30	40	50
$P(n)$	0.003	0.04	0.25	0.41	0.48	0.51	0.71	0.89	0.97

If you have twenty-three guests, it is as likely as not that two of them share a birthday, but if you have forty or more guests, hoping that two of them share a birthday is a pretty good bet.

Pascal's triangle*

The numbers $_nC_k$ have significance not only in counting choices and computation of probabilities, but in many other areas of mathematics. Here is an important example.

Consider the following triangular table of natural numbers: On top of the triangle, we place the number 1. The next row under the top consists of two 1's:

$$1$$
$$1 \qquad 1$$

Each next row is one number longer than the previous one, and is constructed according to the following rule. The first number in a row is always 1. Each subsequent number is obtained as the sum of the two numbers that stand to the left and to the right above it in the previous row. The last number is again 1. Proceeding in this way, we obtain the following (infinite) triangular array of numbers, known as Pascal's triangle (see Figure 15.1).

It is named in honor of the seventeenth century French mathematician Blaise Pascal, who described its properties in a systematic way, although it was known long before in ancient India and Greece. The remarkable fact is that the numbers in Pascal's triangle are in fact the numbers of combinations, as shown in Figure 15.2.

Recall that by convention we have $_0C_0 = \frac{0!}{0!0!} = 1$, $_1C_0 = {}_1C_1 = \frac{1!}{0!1!} = 1$, and also $_nC_0 = {}_nC_n = \frac{n!}{0!n!} = 1$. This takes care of the top two rows and the sides of the triangle.

For the numbers inside of the triangle, the statement holds because the numbers of combinations listed in each row can be constructed recursively according to the same rule that determines the numbers in Pascal's triangle. Namely, for any $n \geq 2$ and any $1 \leq k \leq n - 1$, the following equality holds:

$$_nC_k = {}_{n-1}C_{k-1} + {}_{n-1}C_k.$$

This statement can be checked by direct computation using the explicit formulas for the numbers of combinations (see Question 15.15).

However, we find it more instructive to explain this equality by counting unordered collections of items. Suppose we want to count the collec-

Figure 15.1:   First nine rows of Pascal's triangle.

$$
\begin{array}{ccccccccccccccccccc}
& & & & & & & & & 1 & & & & & & & & & \\
& & & & & & & & 1 & & 1 & & & & & & & & \\
& & & & & & & 1 & & 2 & & 1 & & & & & & & \\
& & & & & & 1 & & 3 & & 3 & & 1 & & & & & & \\
& & & & & 1 & & 4 & & 6 & & 4 & & 1 & & & & & \\
& & & & 1 & & 5 & & 10 & & 10 & & 5 & & 1 & & & & \\
& & & 1 & & 6 & & 15 & & 20 & & 15 & & 6 & & 1 & & & \\
& & 1 & & 7 & & 21 & & 35 & & 35 & & 21 & & 7 & & 1 & & \\
& 1 & & 8 & & 28 & & 56 & & 70 & & 56 & & 28 & & 8 & & 1 & \\
1 & & 9 & & 36 & & 84 & & 126 & & 126 & & 84 & & 36 & & 9 & & 1 \\
\end{array}
$$

Figure 15.2:   Pascal's triangle in terms of combinations.

$$
\begin{array}{ccccccccccccc}
& & & & & & _0C_0 & & & & & & \\
& & & & & _1C_0 & & _1C_1 & & & & & \\
& & & & _2C_0 & & _2C_1 & & _2C_2 & & & & \\
& & & _3C_0 & & _3C_1 & & _3C_2 & & _3C_3 & & & \\
& & _4C_0 & & _4C_1 & & _4C_2 & & _4C_3 & & _4C_4 & & \\
& _5C_0 & & _5C_1 & & _5C_2 & & _5C_3 & & _5C_4 & & _5C_5 & \\
_6C_0 & & _6C_1 & & _6C_2 & & _6C_3 & & _6C_4 & & _6C_5 & & _6C_6 \\
\end{array}
$$

tions of k out of n items, where $1 \le k \le n - 1$. Let us separate one item out of n and call it A. Then the number of all collections of k out of n items is the sum of those collections that contain A and those that do not:

$$
{}_nC_k = \left\{ \begin{array}{lcl} \text{\# collections containing } A & = & {}_{n-1}C_{k-1} \\ & + & \\ \text{\# collections not containing } A & = & {}_{n-1}C_k. \end{array} \right.
$$

The number of collections containing A is the number of collections of $k - 1$ out of $n - 1$ items, ${}_{n-1}C_{k-1}$, because we have to choose $k - 1$ items from $n - 1$ items and then add A. The number of collections not containing A is the number of collections of k out of $n-1$, ${}_{n-1}C_k$, because we have to exclude A from the set of possible items. Therefore, we have the desired equality, ${}_nC_k = {}_{n-1}C_{k-1} + {}_{n-1}C_k$.

The recursive relations in Pascal's triangle provide a source of algorithms for computations involving the numbers ${}_nC_k$ (without computing the factorials!), and play an important role in various fields of mathematics. We will point out just one of the triangle's applications, namely the *binomial theorem*.

Let us do some algebra. Consider the expression $(x + y)^n$, where x and y are two variables, and expand it out. For $n = 2, 3$ we have:

$$
(x + y)^2 = x^2 + 2xy + y^2, \qquad (x + y)^3 = x^3 + 3x^2y + 3xy^2 + y^3.
$$

The coefficients are precisely the second and third rows of Pascal's triangle (if we start counting rows from 0). This is not a coincidence. In fact, the following formula holds for any natural n:

$$
(x + y)^n = {}_nC_0 x^n + {}_nC_1 x^{n-1}y + {}_nC_2 x^{n-2}y^2 + \cdots
$$
$$
+ {}_nC_k x^{n-k}y^k + \cdots + {}_nC_{n-1}xy^{n-1} + {}_nC_n y^n.
$$

Using the summation notation, we can write

$$
(x + y)^n = \sum_{k=0}^{n} {}_nC_k \, x^{n-k}y^k.
$$

This is the statement of the binomial theorem: if you expand out the nth power of a sum of two variables, all coefficients in front of the powers of the variables can be read off the nth row of Pascal's triangle. The reason is that the coefficients satisfy the same recursive relation as the numbers in the triangle (see Question 15.16 for more details). Another way to prove the binomial theorem is to interpret the coefficients in the binomial decomposition as numbers of certain unordered choices (think which). The fact that the numbers of unordered collections of items, the entries of Pascal's triangle, and the binomial coefficients turn out to be the same illustrates the unity of mathematics: an important concept transcends the boundaries of the area from which it originated.

EXERCISES

Question 15.1. Compute the following numbers:

$$_8C_3, \quad _8C_5, \quad _{14}C_8, \quad _{14}P_8.$$

Question 15.2. Which is greater:

$$_{30}P_7 \quad \text{or} \quad _{30}P_{23} \; ?$$
$$_{30}P_7 \quad \text{or} \quad _{30}C_7 \; ?$$
$$_{30}P_7 \quad \text{or} \quad _{30}C_{23} \; ?$$

Question 15.3.   1. How many distinct permutations can be made using the letters of the word TENNESSEE?

   2. How many permutations of three different letters can be made using the letters of the word TENNESSEE?

Question 15.4. A corporation has seven members on its board of directors. In how many different ways can it elect a president, vice-president, secretary, and treasurer, assuming that no director can be elected to two different posts?

Question 15.5. A radio DJ has fifteen records to choose from. In how many ways can she schedule a program containing five songs?

Question 15.6. How many ways are there to choose two red, two pink, and four white roses from a vase that contains eighteen roses, six of each of the three colors?

Question 15.7. Planet Kijij has a law according to which a national flag of a country has to consist of equal bands of three different colors, arranged either horizontally or vertically. The only colors allowed to be used in these flags are purple, yellow, blue, white, and brown. How many different national flags can the planet have?

Question 15.8. A circus master has to choose six horses, two dogs, and three parrots to take on a tour across the country out of a total of ten horses, four dogs, and five parrots. How many different groups of animals can be selected?

Question 15.9. A box contains nine transistors, three of which are defective. If three transistors are selected at random, what is the probability that

1. all are defective?

2. none are defective?

Hint: Use combinations to count the number of possible and desired outcomes.

Question 15.10. In how many ways can a committee of four professors and five students be formed from a group of seven professors and ten students?

Question 15.11. An experiment consists of fifteen consecutive fair coin tosses. Find the probability of getting

1. exactly eleven tails.

2. at least eleven tails. *Hint:* Break it into a union of disjoint events.

3. eleven tails in the first eleven tosses.

Question 15.12. In Example 15.12, find the probability of having at least two bottles of orange juice. What is the sum of this probability with the probability of having at least two bottles of mango juice, computed in Example 15.12 (3)?

Question 15.13. A box holds twenty-five muffins, including ten blueberry muffins, six carrot muffins, five chocolate chip muffins, and four apple muffins. A person randomly chooses three muffins.

1. What is the probability that all three muffins are chocolate chip?

2. What is the probability that none of the muffins is blueberry, and exactly one of them is apple?

Question 15.14. Seven cards are drawn randomly without replacement from a standard deck. What is the probability that five are clubs and two are spades (in any order)?

Question 15.15. Use the formula for $_nC_k$ in terms of the factorials to check the identity

$$_nC_k = {_{n-1}C_{k-1}} + {_{n-1}C_k}$$

that holds for any $n \geq 2$ and any $1 \leq k \leq n-1$.

Question* 15.16. Suppose you have decompositions of the nth and $(n-1)$st power of a sum of two variables,

$$(x+y)^n = \sum_{k=0}^{n} a(n,k)x^{n-k}y^k, \qquad (x+y)^{n-1} = \sum_{k=0}^{n-1} a(n-1,k)x^{n-1-k}y^k,$$

where $a(n,k)$ and $a(n-1,k)$ are the coefficients. Show that for each k such that $1 \leq k \leq n-1$, the following relation holds:

$$a(n,k) = a(n-1,k-1) + a(n-1,k).$$

Hint: Write $\sum_{k=0}^{n} a(n,k)x^{n-k}y^k = (x+y)\sum_{k=0}^{n-1} a(n-1,k)x^{n-1-k}y^k$ and compare the coefficients of the term $x^{n-k}y^k$ on both sides.

Chapter 16

Bayes' law: How to win a car... or a goat

In this chapter we discuss how the occurrence of one event can influence the probability that another event will occur. If one event has no influence on the odds of another, such events are called independent. In the previous two chapters we considered examples of such events, for instance, consecutive coin tosses. Many events in the real world are not independent: for instance, a budget cut influences the probability of an employee getting a raise. We will consider ways to determine the probability that an event will happen given that another event has occurred. Bayes' law of conditional probability provides an efficient method of computing such probabilities. In particular, it can be applied when dealing with false positives in medical tests, in assessing the probability of spotting a rare species, and in many other situations where uncertainty plays a significant role.

MATH

Independent events

Two events are independent if the occurrence of one has no influence on the probability of the other. For example, tosses of a fair coin are independent: no matter what the result of the previous tosses is, each new toss will result in heads or tails with probability 1/2.

Formally speaking, two events A and B are *independent* if and only if the probability of both of them happening at the same time is equal to the product of their probabilities. The favorable outcomes for events A and B happening at the same time are all outcomes that are favorable for both A and B. In mathematical notation, it is the *intersection* of the sets of favorable outcomes for each of the events, and it is denoted by $A \cap B$. The probability $P(A \cap B)$ of independent events A and B happening at the same time is

$$P(A \cap B) = P(A) \cdot P(B).$$

Thus, the probability of getting two heads in two coin tosses is $1/2 \cdot 1/2 = 1/4$.

One way to check if two events A and B are independent is to compute the probabilities $P(A)$, $P(B)$ and $P(A \cap B)$ and check if the formula $P(A \cap B) = P(A) \cdot P(B)$ holds. Many random events are not independent.

Example 16.1. Suppose you draw two cards randomly from a deck. Let A denote the event that the first card is a diamond, and B that the second card is a diamond. Are events A and B independent?

Solution: Clearly, $P(A) = \frac{13}{52} = \frac{1}{4}$. What about $P(B)$, the probability of drawing a diamond from the deck that is already missing a card? In Example 14.8 (2) and (3) in Chapter 14, we have already computed that this probability is the same as for the first card, $P(B) = \frac{1}{4}$. Now we compute $P(A \cap B)$. This is the event that both cards are diamonds, and its probability is:

$$P(A \cap B) = \frac{{}_{13}C_2}{{}_{52}C_2} = \frac{13 \cdot 12}{52 \cdot 51} = \frac{12}{4 \cdot 51} = \frac{1}{17}.$$

Here we used the notation ${}_kC_n$ for the number of unordered collections of k out of n items, which was introduced in Chapter 15.
On the other hand,

$$P(A) \cdot P(B) = (1/4) \cdot (1/4) = 1/16.$$

Therefore, $P(A) \cdot P(B) \neq P(A \cap B)$, and the events are not independent.
□

Conditional probability

Can we have a formula that relates the probabilities of A and B and $A \cap B$, if events A and B are not independent?

To do this we need to introduce the notion of conditional probability. The *conditional probability* (denoted $P(A|B)$) of event A given B is the probability of event A occurring, given that event B has occurred. With this definition, the probability that both events A and B occur is equal to the product of the probability of B times the conditional probability $P(A|B)$:

$$P(A \cap B) = P(A|B) \cdot P(B).$$

Clearly, we can define the conditional probability $P(B|A)$ in a similar way, and obtain another formula for $P(A \cap B)$:

$$P(A \cap B) = P(B|A) \cdot P(A).$$

These formulas hold for any two events A and B. If events A and B are independent, then by definition $P(A|B) = P(A)$, because event B has no influence on the occurrence of A. Similarly, $P(B|A) = P(B)$. In this case both formulas for $P(A \cap B)$ transform into $P(A \cap B) = P(A) \cdot P(B)$.

If the probabilities $P(A)$ and $P(B)$ are nonzero, we can rewrite the above formulas for $P(A \cap B)$ to express $P(A|B)$ and $P(B|A)$:

$$P(A|B) = \frac{P(A \cap B)}{P(B)}, \qquad P(B|A) = \frac{P(A \cap B)}{P(A)}.$$

If furthermore events A and B are independent, plugging in $P(A \cap B) = P(A) \cdot P(B)$ yields $P(A|B) = P(A)$ and $P(B|A) = P(B)$.

Let us reconsider Example 16.1: A is drawing a diamond; B is drawing another diamond (without replacement). Then $(A \cap B)$ is drawing two diamonds in a row, and we computed its probability to be equal to $1/17$. The probabilities $P(A) = P(B) = 1/4$. What is the conditional probability $P(B|A)$? According to the formula, it is

$$P(B|A) = \frac{P(A \cap B)}{P(A)} = \frac{\frac{1}{17}}{\frac{1}{4}} = \frac{4}{17}.$$

We can find this probability directly. If we know that the first card was a diamond, then there are only 12 choices left for the second card to be a diamond, out of 51 cards total:

$$P(B|A) = \frac{12}{51} = \frac{4}{17}.$$

Note that the reverse conditional probability, $P(A|B)$, is the same according to the formula

$$P(A|B) = \frac{P(A \cap B)}{P(B)} = \frac{\frac{1}{17}}{\frac{1}{4}} = \frac{4}{17}.$$

This happens because $P(A|B)$ is the probability of the first card being a diamond, *given* that the second is a diamond; therefore there are only 12 choices for the first card, and we arrive at the same number $P(B|A) = 12/51 = 4/17$.

Bayes' law of conditional probabilities

Now compare the two formulas above for the probability of both events happening:

$$P(A \cap B) = P(B) \cdot P(A|B), \quad \text{and}$$
$$P(A \cap B) = P(A) \cdot P(B|A).$$

From these formulas, we can conclude:

$$P(A) \cdot P(B|A) = P(B) \cdot P(A|B).$$

Assuming that $P(A)$ is nonzero, this allows us to compute the conditional probability $P(B|A)$ in terms of the reverse conditional probability $P(A|B)$, $P(A)$ and $P(B)$:

$$P(B|A) = \frac{P(A|B) \cdot P(B)}{P(A)}, \qquad P(A) \neq 0.$$

This is *Bayes' law* of conditional probabilities, which holds for any two events A and B with positive probabilities.

If A and B are independent, then $P(B|A) = P(B)$ and $P(A|B) = P(A)$, and Bayes' law becomes a trivial identity.

EXAMPLES

Example 16.2. Draw two cards randomly from a deck (without replacement). A is the event that the first card is a diamond ($\diamond$); B, that the second card is a queen (Q). Are events A and B independent?

Solution: Compute the probabilities:

$$P(\diamond) = \frac{13}{52} = \frac{1}{4}, \qquad P(Q) = \frac{4}{52} = \frac{1}{13}.$$

Note that just as in Example 16.1, if we don't know what the first card was, the probability of the second card being a queen is the same as the probability of drawing a queen from a full deck.

To count the number of outcomes with the first card being a diamond and the second a queen we have to consider two cases: (1) the first card is a diamond and not a queen: twelve possibilities; and (2) the first card is the queen of diamonds: one possibility. In the first case, the second card can be any of the four queens; in the second, any of the three remaining queens. In total, we have $12 \cdot 4 + 1 \cdot 3 = 51$ desired outcomes. There are $_{52}P_2 = 52 \cdot 51$ possible outcomes of drawing two cards from the deck without replacement. Therefore, the probability of $A \cap B$ is

$$P(A \cap B) = \frac{51}{52 \cdot 51} = \frac{1}{52} = \frac{1}{4} \cdot \frac{1}{13} = P(A) \cdot P(B).$$

Therefore, events A and B are independent.   □

Example 16.3. Suppose we draw a card and it turns out to be red. What are the odds that the card is a diamond?

Solution: Let A denote getting a diamond and B getting a red card. Then $P(A \cap B)$ is the probability that a card is red and a diamond, which simply means that it is a diamond, $P(A \cap B) = 1/4$. The probability of a card being red is $P(B) = 1/2$. Therefore, the probability of a card being a diamond, given that it is red, is:

$$P(A|B) = P(A \cap B)/P(B) = \frac{1/4}{1/2} = \frac{1}{2},$$

which is intuitively clear.   □

Practice problem 16.4. Compute the conditional probability of the following events:

1. Rolling two dice and getting at least one 6, given that the sum of the outputs is 8.

   Answer: There are only five possibilities: $(2,6),(3,5),\ldots,(6,2)$, which yields: $2/5$.

2. Tossing a fair coin twice and getting exactly one head, given that not all outputs are tails. *Answer:* There are only three possibilities: HH, HT, TH, which yields: $2/3$.

Example 16.5. Rolling three dice, what is the probability of getting a 3 on the first roll if the sum of the three outputs is 6?

Solution: Let A denote the event that the sum of the three outputs is 6, and B the event of getting a 3 on the first roll. We need to find $P(B|A)$. By definition,

$$P(B|A) = \frac{P(A \cap B)}{P(A)}.$$

The outcomes in A are $(2,2,2), (1,1,4)(4,1,1), (1,4,1)$, and the six permutations of $(1,2,3)$, totalling ten outcomes. (Why six? The number of permutations of three elements is $_3P_3 = 3! = 6$.) Then we have $P(A) = 10/6^3 = 5/108$. The outcomes of $(A \cap B)$ are $(3,2,1)$ and $(3,1,2)$, so $P(A \cap B) = 2/6^3 = 1/108$. Then

$$P(B|A) = \frac{\frac{1}{108}}{\frac{5}{108}} = \frac{1}{5}.$$

We could have used Bayes' law to compute the same probability:

$$P(B|A) = \frac{P(A|B) \cdot P(B)}{P(A)}.$$

Then we need to know $P(B)$ and $P(A|B)$. The first is easy: the probability of getting a 3 in one roll of a die is $P(B) = 1/6$. The second is the probability of getting the sum of the three outputs equal to 6, given that the first output is 3. This is the same as the probability of getting the sum of two outputs in two rolls equal to 3. There are two possibilities: $(1,2)$ and $(2,1)$. Therefore, $P(A|B) = 2/36 = 1/18$. Bayes' law gives

$$P(B|A) = \frac{P(A|B) \cdot P(B)}{P(A)} = \frac{\frac{1}{18} \cdot \frac{1}{6}}{\frac{5}{108}} = \frac{1}{5}.$$

$\square$

APPLICATIONS

Bayes' law is an extremely important result for the purposes not only of probability theory, but also for modern statistical methods. An entire sub-field of *Bayesian statistics* exists to provide sophisticated and involved techniques in data analysis. We will see several examples of Bayes' law applied to real-world settings below – as well as one (simplified) application to constructing *spam filters*; see Questions 16.15 and 16.16 in the Exercises. You may also have seen an application of Bayesian methods in a recent election: in the 2008 and 2012 US presidential elections, the psephologist Nate Silver used Bayesian methods to correctly predict the winners in every one of the fifty states (with the sole exception of Indiana in 2008).

Making use of the relevant information

Conditional probability provides a way to estimate the impact of any additional information on the chances of an event happening. It comes in handy when you want to evaluate your chances with the best possible precision, and when the available information is incomplete and only partially relevant. Here is an example.

Example 16.6. Suppose there are two red, three blue, and four white marbles in a bag. The experiment consists of choosing two marbles randomly. Suppose someone had a glimpse of the chosen marbles and knows that none of them are red. You would like to estimate the probability that at least one of the marbles is white. Or the opposite, someone tells you they saw a glimpse of white color. What are the chances that none of the marbles are red?

Solution: From the viewpoint of the theory of conditional probability, the two questions are closely related and we will be able to answer them both at once. Let us denote the event that none of the chosen marbles are red by A, and the event that at least 1 of the marbles is white by B. Then the first question asks for the conditional probability $P(B|A)$, and the second for $P(A|B)$. With the formulas

$$P(A|B) = \frac{P(A \cap B)}{P(B)} \quad \text{and} \quad P(B|A) = \frac{P(A \cap B)}{P(A)},$$

both questions can be answered if we find the probabilities of the events A and B, and $A \cap B$, which is the probability that events A and B will both happen at once.

For all three probabilities, the total number of outcomes is given by the number of unordered choices of two out of nine total marbles in the bag. For event A, the number of favorable outcomes is the number of choices of two out of seven blue and white marbles.

$$P(A) = \frac{_7C_2}{_9C_2} = \frac{\frac{7 \cdot 6}{2}}{\frac{9 \cdot 8}{2}} = \frac{7}{12}.$$

Event B is a union of two disjoint events: both marbles are white, and exactly one marble is white. For the former the number of favorable outcomes is the number of choices of two out of four white marbles; for the latter, it is the number of choices of one out of four white ones times the number of choices of one out of the remaining five non-white marbles. We have

$$P(B) = \frac{{}_4C_2 + {}_4C_1 \cdot {}_5C_1}{{}_9C_2} = \frac{\frac{4 \cdot 3}{2} + 4 \cdot 5}{\frac{9 \cdot 8}{2}} = \frac{6 + 20}{36} = \frac{13}{18}.$$

Similarly, event $A \cap B$ is a union of two disjoint events: both marbles are white, and exactly one marble is white while the other is blue (because none of the marbles can be red). The only difference from the previous computation is that now we can choose the second non-white marble only out of the three blue ones:

$$P(A \cap B) = \frac{{}_4C_2 + {}_4C_1 \cdot {}_3C_1}{{}_9C_2} = \frac{\frac{4 \cdot 3}{2} + 4 \cdot 3}{\frac{9 \cdot 8}{2}} = \frac{6 + 12}{36} = \frac{1}{2}.$$

Now we can compute the conditional probabilities

$$P(A|B) = \frac{\frac{1}{2}}{\frac{13}{18}} = \frac{9}{13} \simeq 0.69, \qquad P(B|A) = \frac{\frac{1}{2}}{\frac{7}{12}} = \frac{6}{7} \simeq 0.86.$$

So, if you saw that none of the marbles were red, there is a pretty good chance (86%) that at least one of them is white. If you saw a white marble, then the chance that none of the marbles are red is approximately 69%.

The conditional probabilities we just found can be calculated directly by counting the favorable and total choices (see Question 16.4).   □

Imperfect correlation

Bayes' law is useful when you want to find a conditional probability $P(B|A)$ and the reverse conditional probability $P(A|B)$ is given or easy to find. Here is an example.

Example 16.7. Suppose you have a box of chocolates with three types of fillings: cherry, almond, and coffee. You know that 1/4 of all chocolates have cherry filling, 1/4 coffee, and 1/2 almond. The chocolates are wrapped in colorful foil according to the following rule: (1) all cherry-filled chocolates are wrapped in red foil; (2) of coffee-filled chocolates, 1/2 are wrapped in yellow and 1/2 in brown; and (3) of almond-filled choco-lates, 1/3 are wrapped in red, 1/3 in yellow, and 1/3 in blue. Given that a chocolate is wrapped in red, what are the chances that it has cherry filling?

Solution: This is a classic example where Bayes' law works best. Denote by A the event that a chocolate is wrapped in red; and by B the event that it has cherry filling. We need to find $P(B|A)$. The reverse probabil-ity $P(A|B)$ is given: if a chocolate holds cherry filling, it is always red, so $P(A|B) = 1$. The probability $P(B)$ is also given: 1/4 of the choco-lates have cherry filling, therefore $P(B) = 1/4$. Let us compute $P(A)$ $= P(\text{wrapped in red})$:

$$P(A) = \frac{1}{4} \cdot 1 + \frac{1}{2} \cdot \frac{1}{3} = \frac{1}{4} + \frac{1}{6} = \frac{5}{12}.$$

Here the first summand accounts for the cherry-filled chocolates wrapped in red and the second summand accounts for the almond-filled chocolates wrapped in red. Finally, Bayes' law gives:

$$P(B|A) = \frac{P(A|B) \cdot P(B)}{P(A)} = \frac{1 \cdot \frac{1}{4}}{\frac{5}{12}} = \frac{3}{5}.$$

$\square$

In this example, you have an easily perceived property (the color of the wrapping) and an important but invisible property (the type of filling). You know how the color of the wrapping is chosen according to the type of filling, but you would like to know the opposite: what can be deduced about the type of filling from the color of the wrapping, and how reliable will the conclusion be? More examples of this kind can be imagined: the quality of a fabric versus the pattern, the color of a car versus the type of engine, and so on.

The tree diagram

To visualize the distribution of probabilities in situations similar to Exam-ple 16.7, we can draw a *tree diagram*. Tree diagrams are useful whenever we have multiple levels of choices and know how the probabilities are distributed between the choices at each level. For example, here we have three kinds of fillings (cherry, almond, coffee) that occur with probabilities $1/4, 1/2, 1/4$, respectively. This information is encoded in the first level of

Figure 16.1: The tree diagram for the chocolate problem.

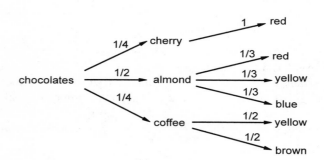

the tree diagram in Figure 16.1. In the next level, we add wrapping colors that can occur for each filling, with their respective probabilities.

The resulting diagram allows us to find probabilities by multiplying coefficients along the "tree branches." For example, to find the probability of a chocolate being wrapped in red, we have to multiply the coefficients along the branches that lead to "red" and add the results for each path: $1/4 \cdot 1 + 1/2 \cdot 1/3 = 5/12$. The probability of a chocolate being wrapped in red *and* having cherry filling equals the product of the probabilities along the path that leads to "red" and passes through "cherry": $1 \cdot 1/4 = 1/4$. Dividing the latter by the former, we get the conditional probability that a chocolate has cherry filling if it is wrapped in red: $\frac{1}{4}/\frac{5}{12} = 3/5$.

Tree diagrams are appealing because they illustrate how probabilities split, and they can be applied in many of the examples considered below. A tree diagram represents the same computation that follows from the Bayes' law formula, so the two methods are essentially the same.

The grid method

Another approach to solving problems such as Example 16.7 is called the *grid method.* This does not involve writing down mathematical equations – yet, it performs exactly the same computations as in Example 16.7. First, draw a table listing the various perceived properties (colors) in the rows, and the important properties (fillings) in the columns – with one extra column and row in which to write the totals along each row and column:

	Cherry	Coffee	Almond	Total
Red				
Yellow				
Brown				
Blue				
Total				

Now suppose we start with a large number of chocolates – say, 600. Based on the given problem, we can then write down the total numbers of cherry, coffee, and almond chocolates – i.e., the numbers in the last row:

	Cherry	Coffee	Almond	Total
Red				
Yellow				
Brown				
Blue				
Total	150	150	300	600

Now given that there are 150 cherry chocolates, we can fill out the entire first column: all cherry chocolates are red, so there must be 150 red (cherry) chocolates, and 0 cherry chocolates wrapped in other colors. Similarly, the column corresponding to coffee reads: $0, 75, 75, 0$; and the column for almond reads: $100, 100, 0, 100$.

We can now complete the table by adding up the numbers in each row and filling in the row sums.

	Cherry	Coffee	Almond	Total
Red	150	0	100	250
Yellow	0	75	100	175
Brown	0	75	0	75
Blue	0	0	100	100
Total	150	150	300	600

At this point we should be able to check that the entries in the last column add up to 600.

Remark. The method works for any initial choice of the number of chocolates. We set it equal to 600, so that even after multiplying by the various proportions, all the entries in the above table will be integers. In fact, this is not necessary: because we will ultimately be dividing one entry by another, choosing any total number will do – even 1. Just note that all entries then have to be scaled proportionately. For instance, if we decide to start with a total of 5 chocolates, then 5/4 are cherry-flavored, 5/4 are coffee-flavored, and 5/2 are almond-flavored. The entries in the various columns of the tables also get rescaled accordingly – leaving the final answers exactly the same. Of course, if we talk about individually wrapped chocolates, fractional numbers can hardly apply directly, but rather have a strictly theoretical meaning.

Equipped with this table, we can solve the question in Example 16.7. Namely, given that a chocolate is wrapped in red, what is the chance that it has cherry filling? To see this, simply examine the first row. There are a total of 250 red chocolates, 150 of which are the favorable outcomes. Thus, we obtain:

$$P(\text{cherry}|\text{red}) = \frac{150}{250} = \frac{3}{5}.$$

The computations involved in filling out the table seem easy to work out – but at the same time they closely follow the calculations according to Bayes' law carried out in Example 16.7. For instance, we studied the probability of a chocolate being wrapped in red, and found that it is 5/12. From the table, we can immediately obtain this answer as well: $P(\text{red}) = 250/600 = 5/12$. □

The advantage of the grid method is that it provides not just the first row but the rest of the table as well. Why is computing the entire table useful? Because now we can solve any kind of conditional probability problem in this setting without performing any additional computations. This reduces the amount of work to be done if there is more than one question to answer. Here are two examples:

1. If you select a yellow-wrapped chocolate, what is the probability that it is coffee? *Answer:* 75/175 = 3/7.

2. If you select a blue-wrapped chocolate, it immediately follows that the chocolate must be almond-flavored.

Disease detection and drug testing

An important application of Bayes' law is in assessing the results of various medical tests. In the next example, we will use Bayes' law to solve problems involving *false positives* in disease detection and drug testing.

Example 16.8. Suppose a test claims to detect people with a rare disease nine times out of ten, and for people without the disease it correctly says they do not have it 95% of the time. Suppose these figures were obtained by having tested a population of 1,000 subjects, of whom only 100 have the disease (this was verified independently of the test). This is the information we have on the reliability of the test.

Now suppose a subject undergoes the test, and the result is positive. Based on the available information, what is the chance that the subject does have the disease?

Solution: It is tempting to assume that the chance is 90%, because the test is said to be 90% reliable on sick patients. In fact, we will see that the probability of the subject having the disease is much lower.

Let D, N, $+$, $-$ denote the events that the subject has the disease, does not have the disease, is tested positive, and is tested negative by the test, respectively. (Note that the diagnosis being positive is one thing, and suffering from the disease is a completely different event.)

Let us use the grid method to solve the problem. For example, let us start with 1,000 patients and draw the table:

	Diseased (D)	Not Diseased (N)	Total
Positive (+)			
Negative (−)			
Total	100	900	1,000

Now fill in the rest of the entries in the first column, then in the second column, and finally add up the row totals. This yields the following table.

	Diseased (D)	Not Diseased (N)	Total
Positive (+)	90	45	135
Negative (−)	10	855	865
Total	100	900	1,000

We conclude that the probability of a patient who tested positive to actually have the disease is $P(D|+) = \frac{90}{135} = \frac{2}{3} = 66.67\%$. □

We could also have solved the problem by using Bayes' law directly. First we write the given information about the test in terms of probabilities. For example, given that a patient has the disease, he is diagnosed to have it nine times out of ten. Thus,

$$P(+|D) = 0.9, \quad \text{and therefore} \quad P(-|D) = 1 - 0.9 = 0.1.$$

Similarly, we know the percent of the subjects who have and do not have the disease:

$$P(D) = 100/1,000 = 0.1, \quad P(N) = 1 - P(D) = 0.9.$$

For the patients who do not have the disease, the test gives the following results:

$$P(-|N) = 0.95, \quad P(+|N) = 1 - 0.95 = 0.05.$$

We would now like to determine the probability that the patient has the disease given that his test is positive. This corresponds to the conditional probability $P(D|+)$. *Attention:* this probability is different from the reverse conditional probability $P(+|D)$, which is claimed to be 90%.

To compute the desired probability, we now use Bayes' law, taking into account the above information:

$$P(D|+) = \frac{P(+|D) \cdot P(D)}{P(+)} = \frac{0.9 \cdot 0.1}{P(+)} = \frac{0.09}{P(+)}.$$

The probability $P(+)$ is the fraction of the subjects who are *diagnosed* to have the disease. It is 90% of the 100 users who actually have the disease, and 5% of the 900 people who do not have the disease. The events of having and not having the disease are disjoint and complementary; therefore the probability of testing positive is the sum of the probability of testing positive while having the disease and the probability of testing positive while being healthy. We have:

$$P(+) = P(+|D) \cdot P(D) + P(+|N) \cdot P(N) = 0.9 \cdot \frac{100}{1,000} + 0.05 \cdot \frac{900}{1,000}$$
$$= 0.09 + 0.045 = 0.135.$$

(*Note:* this is also reflected in the table.) Hence

$$P(D|+) = 0.09/0.135 = 2/3.$$

The chance of the patient having the disease is not as high as 9 in 10; it is closer to 67%. □

Similarly, one can have examples where even fewer subjects in the reference group had the disease to start with – in this case, it may happen that $P(D|+) < 0.5$ – which would mean that a "+" means that it is *more* likely that the patient does *not* have the disease. Of course, $P(D|-)$ is still much smaller (see Question 16.11).

The lesson is: do not confuse a conditional probability $P(A|B)$ with the reverse conditional probability $P(B|A)$; rather, compute one from the other using the grid method or Bayes' law.

Here is another example.

Example 16.9. Suppose you have two tests for disease D: Test A claims to diagnose correctly both diseased and healthy people in 99% of cases and was carried out in a sample of subjects where only 1% had the disease. Test B claims to diagnose correctly both diseased and healthy people in 97% of cases and was carried out in a sample of subjects where only 5% had the disease. Which test is more reliable?

Solution: The problem can be solved either by computing the conditional probabilities by Bayes' law formula, as we do below, or by the grid method (see Question 16.12).

Let us denote the event of having the disease by N and not having it by N; testing positive by + and testing negative by −. We will use indices A and B for probabilities obtained for tests A and B, respectively. To compare reliability, we need to compare two conditional probabilities: $P_A(D|+)$ and $P_B(D|+)$.

We have $P_A(+|D) = 99/100$, $P_A(+|N) = 1/100$, $P_A(D) = 1/100$, and $P_A(N) = 99/100$. To compute $P_A(D|+)$ by Bayes' law, we need to find $P_A(+)$. Proceeding as before, we have

$$P_A(+) = P_A(+|D) \cdot P_A(D) + P_A(+|N) \cdot P_A(N) = \frac{99}{100} \cdot \frac{1}{100} + \frac{1}{100} \cdot \frac{99}{100}.$$

Then, by Bayes' law, we compute the probability that a person has the disease if she tests positively using test A:

$$P_A(D|+) = \frac{P_A(+|D) \cdot P_A(D)}{P_A(+)} = \frac{0.99 \cdot 0.01}{0.99 \cdot 0.01 + 0.01 \cdot 0.99} = \frac{1}{2}.$$

Does it get any better with test B? We have $P_B(+|D) = 97/100$, $P_B(+|N) = 3/100$, $P_B(D) = 5/100$, and $P_B(N) = 95/100$. To compute $P_B(D|+)$ by Bayes' law, we need to find $P_B(+)$. We have

$$P_B(+) = P_B(+|D) \cdot P_B(D) + P_B(+|N) \cdot P_B(N) = \frac{97}{100} \cdot \frac{5}{100} + \frac{3}{100} \cdot \frac{95}{100}.$$

Then by Bayes' law we compute the probability that a person has the disease if she tests positively using test B:

$$P_B(\mathrm{D}|+) = \frac{P_B(+|\mathrm{D}) \cdot P_B(\mathrm{D})}{P_B(+)} = \frac{0.97 \cdot 0.05}{0.97 \cdot 0.05 + 0.03 \cdot 0.95} = \frac{0.0485}{0.077}$$
$$\simeq 0.63.$$

Test B is more reliable.   □

The reason that test B turned out to be more reliable in Example 16.9, even though it claimed a lower percentage of correct diagnoses, is that it was tested on more subjects who had the disease. In general, to have a reliable test for a disease, you need to get high conditional probability $P(+|\mathrm{D})$ on a possibly *large* sample of subjects who have the disease (and of course as small a fraction of mistakes $P(+|\mathrm{N})$ as possible on healthy people).

Estimating the true proportion of diseased people. In the above disease detection problems, the solution involves two steps: (1) estimate the proportion of diseased people *among all people*, and (2) carry out calculations involving Bayes' law. In the first step, it is important to understand that the estimate would have to be obtained separately and independently from the data provided by the test. The efficiency of Bayes' method in estimating reliability of a given medical test depends on the accuracy of this data. Usually hospitals have reliable data on the proportion of patients who had the disease among the patients who exhibited alarming symptoms. To find the proportion of the diseased people among the total population, we have to multiply this number by the proportion of people exhibiting alarming symptoms among the total population. In the present chapter, we assume that the proportion of diseased people out of the total population is given.

Looking for a rare bird?

Another situation where Bayes' law can be useful occurs in observational biology. Sometimes a rare species has a particular visible characteristic that is, however, shared by a small percentage of a more common variety. If you see an animal or a plant with such a characteristic, how certain can you be that you saw a representative of the rare species?

Example 16.10. Suppose that a particular pattern occurs on the wings of 100% of individuals of a rare butterfly species. A very similar pattern, indistinguishable from a distance, occurs in just about 2% of a more common species. You know that only 4 out of 1,000 similar-looking butterflies belong to the rare species. If you spot a butterfly with the characteristic pattern, what is the probability that you saw a representative of the rare species?

Solution: Let us solve the question using the grid method. It is natural in this setting to choose the total number of butterflies equal to 1,000. Then 4 butterflies are rare, and 996 common. We need to compute 2% of 996. We have $9.96 \cdot 2 = 19.92$. This is the "number" of common butterflies having the characteristic pattern. The rest of the common butterflies do not have the pattern: $996 - 19.92 = 976.08$. Now we can fill in the entries in each column, then add up the row sums, to get:

	Rare	Not Rare	Total
Pattern	4	19.92	23.92
No pattern	0	976.08	976.08
Total	4	996	1,000

To answer the question, we compute:

$$P(\text{Rare}|\text{Pattern}) = \frac{4}{23.92} = 0.167 \simeq 17\%.$$

The probability that you have seen a rare butterfly is only 17%. The answer does not sound encouraging, but this makes a quest for finding a genuine representative of a rare species even more exciting.

Note: in this case the table contains decimal numbers; this could have been corrected by starting with 100,000 butterflies instead of just 1,000.   □

Monty Hall problem

Bayes' law can be used to solve many of the tricky questions in probability theory. A well-known example is the Monty Hall problem.

Example 16.11. The question derives its name from an American TV quiz show host, whose shows contained similar puzzles. In the original formulation, a show participant (a player) is given a chance to win a big prize (a car) that is hidden behind one of three closed doors. Behind the other two doors, mock prizes (goats) are hidden. The player is asked to choose a door. Once the door is chosen, the host, who knows which door conceals the car, opens one of the remaining two doors, to reveal a goat. Only two doors remain closed now, and behind one of them is the car. The host offers the player another chance: to switch to the other closed door, or to stay with her first choice. Should the participant switch or stay? Does it make a difference?[1]

Figure 16.2: The Monty Hall problem.

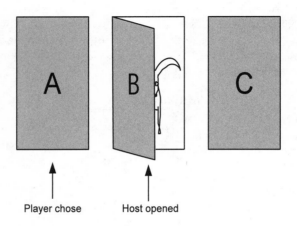

Solution: Let us denote the three doors by letters A, B, and C. Suppose the player has chosen door A, and the host opened B to reveal a goat. The doors A and C are closed, and the car is behind one of them. Let us denote the probabilities that the car is behind door A and door C as $P(A)$ and $P(C)$, respectively. Let us denote by $P(\text{op } B)$ the probability that the host opened door B. To decide if the player should switch, we have to compare the conditional probabilities that the car is behind A or C, given that the host opened B. These are the following probabilities:

$$P(A|\text{op } B) \quad \text{and} \quad P(C|\text{op } B).$$

We will use Bayes' law to compute them both. We have

$$P(A|\text{op } B) = \frac{P(\text{op } B|A) \cdot P(A)}{P(\text{op } B)}, \quad P(C|\text{op } B) = \frac{P(\text{op } B|C) \cdot P(C)}{P(\text{op } B)}.$$

---

[1]This problem has been referenced in many sources; in particular, it was mentioned briefly in the movie *21*. You can find the relevant video clip online.

Assuming the game is fair, the probability of the car being hidden behind each of the three doors is one-third: $P(A) = P(B) = P(C) = 1/3$. The probability that the host opens door B if the player chose A and the car is behind A is one-half, because behind both B and C there are goats and the host can decide to open either one of them with equal chance. This means that $P(\text{op } B|A) = 1/2$. The probability that the host opens B if the car is behind C is 1, because it is the only door that conceals a goat and is not already chosen by the player: $P(\text{op } B|C) = 1$. If the car is behind B, the host won't open it: $P(\text{op } B|B) = 0$. The tree diagram in Figure 16.3 shows the chances that the host opens door B or C, if he knows that the car is behind each of the three doors.

Figure 16.3: Tree diagram assuming the player chose door A.

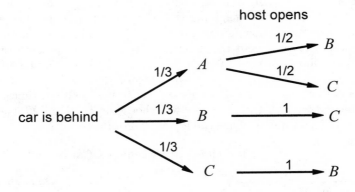

To find the probability of the host opening door B, we split it into the sum of probabilities of three disjoint events when the car is behind A, B, or C:

$$P(\text{op } B) = P(\text{op } B|A) \cdot P(A) + P(\text{op } B|B) \cdot P(B) + P(\text{op } B|C) \cdot P(C)$$
$$= \frac{1}{2} \cdot \frac{1}{3} + 0 \cdot \frac{1}{3} + 1 \cdot \frac{1}{3} = \frac{1}{2}.$$

In the tree diagram this corresponds to summing up the products of coefficients along the paths leading to B. Now by Bayes' law (or by multiplying the probabilities along the paths passing through A and C, respectively), we have

$$P(A|\text{op } B) = \frac{P(\text{op } B|A) \cdot P(A)}{P(\text{op } B)} = \frac{\frac{1}{2} \cdot \frac{1}{3}}{\frac{1}{2}} = \frac{1}{3},$$

$$P(C|\text{op } B) = \frac{P(\text{op } B|C) \cdot P(C)}{P(\text{op } B)} = \frac{1 \cdot \frac{1}{3}}{\frac{1}{2}} = \frac{2}{3}.$$

Therefore, given that the host opened door B, the probability that the car is behind C is twice as high as the probability that the car is behind A. The player should always switch to the remaining closed door to improve her chances of winning the prize. □

Let us see how the grid method works in this case. First we need to figure out the total number we will be counting here. It is not the number of doors or cars or goats or players. Upon contemplation, we realize that it should be the total number of experiments being carried out. (The player can play only once, but imagining that she could play many times, with different outcomes, allows us to evaluate the probability.) Suppose the game show host carries out the experiment thirty times and each time the contestant chooses door A. We can assume that the car is behind doors A, B, C 10 times each. Now if the car is behind door B (or door C), then the host will choose door C (or door B, respectively) all ten times. If the car is behind door A, then the host will choose doors B and C five times each. In other words, we have the following grid:

	Car behind door A	Car behind door B	Car behind door C	Total
Host opens door A	0	0	0	0
Host opens door B	5	0	10	15
Host opens door C	5	10	0	15
Total	10	10	10	30

Now given that the host has opened door B, we compute the probability that switching is desirable to be: $10/15 = 2/3$ (as opposed to not switching, which is desirable only in $5/15 = 1/3$ of the cases). Similarly, given that the host has opened door C, we compute that switching is helpful in two-thirds of the cases, and not switching only in one-third of the cases. Therefore the participant should switch. □

We were able to reach the answer by a relatively simple sequence of computations based on Bayes' law, or on the grid method. However, the conclusion is far from being intuitively obvious. Indeed, this conclusion, given by Marilyn vos Savant in reply to a reader's question in her column "Ask Marilyn" in *Parade* magazine in 1990, started a long and heated discussion among both specialists and the general public. The result seemed to be so counter-intuitive that even many professional mathematicians initially argued that switching or not should make no difference to the player's chances of winning the prize. The grid method as presented above may help persuade the skeptical: before the game show host opens a door, the car is equally likely to be behind any of the three doors (observed from the column sums). But once the host opens a door, we are restricted to one particular row of the table, and the rows are visibly inequivalent, which makes the conditional probabilities different and shows that switching is indeed preferable.

Vos Savant added a further explanation: imagine there were a million doors, with goats behind all but one of them. Then if a player initially chooses door 1 and the host opens all remaining doors except door number $777,777$, should the player switch to that door? Most people agreed that in this case switching is preferable.

Eventually, with the help of theoretical arguments and numerical simulations, it was universally accepted that the player should always switch. But the deceptive simplicity of the question gave rise to multiple modified versions of it that continue to attract widespread interest. Here is one.

Example 16.12. As before, suppose that there is a car behind one of the doors A, B, or C, and goats behind the two remaining doors. Suppose that the player chooses door A. Imagine that, at this moment, a heavy truck passes by, causing the building to shake and door B to open and reveal a goat. Given a chance to switch to door C, should the player switch or stay with her initial choice? Does it make a difference?

Solution: This innocent-looking modification turns out to be crucial and completely changes the rules of the game. Let us use Bayes' law to compare the player's chances to win if she switches and if she stays. As before, $P(A) = P(B) = P(C) = 1/3$ will denote the probabilities of the car being hidden behind the respective doors, with no additional information given. The event (ac B) stands for door B opening accidentally. We should compare the conditional probabilities $P(A|\text{ac } B)$ and $P(C|\text{ac } B)$. They are given by the formulas

$$P(A|\text{ac } B) = \frac{P(\text{ac } B|A) \cdot P(A)}{P(\text{ac } B)}, \qquad P(C|\text{ac } B) = \frac{P(\text{ac } B|C) \cdot P(C)}{P(\text{ac } B)}.$$

It is hard to estimate the probability of door B opening because of the

Figure 16.4: Modified Monty Hall.

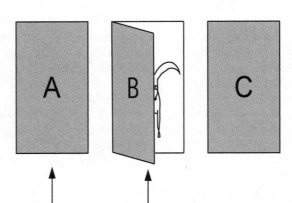

passing truck. It might depend on the quality of the building and the properties of the door, and on the probability of the truck passing by at the right moment. But one thing we can be sure about is that this probability does not depend on whether the car is behind door A or door C. In formulas, this means that

$$P(\text{ac } B|A) = P(\text{ac } B|C).$$

Because we also have $P(A) = P(C)$, the conditional probabilities we want to compare are the same:

$$P(A|\text{ac } B) = \frac{P(\text{ac } B|A) \cdot P(A)}{P(\text{ac } B)} = \frac{P(\text{ac } B|C) \cdot P(C)}{P(\text{ac } B)} = P(C|\text{ac } B).$$

Switching doors in this case does not make any difference to the player's chances of winning. $\qquad\square$

A small change in the framing of the question led to a completely different conclusion because of the different amount of information contained in the event of opening door B in the two cases. In the Monty Hall problem, the host *knows* where the car is and this knowledge informs his decision to open door B. In the modified version, the passing truck, the shaking building, and door B itself *know nothing* about the location of the car. The only information we get in this case is that the car is not behind door B. This makes the remaining two options equally probable.

The same mathematical problem is given a nice fictional twist in the novel *Sweet Tooth* by the British writer Ian McEwan.[2] A jealous husband knows his wife is with her lover in one of three hotel rooms. Just when he is ready to take his chance and break into one of the rooms, another door opens and a family with small children comes out. Should the husband switch to the third door, or stay with his initial choice? You can compare the conditional probabilities. Or you can read the book.

EXERCISES

Question 16.1. Are the following pairs of events A and B independent? Justify your answer by computing the probabilities $P(A)$, $P(B)$ and $P(A \cap B)$.

1. Toss three fair coins. A: the first is a head. B: there are exactly two heads.

2. Draw two cards from a deck without replacement. A: the first card is a number from 2 to 10; B: the second card is a king or an ace.

3. Roll two fair dice. A: the sum of the outputs is 7. B: the first output is odd.

Question 16.2. Suppose I roll two fair dice. For the events A and B given below, find the probabilities $P(A)$, $P(B)$, and $P(A \cap B)$. Are the events independent?

1. Event A: the sum of the outputs is 8. Event B: both numbers are even.

2. Event A: the sum of the outputs is even. Event B: the first output is 5.

3. Event A: both outputs are even. Event B: the first output is greater than 3.

Question 16.3. Use the definition of the conditional probability $P(A|B) = P(A \cap B)/P(B)$ to answer the following questions:

1. Suppose I roll two dice, and the sum is odd. Given this, what are the chances that the sum is 11?

---

[2]Ian McEwan, *Sweet Tooth*, Anchor Books, 2012.

2. Suppose a bag has 100 red marbles, 50 green marbles, and 25 blue marbles. Apart from their color, the marbles are indistinguishable and mixed randomly in the bag. If I draw two marbles randomly without replacement from the bag and neither of them is green, what are the odds that they are both red?

Question 16.4. In the conditions of Example 16.6, compute the conditional probabilities $P(A|B)$ and $P(B|A)$ explicitly by counting the favorable and total outputs in each case.

Question 16.5. A bookshelf has five books in French, five in Spanish, and five in German. Two books are picked at random. Event A: both books are in French. Event B: none of the books are in Spanish. Find the probabilities $P(A)$, $P(B)$, $P(A|B)$, $P(B|A)$. Are the events independent?

Question 16.6. There are three red, five yellow, and four white carnations in the vase. Two are chosen randomly. Event A: none of them are red. Event B: at least one is yellow. Find the probabilities $P(A)$, $P(B)$, $P(A \cap B)$, $P(A|B)$, and $P(B|A)$. Are the events independent?

Question 16.7.   1. Which is more likely: drawing two cards without replacement from a deck and getting two clubs, or getting four heads in four fair coin tosses?

2. What is the conditional probability $P(A|B)$ of the following event: roll three fair dice. Event A: the second die yields 2. Event B: the sum of the outputs is 3.

Question 16.8. A criminal investigation is conducted in a three-story building that has four windows on each floor. It is known that on the night of the crime, exactly two windows in the building were lit. The first witness testifies that none of the third floor windows were lit. The second witness testifies that at least one window on the first floor was lit. The suspect states that the two leftmost windows on the first floor were not lit. What is the probability that the suspect is telling the truth, given

1. the testimony of the first witness.

2. the testimony of the second witness.

3. both testimonies combined.

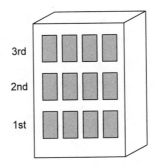

Question 16.9. Suppose an entire school goes on a picnic – as many boys as girls. The boys all wear jeans; a third of the girls wear skirts, and the rest wear jeans. Given that a randomly picked student is wearing jeans, what are the odds that the student is a girl? *Hint:* Use Bayes' law.

Question 16.10. A Toyota factory manufactures twice as many Priuses as Corollas. Priuses are painted red, blue, or silver in equal numbers; Corollas are painted white, black, red, or blue in equal numbers. Given that a car manufactured in this factory is red, find the probability that it is a Prius. *Hint:* Use Bayes' law.

Question 16.11. In the conditions of Example 16.8, find the probability that the subject has the disease, given that the subject tested negative.

Question 16.12. Solve Example 16.9 using the grid method. Check that every probability computed in Example 16.9 also can be computed using the two tables (for the two tests) which you have obtained.

Question 16.13. Suppose a new test for a rare disease claims to detect it among diseased people 99% of the time. Among the healthy people tested, the test is negative 95% of the time. Suppose the disease statistically occurs in 25 out of 1,000 people. If a new patient tests positively, what is the probability that she or he has the disease?

Question 16.14. Suppose there are two competing tests in the market for detecting a particular disease: one (call it test A) claims to correctly diagnose both diseased and healthy people 99% of the time, whereas the other (test B) does so only 95% of the time. We know independently that 99% of the subjects on whom test A was carried out were healthy, and 94% of the subjects for test B were healthy. Given this, which test is more reliable (has higher $P(D|+)$)?

Question 16.15. Here is a simplified version of a *Bayesian spam filter*, used by email services to classify whether or not a particular email is genuine or spam. You may have received emails claiming to be from the prince of some country and which promise a lot of money if you send him your social security number, date of birth, bank account number, and so on.[3]

Thus, let us say your email software searches all incoming emails for occurrences of the word "prince." It is known that roughly 1% of the emails you get every week are spam. In spam emails, the word "prince" occurs 7% of the time on average, and in regular emails it occurs 0.01% of the time. Now suppose you get an email containing the word "prince." What is the probability that this email is spam? (The software will then classify the email as spam or not, randomly, with this probability.) Do not worry; this problem is only a simplified version of the actual spam filter in your email service.

Question 16.16. An email software searches all incoming emails for the word "lottery." It is known that roughly 2% of incoming emails are spam. In spam emails, the word "lottery" occurs 20% of the time on average. In regular emails, it occurs 0.02% of the time. Now given an incoming email which contains the word "lottery," what is the probability that it is spam? Given an email which does not contain "lottery," what is the probability that it is genuine?

Question 16.17. Recall from this chapter the vos Savant thought experiment where there are one million doors instead of three in the original Monty Hall problem. We now consider another variant of the Monty Hall problem along similar lines. Namely, suppose there are five doors instead of three, and there is a car behind one of them. Once the player chooses a door, the host opens two other doors instead of one. Would it then be in the player's best interests to switch? In this question we carry out a step-by-step analysis using the grid method.

1. Let us denote the five doors by A through E. Suppose the player has chosen door A. What are the probabilities $P(A), P(B), \ldots$ of the car being behind door $A, B, \ldots$?

2. If the car is behind door A, then the host is free to choose any two of the remaining four doors – this can be done in $_4C_2 = 6$ ways: BC, BD, BE, CD, CE, DE. So let us suppose there are $5 \cdot {}_4C_2 =$

---

[3]Please don't! With 99.999% probability, it is a scam.

30 times that this experiment takes place, and in all of them the participant chooses door A, but the car is behind doors $A, B, \ldots, E$ in six cases each.

Now, in the six cases when the car is behind door A, the host can choose any of the six combinations of two doors with equal probability. Thus, the host would open BC in one of the six experiments, BD in one, and so on. We obtain the following table:

Host opens door:	Car behind door A	Car behind door B	Car behind door C	Car behind door D	Car behind door E	Total
BC	1					
BD	1					
BE	1					
CD	1					
CE	1					
DE	1					
Total	6	6	6	6	6	30

3. Fill in the second column. Clearly if the car is behind door B, then the host cannot open any pair of doors with B in them. How many times out of 6 total would the host open one of the pairs of doors which does not include A or B?

4. Proceeding in a similar way, fill in the rest of the table. What are the row sums?

5. Now suppose that the host chose to open doors B and E after the participant chose door A. Is it in the interest of the participant to switch to *either* of the remaining doors? Does the answer depend on which pair of doors the host opens?

Question* 16.18. *General case of the Monty Hall problem.* Suppose that the game show contains n doors labeled $D_1, \ldots, D_n$, and behind exactly one of these doors is a car; the other doors have goats. Also suppose that the participant chooses door D_1, say - and the host opens some k other doors. We will assume that $1 \leq k \leq n - 2$, so that there is at least one other door left closed after the host has opened k doors.

1. We will not actually draw a grid, but will visualize it. First, what should be the dimensions (number of rows and columns) of the grid? *Hint:* The previous question can help you guess.

2. Now suppose the host repeats this experiment $n \cdot {}_{n-1}C_k$ times - i.e., ${}_{n-1}C_k$ times the number of doors the car can be behind. If the car is behind door D_1, then the host is equally likely to open any k of the remaining $n-1$ doors – which means opening each such set of k doors exactly once among the ${}_{n-1}C_k$ times.

3. If on the other hand the car is behind door D_i for some $2 \leq i \leq n$, then the host cannot open D_1 or D_i. How many sets of k doors is the host allowed to open?

   What is the average number of times the host would open one of these sets of doors? To do this computation, one has to divide one number of combinations by another. Check that the answer is: the host opens each set of k doors not containing D_1, D_i an average of $\frac{n-1}{n-1-k}$ times.

4. Fill in the given grid. The only entries in the grid (except for the "Total" row and column) should be 1, 0, or $\frac{n-1}{n-1-k}$.

5. Finally, suppose the host has opened some *fixed* set of k doors. Using the grid, given that the host has opened these doors, is it in the interest of the participant to switch from door D_1 to any of the doors that remain closed?

Chapter 17

Statistics: Babe Ruth and Barry Bonds

In this chapter and the next, we discuss the basics of data analysis and statistics. In today's world, data is everywhere. The world is increasingly reliant on understanding how observed quantities depend on one another, and on harnessing this understanding to make informed decisions.

In the previous chapter, you have already seen instances of such decision making using the available information or data: in Example 16.9 we had to decide which of two competing tests is more reliable in detecting disease. The Monty Hall problem is another example of decision making using incomplete information. In the present chapter, we discuss some of the fundamental and most widely used tools of data analysis.

MATH

Statistics is the science of analyzing data. In this book, we will assume that data comes in the form of the values that some variable – X, say – takes as it applies to all objects in a set. The set of objects may be the students in a class, with X denoting their heights, weights, or scores on a test. Or the set can be days in February, with X denoting the amount of snowfall on the Yale campus on a particular day. In all examples we consider in this book, the values of X will be real numbers. For each such dataset, we define the following notions:

Definition. Suppose we are given a dataset of values $x_1, x_2, \ldots, x_n$ (which we will call *observations*) assumed by a variable X, where X takes real values. For each such dataset, we define the following notions:

1. The *mean* or *average* value of the dataset is denoted by $\overline{x}$ (read "x bar"), and defined to be

$$\overline{x} = \frac{x_1 + x_2 + \cdots + x_n}{n} = \frac{1}{n} \sum_{i=1}^{n} x_i.$$

2. When n is odd, the *median* of $x_1, x_2, \ldots, x_n$ is obtained by arranging the observations in increasing order and choosing the middle value among them. If the number of observations n is even, then choose the two "middle values" after arranging the x_i in increasing order, and define the median to equal their mean.

3. The *mode* of the observations $x_1, x_2, \ldots, x_n$ is the most frequently occurring value among them, if it exists.

Practice problem 17.1. Suppose that score results on a placement test in a class of 10 students are as follows:

$$78, \ 70, \ 82, \ 64, \ 99, \ 73, \ 92, \ 56, \ 70, \ 76.$$

Find the mean, median, and mode of the observations.

Answer: Mean 76, median 74.5, mode 70.

Now we are ready to introduce more advanced notions commonly used in statistical analysis, the *variance* and *standard deviation*. The *variance* of the dataset indicates how much data points differ from the mean. It is denoted by σ_X^2 and defined as follows:

$$\sigma_X^2 = \frac{(x_1 - \overline{x})^2 + (x_2 - \overline{x})^2 + \cdots + (x_n - \overline{x})^2}{n}.$$

To simplify the formula, we can use the sum notation, just as we did for the geometric series in Chapter 9. Then we can write

$$\sigma_X^2 = \frac{1}{n} \sum_{i=1}^{n} (x_i - \overline{x})^2.$$

The *standard deviation* is the square root of the variance, denoted by σ_X.

In some cases it is more convenient to use an alternate formula for the variance:

$$\sigma_X^2 = \frac{1}{n}(x_1^2 + \cdots + x_n^2) - \overline{x}^2.$$

We will explain why this formula is equivalent to the definition of σ_X^2 at the end of this chapter.

Example 17.2. For the data given in Practice problem 17.1, find the variance and the standard deviation.

Solution: We know that $\overline{x} = 76$. Let us write down the list of differences between the data values and $\overline{x}$:

$$2, \ -6, \ 6, \ -12, \ 23, \ -3, \ 16, \ -20, \ -6, \ 0.$$

Then we can find the variance by summing up the squares of these numbers:

$$\sigma_X^2 = \frac{1}{10}(4 + 36 + 36 + 144 + 529 + 9 + 256 + 400 + 36 + 0)$$
$$= \frac{1,450}{10} = 145.$$

The standard deviation is $\sigma_X = \sqrt{145} \simeq 12.04$. □

Often data comes in the form of more than one measurement for each object: for example, scores on the first and the second midterm exams in a class of students, or the systolic and diastolic blood pressure numbers for the patients in a hospital. In this case to each object we associate two variables, say, X and Y, and the data is called *bivariate*.

Given bivariate data $(x_1, y_1), \ldots, (x_n, y_n)$ for two random variables (X, Y), an important characteristic is their *covariance*. It is defined to be the average of the products of the differences between each component in a bivariate observation and its respective mean:

$$\text{Cov}(X, Y) = \frac{1}{n} \sum_{i=1}^{n} (x_i - \overline{x})(y_i - \overline{y}).$$

Note that the variance of a dataset $x_1, \ldots, x_n$ can be interpreted as the covariance of the dataset with itself: if we substitute x_i for all y_i and $\overline{x}$ for $\overline{y}$ in the covariance formula, we will obtain the variance σ_X^2:

$$\text{Cov}(X, X) = \frac{1}{n} \sum_{i=1}^{n} (x_i - \overline{x})(x_i - \overline{x}) = \frac{1}{n} \sum_{i=1}^{n} (x_i - \overline{x})^2 = \sigma_X^2.$$

It is natural to ask if an alternate formula for the covariance exists, just as one did for the variance (see above). Here is the answer:

$$\text{Cov}(X, Y) = \frac{1}{n}(x_1 y_1 + \cdots + x_n y_n) - \overline{x} \cdot \overline{y} = \frac{1}{n} \sum_{i=1}^{n} x_i y_i - \overline{x} \cdot \overline{y}.$$

The formula is obtained using calculations similar to those for the variance; see Question 17.6.

For bivariate data X, Y we can also introduce their *correlation (coefficient)* $\rho_{X,Y}$. It is defined as

$$\rho_{X,Y} = \frac{\text{Cov}(X, Y)}{\sigma_X \sigma_Y}.$$

The quantity $\rho_{X,Y}$ is also known by many other names in the literature, in particular, Pearson's r or the Pearson product-moment correlation coefficient.

The above quantities are the first to be computed in a data analysis problem. They constitute useful summary statistics of the data.

APPLICATIONS

Statistics begins with data that might come in all shapes and forms. For instance, the heights of students in a classroom can be collected in the form of a list of integer numbers of inches. The data results one obtains from an image search for "basketball" are in the form of pictures. Analyzing brain activity, or who is friends with whom on each day of 2013, involves studying data in the form of neural networks or "friendship graphs," respectively. Today, data analysis is an extremely important subject for most large organizations, and even machines are programmed to "learn" about the data, to analyze it, and to detect patterns in it. Of course, data analysis was around long before the advent of companies like IBM, Google, or Facebook – even before computers were around. For instance, governments of nations and other political units have been collecting and analyzing census data on the ages, occupations, and other attributes of populations for thousands of years.

In this book, we will restrict our attention to studying quantitative data in the form of lists of values, such as heights of students in a classroom. We now discuss how quantities such as the mean, median, variance, and correlation allow us to draw meaningful conclusions about the observations in the data.

Example 17.3. The following dataset lists the number of home runs by year which were hit by Babe Ruth when he played for the New York Yankees from 1920 to 1934:[1]

$$54, \; 59, \; 35, \; 41, \; 46, \; 25, \; 47, \; 60, \; 54, \; 46, \; 49, \; 46, \; 41, \; 34, \; 22.$$

Find the mean, median, and mode annual number of home runs Babe Ruth hit as a Yankee. Also find the variance and standard deviation.

Solution: It is easy to check that

$$x_1 + \cdots + x_{15} = 54 + 59 + \cdots + 22 = 659.$$

Therefore the average number of home runs hit by Babe Ruth per year as a Yankee is $\overline{x} = 659/15 = 43.93 \simeq 44$.

The median of the above dataset can be obtained by arranging the fifteen numbers in increasing order. Then the eighth number is 46, and there are precisely seven numbers before or after it. Thus the median number of home runs is 46.

Next, note that the number 46 appears thrice in the above dataset, and no other number occurs as frequently. Therefore the mode is also 46.

Finally, we compute that $x_1^2 + \cdots + x_{15}^2 = 30,723$, so the variance and standard deviation are, respectively,

$$\sigma_X^2 = \frac{1}{15}(30,723) - 43.93^2 = 118.3551, \qquad \sigma_X = \sqrt{118.3551} = 10.879.$$

$\square$

Example 17.4. At the time we wrote this book, there were only six tennis players in history with more combined Grand Slam titles than Serena Williams (34): Margaret Court (64), Martina Navratilova (59), Billie Jean King (39), Margaret duPont (37), Louise Clapp (35), and Doris Hart (35).[2] Find the mean and median numbers of Grand Slam titles won by these six players.

Solution: It is easy to check that the mean is $269/6 = 44.833 \simeq 45$ titles, while to find the median, we arrange the data in a non-decreasing order:

$$35, \quad 35, \quad 37, \quad 39, \quad 59, \quad 64.$$

Then the median is the average of the third and fourth numbers on the list, i.e., 38 titles. $\square$

---

[1]Source: http://www.baseball-reference.com.

[2]Source for this and the following example: http://en.wikipedia.org.

Robustness of the median. Typically, both the mean and median are supposed to represent *measures of central tendency* – i.e., a statistic or quantity which is "central" in a given dataset. In Example 17.4, the mean was 44.833, which is not really "central" to the dataset: it is higher than all but two of the observations. Why does this discrepancy occur? The reason comes from the two highest values: 64 and 59 are far larger than the other four values. These two largest values are called *outliers*, lying far away from the remaining dataset and distorting the average value unreasonably. On the other hand, the median is 38, which seems to be more "central" or closer to the majority of the observations. This is true in general: the median is a more robust measure of central tendency than the mean – i.e., if we replace a few of the observations by outliers, then the mean can change quite a bit, whereas the median remains essentially the same.

Practice problem 17.5. In the 1960s, the Beatles recorded the following number of songs per year:

$$4, \ 3, \ 20, \ 66, \ 41, \ 37, \ 16, \ 30, \ 45, \ 42.$$

Compute the mean and median number of songs they recorded per year from 1960 to 1969.
Answer: Mean $= 304/10 = 30.4$, median $= (30 + 37)/2 = 33.5$.

As we discussed above, the mean and median describe a "central value" among the observations. What is the significance of the variance? It measures the *spread* of the data. Here is an example of a typical interpretation of the variance.

Example 17.6. The following dataset lists the number of home runs, by year, which were hit by Barry Bonds when he played for the San Francisco Giants from 1993 to 2007:[3]

$$46, \ 37, \ 33, \ 42, \ 40, \ 37, \ 34, \ 49, \ 73, \ 46, \ 45, \ 45, \ 5, \ 26, \ 28.$$

Find the mean and median annual number of home runs hit by Barry Bonds as a Giant. Also find the variance and standard deviation – compare them with Babe Ruth's statistics from Example 17.3.

Solution: The average for Bonds is $586/15 = 39.07 \simeq 39$, while the median is computed by arranging the above scores in increasing order and

---

[3]Source: http://www.baseball-reference.com.

considering the eighth ranked score: 40. (We do not consider the mode here, as observation reveals that three scores – 37, 45, and 46 – occur the most frequently, i.e., twice each.)

To compute the variance, we will use the second formula:

$$\sigma_X^2 = \frac{1}{n}(x_1^2 + \cdots + x_n^2) - \overline{x}^2.$$

Adding up the squares of the scores, we get $25,844$. Therefore the variance and standard deviation equal, respectively,

$$\sigma_X^2 = \frac{1}{15}(25,844) - 39.07^2 \simeq 196.47, \qquad \sigma_X = \sqrt{196.47} = 14.017.$$

□

Interpretation. Note that the two time periods under consideration over which the two players played for a single team are equal in length: 15 years. The comparison between the means (or, equivalently, the totals if we multiply by 15) indicates that Ruth hit more home runs than Bonds on average during these two 15-year periods. Note that the mean is a *summary statistic*, so it does not imply that Ruth hit more home runs than Bonds every year in their respective times.

Also note that the variance – i.e., spread – of the scores for Bonds is greater than that for Ruth over the 15-year time period. This indicates that Ruth's home run numbers were prone to less *fluctuation* or *volatility* than those of Bonds during that time.

Is there a way that we might have suspected whose scores had greater variance merely by examining the data? Yes, and here is how: the variance and standard deviation measure the spread of the data, i.e. its deviation from the mean. Therefore outlier values (on either side of the mean) would add to the spread, because in computing the variance we add their squared deviations from the mean. The observations for Barry Bonds include a 73 and a 5, both of which are very far removed from the mean. (The 5 was because he played only 14 games in 2005.) On the other hand, Babe Ruth's highest and lowest scores are 60 and 22, respectively, which are less spread out. These observations already suggested that Bonds had a greater variance (and standard deviation) than Ruth.

The above examples were relatively simple to interpret. However, real-world examples can get significantly more involved. In those cases, data analysis and interpretations often vary from person to person and can be the subject of endless debates and discussions – not just among sports enthusiasts, but also in making policy decisions based on real-world data in government and industry, or even in deciding which stocks and shares to buy or sell at any given moment in financial market trading.

Also note that interpretations may be drawn from data analysis in settings that do not make sense in the real world. For instance, we could have compared the variances of the Beatles song numbers in Practice problem 17.5 and of Babe Ruth's scores from Example 17.3. While this may provide information about who was less volatile in their own field, the two fields are quite unrelated and the two quantities being measured are quite different from each other, rendering such comparisons not very meaningful.

Statistical significance and three-sigma limits

Often in measuring a quantity, we encounter fluctuations due to a variety of causes. How can we tell if a particular observation deviates from the expected amount, or mean, due to a non-random cause? Usually we say that observations that are sufficiently far removed from the mean are *statistically significant* – i.e., their error or deviation is not merely due to the underlying randomness or measurement errors.

A way to make this precise is as follows: given observations $x_1, \ldots, x_n$ for a variable or observed quantity X, determine for an observation x_i the multiple of σ_X by which it differs from the mean $\overline{x}$. For the purposes of this chapter, let us define the *statistical significance* (ss) of an observation x_i to be:

$$\mathrm{ss}(x_i) = \frac{x_i - \overline{x}}{\sigma_X}.$$

Typically, observations with statistical significance above 1 or below -1 are considered to be significant.

Example 17.7. Consider the home run totals of Barry Bonds in Example 17.6. Compute the statistical significance of the first, highest, and lowest observations.

Solution: We recall that $\overline{x} = 39.07$ and $\sigma_X = \sqrt{196.47} \simeq 14.02$. Using this,

$$\mathrm{ss}(x_1) = \frac{46 - 39.07}{14.02} \simeq 0.49, \qquad \mathrm{ss}(x_{\max}) = \frac{73 - 39.07}{14.02} \simeq 2.42,$$

$$\mathrm{ss}(x_{\min}) = \frac{5 - 39.07}{14.02} \simeq -2.43.$$

Thus the maximum and minimum observations are statistically significant.
□

You may have heard the term *three-sigma limits*. In fact, this corresponds to entries from a distribution of scores or observations whose statistical significance is either more than 3 or less than -3. These are outliers that fall well outside the expected ranges of acceptable or likely observations, and constitute extremely rare events. For example, in financial transactions, computer programs are automatically trained to detect three-sigma events and to alert the right people or entities in order to further investigate what caused such events to occur.

Bivariate data analysis

We now discuss how to analyze data involving two variables, X and Y. The added component is to calculate measures of dependency between the variables, i.e., how a change in one variable correlates with a change in the other. The covariance and correlation coefficient are two such measures.

Example 17.8. A classic example of correlation is of heights and weights. Compute the means, variances, covariance, and correlation between the heights H and weights W of five people.

Index	Height (inches)	Weight (pounds)
1	70	150
2	74	164
3	62	121
4	65	128
5	70	140

Solution: To compute all of these quantities can be quite time-consuming. The task is simplified by using a computer or scientific calculator; the easiest way to perform this kind of computation, though, is to use a spreadsheet. We present here one way to do the calculations systematically. First, we expand the previous table by introducing additional columns:

Index (i)	Height h_i	Weight w_i	h_i^2	w_i^2	$h_i w_i$
1	70	150			
2	74	164			
3	62	121			
4	65	128			
5	70	140			
Total					

We then fill out the individual columns of the table (except for the last entries in each column). Finally, we add the entries in each column to fill out the last row of totals. This yields the following table.

Index (i)	Height h_i	Weight w_i	h_i^2	w_i^2	$h_i w_i$
1	70	150	4,900	22,500	10,500
2	74	164	5,476	26,896	12,136
3	62	121	3,844	14,641	7,502
4	65	128	4,225	16,384	8,320
5	70	140	4,900	19,600	9,800
Total	341	703	23,345	100,021	48,258

Now use the formulas to compute all of the desired quantities (with $n = 5$):

$$\overline{h} = \frac{1}{5} \sum_{i=1}^{5} h_i = \frac{341}{5} = 68.2,$$

$$\overline{w} = \frac{1}{5} \sum_{i=1}^{5} w_i = \frac{703}{5} = 140.6,$$

$$\sigma_H^2 = \frac{1}{5} \sum_{i=1}^{5} h_i^2 - \overline{h}^2 = \frac{23,345}{5} - 68.2^2 = 17.76,$$

$$\sigma_W^2 = \frac{1}{5} \sum_{i=1}^{5} w_i^2 - \overline{w}^2 = \frac{100,021}{5} - 140.6^2 = 235.84,$$

$$\text{Cov}(H, W) = \frac{1}{5} \sum_{i=1}^{5} h_i w_i - \overline{h} \cdot \overline{w} = \frac{48,258}{5} - 68.2 \cdot 140.6 = 62.68,$$

$$\rho_{H,W} = \frac{\text{Cov}(H, W)}{\sigma_H \sigma_W} = \frac{62.68}{\sqrt{17.76 \cdot 235.84}} = 0.9685.$$

$\square$

Interpretation. The correlation coefficient $\rho_{X,Y}$ is a measure of the association of the two variables X and Y, or of their sets of observations $(x_1, \ldots, x_n)$ and $(y_1, \ldots, y_n)$. It can be shown that $\rho_{X,Y}$ is always between -1 and 1 for any two sets of observations (this is called the *Cauchy-Schwarz inequality*). Moreover, $\rho_{X,Y} = 1$ if and only if there exist constants $m > 0$ and c such that $y_i = mx_i + c$ for all i (i.e., the variables X, Y are in a perfect increasing linear relationship). Similarly, $\rho_{X,Y} = -1$ if and only if there exist constants $m < 0$ and c such that $y_i = mx_i + c$ for all i (i.e., the variables X and Y are in a perfect decreasing linear relationship).

To understand the meaning of $\rho_{X,Y}$ better, let us look again at the formula:

$$\rho_{X,Y} = \frac{\operatorname{Cov}(X,Y)}{\sigma_X \sigma_Y} = \frac{\sum_{i=1}^{n}(x_i - \overline{x})(y_i - \overline{y})}{\sqrt{\sum_{i=1}^{n}(x_i - \overline{x})^2}\sqrt{\sum_{i=1}^{n}(y_i - \overline{y})^2}}.$$

If $(x_i - \overline{x})$ is very close to $(y_i - \overline{y})$ for all i, then clearly the numerator is almost equal to the denominator, and the correlation $\rho_{X,Y}$ is very close to 1. That means that the deviations of both the X and Y datasets from their respective means behave the same way with respect to the data points: if it is big and positive for X, it is also big and positive for Y. In other words, the datasets for X and Y have the same pattern of distribution along their data points: if X is greater than average for a given point, than so is Y, and by the same amount. If we mark the points with coordinates (x_i, y_i) for all i on a plane, small X-coordinates will be paired with small Y-coordinates, and large X-coordinates – with large Y-coordinates. The data points will be scattered along a line with a positive slope (see Figure 17.1).

Figure 17.1: A bivariate dataset (X, Y) with a strong positive correlation.

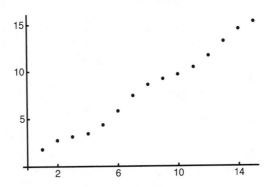

On the other hand, if $(x_i - \overline{x})$ is close to $-(y_i - \overline{y})$ for all i, then the numerator is almost the negative of the denominator, and $\rho_{X,Y}$ is very close to -1. This means that when X is greater than average, Y is less than average by the same amount. This indicates a negative correlation (see Figure 17.2).

Figure 17.2: A bivariate dataset (X, Y) with a strong negative correlation.

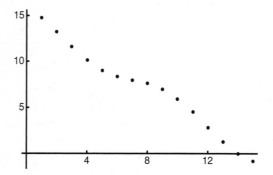

If the differences $(x_i - \overline{x})$ and $(y_i - \overline{y})$ do not follow any pattern, but can have the same or different signs and magnitudes depending on the data point, then the numerator gets closer to zero because many terms will have different signs and cancel each other out. This happens sometimes when X is greater than average while Y is less, and sometimes they both are greater or less than average at the same time. In this case the two datasets have a small or no correlation (see Figure 17.3).

Figure 17.3: A bivariate dataset (X, Y) with no correlation.

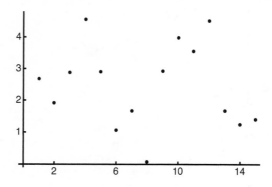

In Example 17.8, we saw that $\rho_{H,W} \simeq 0.97$. Therefore the heights and weights are in a relationship that is very close to being an increasing linear one.

Correlation and causality

Note that while the analysis of data reveals trends and associations between the variables or phenomena under consideration, it does not explain the underlying causal mechanism. There is a well-known phrase that summarizes this principle succinctly: *correlation does not imply causation.* For instance, consider the two sets of home run totals by Babe Ruth (R) and Barry Bonds (B) in Examples 17.3 and 17.6, respectively. If we denote these observations in chronological order by $r_1, \ldots, r_{15}$ for Ruth and $b_1, \ldots, b_{15}$ for Bonds, then

$$\bar{r} = 43.93, \quad \sigma_{\mathrm{R}}^2 = 118.3551, \quad \bar{b} = 39.07, \quad \sigma_{\mathrm{B}}^2 = 196.47, \quad \sum_{i=1}^{15} r_i b_i = 26,885.$$

Therefore, $\rho_{\mathrm{R,B}} = \dfrac{26,885/15 - 43.93 \cdot 39.07}{\sqrt{118.3551 \cdot 196.47}} \simeq 0.498.$

This is a significant degree of positive correlation. However, it does not imply that the performances of Babe Ruth and Barry Bonds were in some way related, or that one caused the other!

On the other hand, consider the heights and weights of a group of people considered in Example 17.8. We have computed that the correlation is pretty high, $\rho_{H,W} = 0.9685$. In this case, it is clear that taller people naturally tend to weigh more, so there is an obvious causal relation.

A high correlation between two observations might mean that there are additional underlying factors that explain why the two variables are correlated. For instance in the Ruth and Bonds examples, it may be that the hitting careers of baseball players follow similar trends in terms of physical ability and mental agility, and both players conform to these trends. Certainly there is a similarity, or correlation, in their performances over time, even if there is no causal relationship between them.

A well-known example of two datasets with clear correlation is the volume of ice cream sales versus the number of drownings in a given area. However, it is unlikely that eating ice cream can cause people to drown. Most probably, the common cause for the numbers in both datasets to rise would be the hot weather.

Some examples are tricky: it is known that the age of a woman at the birth of her first child is positively correlated with the number of years she spent at school.[4] One might argue that being busy studying and developing a career can cause women to postpone having children. Or one might argue that having a child early can prevent a woman from continuing her education. Or is there a third factor that causes both tendencies? In most situations, it is unwise to completely disregard a high correlation between two datasets; rather, try to find the reason for it.

Proof of the alternate formula for the variance*

We will conclude this chapter by giving a proof of the formula for the variance we presented in the beginning of this chapter:

$$\sigma_X^2 = \frac{1}{n}(x_1^2 + \cdots + x_n^2) - \overline{x}^2.$$

Let us see why this formula holds. By definition and the distributive law,

$$\sigma_X^2 = \frac{1}{n}\sum_{i=1}^{n}(x_i - \overline{x})^2 = \frac{1}{n}\sum_{i=1}^{n}(x_i^2 - 2x_i\overline{x} + \overline{x}^2)$$

$$= \frac{1}{n}(x_1^2 + \cdots + x_n^2 - 2\overline{x}x_1 - \cdots - 2\overline{x}x_n + \overline{x}^2 + \cdots + \overline{x}^2)$$

$$= \frac{1}{n}(x_1^2 + \cdots + x_n^2) - \frac{1}{n}\cdot 2\overline{x}(x_1 + \cdots + x_n) + \frac{1}{n}(\overline{x}^2 + \cdots + \overline{x}^2),$$

where the last two equalities follow by rearranging the terms in the summation. Now observe that the second term is precisely $-2\overline{x}$ times the average of the x_i, which equals $-2\overline{x}^2$. Similarly, the third term is precisely $n\overline{x}^2/n = \overline{x}^2$. Therefore the variance equals

$$\sigma_X^2 = \left(\frac{1}{n}\sum_{i=1}^{n}x_i^2\right) - 2\overline{x}^2 + \overline{x}^2 = \frac{1}{n}\sum_{i=1}^{n}x_i^2 - \overline{x}^2,$$

as claimed above. The advantage of this formula over the definition is that the variance is easy to compute even by hand: take the average of the observations x_i and also of their squares x_i^2. The first average equals

---

[4]Source: http://www.demographic-research.org/Volumes/Vol2/3/html/2.htm.

the mean $\overline{x}$, and subtracting its square from the second average yields the variance σ_X^2.

EXERCISES

Question 17.1. The exam scores in a class are: $80, 65, 71, 85, 76, 72, 50$. Find the mean, median, and variance of these scores. Would you say that any of the students has significantly over- or underperformed in the test? Also compute the statistical significance of such scores.

Question 17.2. In "The Hunger Games",[5] an organization called the Capitol rules a country and collects taxes from its thirteen constituents called Districts. Suppose that one year, the monetary values of these taxes were (in thousands of coins):

$$100, \ 75, \ 50, \ 40, \ 61, \ 42, \ 65, \ 26, \ 34, \ 29, \ 32, \ 24, \ 0.$$

1. Find the mean and median tax collected per District, as well as the standard deviation of this dataset.

2. Which Districts paid a significantly different tax amount than the others to the Capitol? Are there any values outside the three-sigma limits? (The Capitol may draw its own conclusions about the political loyalties of the underpaying districts.)

   Hint: For this part you do not need to perform thirteen computations; just start from the highest and lowest values and "work your way in" until you cross the significance thresholds of 1 and -1.

Question 17.3. From 1986-1993 and 1995-1998, Michael Jordan played a full season of 78-82 basketball games per year for the Chicago Bulls in the NBA (in fact he played fewer than twenty NBA games for the Bulls in other years). In those ten years, the number of three-pointers he scored per year are given as follows:[6]

$$12, \ 7, \ 27, \ 92, \ 29, \ 27, \ 81, \ 111, \ 111, \ 30.$$

1. Find the mean and median number of three-pointers Jordan scored per year during this time, as well as the standard deviation.

---

[5]Suzanne Collins, *The Hunger Games*, Scholastic Press, 2008.
[6]Source: http://www.basketball-reference.com.

2. Which of these numbers are statistically significant?

Question 17.4. Continuing with the information from Question 17.2, suppose in a later year of much political unrest the tax collections from the thirteen Districts listed in the same order are as follows:

$$80, \ 64, \ 45, \ 41, \ 50, \ 45, \ 59, \ 21, \ 32, \ 30, \ 25, \ 0, \ 0.$$

1. Find the mean and median taxes collected per district, as well as the standard deviation of this dataset.

2. Which Districts paid a significantly different tax amount from the others to the Capitol? Compute their significance. Are there any outside the three-sigma limits?

3. Find the correlation between the tax collections from both years. (*Note:* How much the correlation differs from 1 should give an idea of the effect of political unrest on tax collections.)

Question 17.5. In two midterms of a class, the students have the following scores (listed in alphabetical order by student name):

$$\text{Midterm } 1 : 45, \ 35, \ 34, \ 41, \ 50$$
$$\text{Midterm } 2 : 33, \ 36, \ 40, \ 43, \ 48$$

1. Find the mean and median scores, as well as the standard deviation, for both midterms.

2. Find the correlation between the scores from both midterms.

Question* 17.6. Prove the formula for the covariance, mentioned in this chapter. Namely, if observations $(x_1, y_1), \ldots, (x_n, y_n)$ were made for the variables X and Y, then show that

$$\text{Cov}(X, Y) = \frac{1}{n}(x_1 y_1 + \cdots + x_n y_n) - \overline{x} \cdot \overline{y} = \frac{1}{n} \sum_{i=1}^{n} x_i y_i - \overline{x} \cdot \overline{y}.$$

Chapter 18

Regression: Chasing connections in big data

In the previous chapter, we saw how to obtain measures of central tendency, spread, and association from univariate and bivariate data. In this chapter, we will learn how to model the relationship between two variables based on the available data.

MATH

The main focus of this chapter is *regression*, a method of data analysis that is used to describe how different sets of data are related to one another. It is widely applied in finance and in biological and social sciences. We begin with *simple linear regression*, where observations for two variables X and Y are expected to be (approximately) linearly related. Suppose our collected observations are $(x_1, y_1), (x_2, y_2), \ldots, (x_n, y_n)$. We are interested in fitting the data into a linear relationship:

$$Y = \alpha + \beta X,$$

where α and β are real numbers which we have to select. Most often it is not the case that any one line will pass through all of the data points and fit the data perfectly. Indeed, the only way this could happen is if the relationship between Y and X is exactly linear and the data points are measured without any errors (in which case there is perfect correlation: $\rho_{X,Y} = 1$ or -1). Otherwise we would like to construct a linear relationship (meaning that we have to select α and β) that is the best possible in a precise sense, which we are going to describe.

Figure 18.1: Line of best fit for bivariate data.

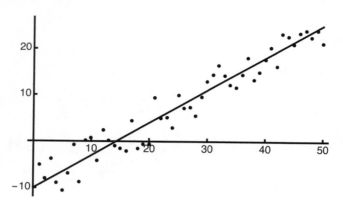

Figure 18.1 depicts a typical dataset with the X-values along the horizontal axis, and the Y-values along the vertical axis. The data points (x_i, y_i) are marked on the graph. It is clear that the line in Figure 18.1 fits the dataset better than, say, the same line moved by 10 units up or down. To make this precise, we say that the *residual errors* – the differences – between the true value y_i and the value $\alpha + \beta x_i$, predicted by the linear model, will increase in magnitude if we move the line up or down or change its slope.

The goal of this chapter is to explain how to obtain the "line of best fit" for bivariate data. We use the method of *least squares*, which ensures that the sum of squares of the residual errors is the least possible. More precisely, suppose we use a linear model $Y = \alpha + \beta X$. Then the predicted value of Y at an observation x_i, denoted by $\widehat{y}_i$, is given by $\widehat{y}_i = \alpha + \beta x_i$. On the other hand, the observed value is y_i. Thus, the residual error ε_i in prediction at x_i equals

$$\varepsilon_i = y_i - \widehat{y}_i = y_i - (\alpha + \beta x_i).$$

The method of least squares aims to choose α and β such that the sum of squares of errors (SSE)

$$SSE = \varepsilon_1^2 + \cdots + \varepsilon_n^2 = \sum_{i=1}^{n}(y_i - \alpha - \beta x_i)^2$$

is minimized.

The method of least squares has several advantages. Most importantly, for all datasets with at least two distinct X-values, there is a choice of α, β such that the minimum is achieved. This choice of α and β is *unique*, and there are simple formulas to compute these values:

$$\beta_{LS} = \frac{\mathrm{Cov}(X,Y)}{\sigma_X^2} = \rho_{X,Y}\frac{\sigma_Y}{\sigma_X} = \frac{\frac{1}{n}\sum_{i=1}^n x_i y_i - \overline{x}\cdot\overline{y}}{\frac{1}{n}\sum_{i=1}^n x_i^2 - \overline{x}^2},$$

$$\alpha_{LS} = \overline{y} - \beta_{LS}\overline{x}.$$

For the definitions of $\mathrm{Cov}(X,Y)$, σ_X, $\rho_{X,Y}$ see Chapter 17. In the above formulas "LS" stands for "least squares." Throughout this chapter we will denote by α_{LS}, β_{LS} the coefficients of the regression model which minimize the sum of squares of errors (SSE). These formulas for α_{LS}, β_{LS} admit a simple verification using only the method of completing the square, discussed in Chapter 1. We will carry out this computation at the end of the chapter and in Question 18.7.

It is also easy to compute the SSE corresponding to $\alpha = \alpha_{LS}, \beta = \beta_{LS}$:

$$SSE_{LS} = n\left(\sigma_Y^2 - \frac{\mathrm{Cov}(X,Y)^2}{\sigma_X^2}\right) = n\left(\sigma_Y^2 - \beta_{LS}^2\sigma_X^2\right).$$

(See Question 18.7 for the computation.) This is the minimum possible value of SSE across all values of α, β.

Moreover, while the sum of the squares of the residual errors achieves its minimum with the least squares approximation, the residual errors themselves sum up to zero. This property of the least squares method will be verified in Question 18.8.

It is remarkable that the method of least squares works not only for a linear model $Y = \alpha + \beta X$, but for *any* model of the form $f(Y) = \alpha + \beta g(X)$ for arbitrary dependences $f(Y)$ and $g(X)$ that assign a real number to every input. Later in this chapter we provide the formulas for α_{LS}, β_{LS}, and SSE_{LS} for general f and g.

The main application of the least squares model is in *prediction* of the anticipated value of one of the correlated variables. Namely, given any value of $X = x$, we predict the value of Y to be: $\widehat{y} = \alpha_{LS} + \beta_{LS}x$. This yields the best possible way to estimate Y based on its correlation with X. In particular, at $X = \overline{x}$, a quick calculation using the definition of α_{LS} shows that the predicted Y-value equals

$$\widehat{y} = \alpha_{LS} + \beta_{LS}\overline{x} = (\overline{y} - \beta_{LS}\overline{x}) + \beta_{LS}\overline{x} = \overline{y}.$$

In other words, the line of best fit passes through the mean values $(\overline{x}, \overline{y})$ of the data.

APPLICATIONS

Applying the method of least squares requires us first to carry out the same computations as in the previous chapter. We now discuss the example of heights and weights from Chapter 17.

Example 18.1. Consider the data presented in Example 17.8. Find the line of best fit for this data for weight as a linear function of height. Compute the sum of squares of errors for this line and for the alternate linear model $W = 3H - 65$.

Solution: We chose this particular alternate linear model because it also predicts values for Y that are close to the actual values in Example 17.8. We will see now how the sums of squares of errors for these two linear models compare.

First, the calculations in Example 17.8 immediately yield:

$$\beta_{LS} = \frac{\mathrm{Cov}(H, W)}{\sigma_H^2} = \frac{62.68}{17.76} \approx 3.529,$$

$$\alpha_{LS} = \overline{w} - \beta_{LS}\overline{h} = 140.6 - 3.529 \cdot 68.2 \approx -100.078.$$

To compute SSE_{LS}, we use the formula stated in the Math section above:

$$SSE_{LS} = n(\sigma_W^2 - \beta_{LS}^2 \sigma_H^2) = 5(235.845 - 3.529^2 \cdot 17.76) \approx 73.3239.$$

We now compute the errors and their sum of squares for the alternate linear model $W = 3H - 65$. In the same table, we also carry out the same computation for the least squares model.

The table in Figure 18.2 verifies that the residual least squares errors add up to zero, as claimed. It also verifies that the sum of squares of errors for the alternate hypothesis ($= 103$) is indeed larger than the sum of squares of least squares errors ($= 73.123$). We point out that the value of SSE_{LS} obtained in the table is 73.123, and it agrees with the value of $SSE_{LS} = 73.3239$ obtained by the formula to the nearest integer. (The discrepancy is an error of computation: if we wanted our estimates to agree to more decimal places, we should have used more decimal places in all of our computations.) □

Figure 18.2: Errors for the least squares and the alternate linear model.

Index	Height (inches)	Weight (pounds)	$\alpha_{LS} + \beta_{LS}h_i$	$\varepsilon_i =$ LS-error	ε_i^2	$3h_i - 65$	$\delta_i =$ error	δ_i^2
1	70	150	146.952	3.048	9.290	145	5	25
2	74	164	161.068	2.932	8.597	157	7	49
3	62	121	118.720	2.280	5.198	121	0	0
4	65	128	129.307	−1.307	1.708	130	−2	4
5	70	140	146.952	−6.952	48.330	145	−5	25
Total				0	73.123		5	103

Fitting other curves

The method of least squares is generally applicable to more than merely linear regressions.

Example 18.2. A certain college found that the number of student organizations increased with the student populations, as follows:

# Students (thousands) S	8	10	11	12	15
# Organizations O	20	31	39	50	72

To predict the number of student organizations as it depends on the number of students, the college chose to use the following quadratic model: $O = \alpha + \beta S^2$.

The next year's incoming class size would make the total student population $16,000$. For how many student organizations should the college plan to allocate its resources? Also compare the sum of squares of errors for this estimate with that for $O = S^2/3 = 0 + \frac{1}{3}S^2$.

Solution: For this problem, we cannot directly use linear regression. Instead, we first introduce a new variable $V = S^2$, and also define $v_i = s_i^2$. Then the regression model under consideration is $O = \alpha + \beta S^2 = \alpha + \beta V$; i.e., we have converted the problem into a linear regression problem. Therefore from above, the least squares coefficients and sum of squares of least squares errors are given by:

$$\alpha_{LS} = \bar{o} - \beta_{LS} \cdot \bar{v}, \qquad \beta_{LS} = \frac{\text{Cov}(V, O)}{\sigma_V^2}, \qquad SSE_{LS} = n\left(\sigma_O^2 - \beta_{LS}^2 \sigma_V^2\right).$$

Now we write out a table with everything needed to carry out the data analysis:

Index (i)	s_i	o_i	$v_i = s_i^2$	v_i^2	o_i^2	$v_i o_i$	$\varepsilon_i = o_i - s_i^2/3$	ε_i^2
1	8	20	64	4,096	400	1,280	-1.33	1.77
2	10	31	100	10,000	961	3,100	-2.33	5.44
3	11	39	121	14,641	1,521	4,719	-1.33	1.77
4	12	50	144	20,736	2,500	7,200	2	4
5	15	72	225	50,625	5,184	16,200	-3	9
Total		212	654	100,098	10,566	32,499		22

Using the table, we now compute:

$$\bar{v} = \frac{1}{5}\sum_{i=1}^{5} v_i = \frac{654}{5} = 130.8,$$

$$\bar{o} = \frac{1}{5}\sum_{i=1}^{5} o_i = \frac{212}{5} = 42.4,$$

$$\sigma_V^2 = \frac{1}{5}\sum_{i=1}^{5} v_i^2 - (\bar{v})^2 = \frac{100,098}{5} - 130.8^2 = 2,910.96,$$

$$\sigma_O^2 = \frac{1}{5}\sum_{i=1}^{5} o_i^2 - \bar{o}^2 = \frac{10,566}{5} - 42.4^2 = 315.44,$$

$$\mathrm{Cov}(V,O) = \frac{1}{5}\sum_{i=1}^{5} v_i o_i - \bar{v}\cdot\bar{o} = \frac{32,499}{5} - 130.8 \cdot 42.4 = 953.88,$$

$$\beta_{LS} = \frac{\mathrm{Cov}(V,O)}{\sigma_V^2} = \frac{953.88}{2,910.96} \approx 0.3277,$$

$$\alpha_{LS} = \bar{o} - \beta_{LS}\bar{v} = 42.4 - 0.3277 \cdot 130.8 = -0.46316,$$

$$SSE_{LS} = n(\sigma_O^2 - \beta_{LS}^2\sigma_V^2) = 5(315.44 - 0.3277^2 \cdot 2,910.96) = 14.199.$$

Thus we conclude that (a) the least squares quadratic model is

$$O = 0.3277V - 0.46316 = 0.3277S^2 - 0.46316,$$

(b) the competing model is $O = 0.3333S^2 = S^2/3$, which is reasonably close to the least squares model, (c) the sum of squares of errors for the competing model is 22, which is strictly greater than the least squares model $SSE_{LS} = 14.199$, as expected.

Finally, we compute the next year prediction using the least squares model. Suppose the student population is to be $16,000$, so that $s = 16$. Then the predicted value using the least squares model is

$$\hat{o} = 0.3277v - 0.46316 = 0.3277s^2 - 0.46316 = 83.428 \approx 83.$$

Thus, the college should plan for about 83 student organizations in the coming year.   □

Derivation of the least squares coefficients*

We end this chapter by deriving the formulas for the least squares coefficients for the linear regression model $Y = \alpha + \beta X$. Following the main idea of the least squares method, we want to choose α_{LS} and β_{LS} so as to minimize the sum of squares of errors

$$SSE = \sum_{i=1}^{n} \varepsilon_i^2 = \sum_{i=1}^{n} (y_i - \alpha - \beta x_i)^2.$$

As mentioned earlier in this chapter, the only mathematical tool we will need is completing the square. To apply this tool, we first use the distributive law and rearrange the terms in the summation:

$$SSE = \sum_{i=1}^{n} ((y_i - \beta x_i) - \alpha)^2 = \sum_{i=1}^{n} (y_i - \beta x_i)^2 + \alpha^2 - 2\alpha(y_i - \beta x_i)$$

$$= n\alpha^2 - 2\alpha \sum_{i=1}^{n} (y_i - \beta x_i) + \sum_{i=1}^{n} (y_i - \beta x_i)^2.$$

Now recall that $y_1 + \cdots + y_n = n\overline{y}$, and similarly, $x_1 + \cdots + x_n = n\overline{x}$. Therefore the SSE in the equation above simplifies to

$$SSE = n\alpha^2 - 2\alpha \cdot (n\overline{y} - \beta \cdot n\overline{x}) + \sum_{i=1}^{n} (y_i - \beta x_i)^2$$

$$= n\left(\alpha^2 - 2\alpha(\overline{y} - \beta\overline{x})\right) + \sum_{i=1}^{n} (y_i - \beta x_i)^2.$$

The first two terms are part of a squared expression, and to complete the square we add and subtract $n(\overline{y} - \beta\overline{x})^2$, to get:

$$SSE = n\left(\alpha^2 - 2\alpha(\overline{y} - \beta\overline{x})\right) + n(\overline{y} - \beta\overline{x})^2 - n(\overline{y} - \beta\overline{x})^2 + \sum_{i=1}^{n} (y_i - \beta x_i)^2$$

$$= n\left(\alpha - (\overline{y} - \beta\overline{x})\right)^2 + A_\beta,$$

where $A_\beta = \sum_{i=1}^{n} (y_i - \beta x_i)^2 - n(\overline{y} - \beta\overline{x})^2$ is formed by the last two terms, and does not depend on α. Therefore if we want to choose α such that SSE is minimized, then the first term on the right-hand side, i.e., $(\alpha - (\overline{y} - \beta\overline{x}))^2$, should be as small as possible. Because it is a perfect square, the smallest it can get is zero, which is achieved precisely when $\alpha_{LS} = \overline{y} - \beta\overline{x}$. Therefore, the least squares coefficients must be related by the formula

$$\alpha_{LS} = \overline{y} - \beta_{LS}\overline{x},$$

as claimed.

To find β_{LS} and complete the derivation of the formulas, one has to consider the value A_β, and find the value of β that minimizes it. This analysis, together with the computation of SSE_{LS}, is reserved for the exercises; see Question 18.7.

Finally, we consider the general regression model

$$f(Y) = \alpha + \beta g(X),$$

where $f(Y)$ and $g(X)$ are some fixed rules assigning a number to each given input. For instance, choosing $f(Y) = Y$ and $g(X) = X$ yields the usual simple linear regression model, while $f(Y) = Y$ and $g(X) = X^2$ was considered in Example 18.2. However, f and g can be any rules: for instance, we may also decide to use $g(X) = e^X$ or $g(X) = \log X$ if the data suggests an exponential or logarithmic dependency between Y and X. We will indeed see such an example in Question 18.4.

Having assumed such a model, and given bivariate data in the form of observations $(x_1, y_1), \ldots, (x_n, y_n)$, we introduce new variables $W = f(Y)$ and $V = g(X)$. Correspondingly, let $w_i = f(y_i)$ and $v_i = g(x_i)$. Then the model for X, Y translates into a simple linear regression model:

$$W = \alpha + \beta V.$$

For this model, we know from above that the least squares coefficients and sum of squares of least squares errors are given by:

$$\beta_{LS} = \frac{\text{Cov}(V, W)}{\sigma_V^2} = \frac{\text{Cov}(g(X), f(Y))}{\sigma_V^2}, \qquad \alpha_{LS} = \overline{w} - \beta_{LS}\overline{v},$$

$$SSE_{LS} = n(\sigma_W^2 - \beta_{LS}^2 \sigma_V^2).$$

Here, $\overline{w} = (w_1 + \cdots + w_n)/n = (f(y_1) + \cdots + f(y_n))/n$ is the average of the quantities $f(y_1), \ldots, f(y_n)$, and σ_W^2 is their variance; similarly for $\overline{v}$ and σ_V^2. In this way we can use the above formulas to obtain least squares coefficients for a large class of regression models.

EXERCISES

Question 18.1. A social network wants to estimate the number of photos uploaded per user on the network. To do this, an analyst counts the number of all uploaded photos on the social network. She finds that this number P is approximately a linear function of the number of users U in the network:

$$P = \alpha + \beta U.$$

Given the following data, find the least squares estimates for α_{LS}, β_{LS}, and SSE_{LS}. Also estimate how many uploaded photos the network will have when the number of users is 20 million.

# Users (millions) U	8	10	11	12	15
# Photos (millions) P	150	170	183	195	231

Question 18.2. Babe Ruth played for six seasons for the Boston Red Sox.[1] In these six years, he played 5, 42, 67, 52, 95, 130 games, respectively, in which he hit 0, 4, 3, 2, 11, 29 home runs, respectively. Assuming that the number of home runs is approximately a linear function of the number of games played, compute the least squares coefficients and sum of squares of errors for this linear regression model. Based only on this model, how many home runs would you estimate Babe Ruth would have hit if he had played 70 games for the Red Sox during some season?

Question 18.3. A scientist placed a certain amount of an unknown radioactive substance in a lead container. When she opened the container ten years later, she estimated that 1 gram of the material remained. Five years later the container contained 0.4 gram of the substance.

1. Estimate both the half-life and initial amount of the radioactive substance. As in Chapter 8, assume that the equation for the amount of radioactive substance is $A(t) = A(0)e^{\ln(1/2)t/T_{1/2}}$, where $T_{1/2}$ is the half-life and t the time since the initial time point. Thus,

$$\ln A(t) = \ln A(0) + \frac{\ln(1/2)}{T_{1/2}}t.$$

This is a linear regression model with unknown coefficients $\alpha = \ln A(0)$ and $\beta = \frac{\ln(1/2)}{T_{1/2}}$, so the goal is to compute the least squares estimates of these quantities, and from them deduce $A(0)$ and $T_{1/2}$.

---

[1]Source: http://www.baseball-reference.com.

2. Estimate the amount of the radioactive substance a further ten years ahead.

3. Verify that the sum of squares of errors for the least squares model, SSE_{LS}, is zero. Can you think of a reason why this should be the case?

Question 18.4. The population of a certain planet is assumed to be growing exponentially. The population in billions, and the number of years since the time it was initially measured, are given in the table below:

# Years since initial measurement t	13	25	32	38
# Population (billions) $P(t)$	5	6	6.5	7

Estimate the population at the time of the initial measurement and 50 years after initial measurement, as well as the relative growth rate. As in the previous question, assume that $P(t) = P(0)e^{kt}$, so that $\ln P(t) = \ln P(0) + kt$. This is a simple linear regression model with $\alpha = \ln P(0)$ and $\beta = k$.

Question 18.5. I have a cubical lump of clay with edge length 1 meter whose weight I would like to determine. Unfortunately, my scale can only accommodate much smaller amounts of clay. I also possess an adjustable metal frame which allows me to cut off small cubes of clay but whose weight itself I do not know. Using this frame, I make small cubes of clay and measure their weights together with the metal frame. Thus my model is

$$W = \alpha + \beta L^3,$$

where W is the weight (in grams) of the cube of clay with an edge of length L. Here, α denotes the weight of the adjustable metal frame (also included in W), and β the density of the clay. My observations are as follows:

Edge length of cube (cm) L	5	7	8	10
Weight of cube (g) W	180	503	750	1,410

Estimate the weight of the metal frame and the density of the clay. Using this data, estimate the weight of the original cubical lump of clay.

Question 18.6. Suppose we are given bivariate data $(x_1, y_1), \ldots, (x_n, y_n)$, for which we are interested in computing the best fitting *horizontal* line. This corresponds to the constant regression model $Y = \alpha$ (regardless of the X-value).

1. Can you guess the value of α that produces the best fitting horizontal line?

2. Assuming that the predicted value of Y is $\widehat{y}_i = \alpha$ for every i, write down an expression for the sum of squares of errors SSE, as it depends on α.

3. Find the value of α that minimizes the expression for SSE. Is it the same as your guess?

Question* 18.7. Consider a general model $f(Y) = \alpha + \beta g(X)$, where f and g are general dependencies and α and β are unknown constants. Denote $W = f(Y)$ and $V = g(X)$; correspondingly, let $w_i = f(y_i)$ and $v_i = g(x_i)$. Then the model for X, Y translates into a simple linear regression model $W = \alpha + \beta V$.

We claimed above that the least squares coefficients and sum of squares of errors are given by:

$$\alpha_{LS} = \overline{w} - \beta_{LS}\overline{v}, \qquad \beta_{LS} = \frac{\text{Cov}(V, W)}{\sigma_V^2}, \qquad SSE_{LS} = n(\sigma_W^2 - \beta_{LS}^2 \sigma_V^2).$$

In the chapter we showed the formula for α_{LS}. Now show that the other two equations hold as well.

Question* 18.8. (Same setting as in the previous question.) Recall that the sum of squares of the errors (SSE) is minimized by this choice of $\alpha = \alpha_{LS}$ and $\beta = \beta_{LS}$. Now show that the sum of the residual errors themselves is zero. In other words, the least squares residual errors cancel each other out:

$$\sum_{i=1}^{n} \varepsilon_i = \sum_{i=1}^{n} (w_i - \alpha_{LS} - \beta_{LS} v_i) = 0.$$

Question* 18.9. (Same setting as in Questions 18.7 and 18.8.) We now discuss what happens when there are only two unequal data points, as in Question 18.3. In this case, show that the line of best fit $W = \alpha + \beta V$ is in fact the unique line that passes through both data points.

Hint: There are at least two ways to solve this problem. One way is to directly verify that at the points v_1 and v_2, the least squares predicted values of W are w_1, w_2, respectively. Another way is to verify that the sum of squares of errors is zero: $SSE_{LS} = \varepsilon_1^2 + \varepsilon_2^2 = 0$. Then both individual residual errors $\varepsilon_1 = w_1 - \widehat{w_1}, \varepsilon_2 = w_2 - \widehat{w_2}$ are zero, i.e., there is no deviation between the predicted value and the true value at both data points.

The bigger story

One of the greatest mathematicians of the twentieth century, Israel Moiseevich Gelfand, once remarked, "When we think about music, we do not divide it into specific areas as we often do in mathematics. If we ask a composer what his profession is, he will answer, 'I am a composer.' He is unlikely to answer, 'I am a composer of quartets.' Maybe this is the reason why when I am asked what kind of mathematics I do, I just answer, 'I am a mathematician.'"[1]

We started the book hoping to provide an accessible overview of several branches of mathematics with an emphasis on their applications. Assuming that the book might be the reader's last experience with mathematics, we tried to cover quite a variety of topics. Now, after having discussed many topics in detail, we would like to emphasize that the diverse chapters you've read are in fact parts of a bigger story.

Even though we tried to keep the exposition light and simple, and stayed mostly at the surface of the mathematical concepts presented in the book, the connections between different chapters appear everywhere. We introduced the number e in Chapter 5 as an outcome of the compound interest model. It appears again as the base of the natural logarithm in Chapters 6, and as the base of the exponential growth and decay models in Chapter 8. The rational numbers, first discussed in Chapter 4, find another incarnation in Chapter 10 as sums of finite or infinite geometric series. Later, in Chapter 11, the geometric series play an important role in explaining how first-order approximations work. The method of completing the square, introduced in Chapter 1, is the main tool in the computation of the regression coefficients in Chapter 18. If we were to attempt a more profound treatment of the subject, the ties between its

---

[1] I.M. Gelfand, *Mathematics as an Adequate Language*, in P. Etingof, V.S. Retakh, I.M. Singer (editors), *The Unity of Mathematics: In Honor of the Ninetieth Birthday of I.M. Gelfand*, Progress in Mathematics, Birkhäuser, 2005.

parts would become even more evident.

Because of the ties, mathematics admits of multiple approaches. You probably noticed that most of the topics considered in this book do not typically appear in high school or introductory college curricula. However, these topics are situated just one step away (in different directions) from high school algebra. None of them, at least at the elementary level, requires any knowledge of calculus.

The tradition to follow elementary algebra with some planar geometry, trigonometry, differentiation, and integration, in this order, and never to deviate from this line, was established long ago. It was never seriously challenged because it suits future mathematicians and science and engineering students, who need these technical skills and after mastering them will immediately continue on to more exciting topics. But, as you have just seen, the house of mathematics has multiple entrances. You can start with modular arithmetic, probability theory, geometric series, or exponents and logarithms. Even if you do not intend to use mathematics professionally, the subject allows and invites you to glimpse some of its precious treasures.

We made an effort to include many easily approachable topics in the book, but some of the potentially interesting subjects were necessarily left out. Perhaps this book will encourage you to investigate them further.

1. *Eulerian paths and circuits.* Imagine a square with one diagonal. You can easily trace it, passing through each edge exactly once. But if you add another diagonal to the square, such tracing becomes impossible. The graphs that admit tracing while visiting each edge exactly once are called Eulerian. Finding out if a graph is Eulerian might seem nothing but an amusing puzzle (one of the famous examples is the problem of finding a path that goes exactly once over each of the seven bridges of Königsberg), but in fact these questions lead to important theorems in graph theory. A good starting point is *Introductory Graph Theory*, by Gary Chartrand, which contains many accessible exercises.[2]

2. *Elements of knot theory.* Tying knots can be practiced with a piece of string, and yet the study and classification of knots in mathematics gives rise to many deep connections with graph theory, abstract algebra, and more advanced topics. There are many introductory texts

---

[2]Gary Chartrand, *Introductory Graph Theory*, Dover Books on Mathematics, 1984.

on the subject, for example, *The Knot Book* by Colin C. Adams, which assumes no further knowledge beyond high school algebra.[3]

3. *Compact surfaces and their classification.* A football and a doughnut are real-world examples of compact surfaces that are clearly different, while a mug with a handle is roughly the same as a doughnut (each has one handle). If you start to think about the difference between a football and a doughnut, and how to formalize it, you arrive at the classification of closed connected surfaces, the topic that gave rise to modern algebraic topology. For an introduction to the world of surfaces, see, for example, *The Shape of Space* by Jeffrey R. Weeks.[4]

4. *Symmetry on the plane, in space, and beyond.* A square is symmetric with respect to certain reflections and rotations (how many?), and so is a cube. In general, symmetric objects on a plane and in space are related to another important notion, that of a group. For a detailed discussion of the properties of the regular polyhedra (cube, tetrahedron, etc.), see *Euler's Gem: The Polyhedron Formula and the Birth of Topology*, by David S. Richeson.[5] An excellent introduction to the ideas of symmetry is *The Symmetries of Things*, by John H. Conway, Heidi Burgiel, and Chaim Goodman-Strauss.[6] An accessible essay on symmetry and its relation to algebra can be found in *The Equation That Couldn't Be Solved: How Mathematical Genius Discovered the Language of Symmetry*, by Mario Livio.[7]

Readers with a taste for more abstract and technical methods might consider continuing their studies in the direction of calculus of real-valued functions, or in the direction of linear algebra. Mastering these subjects, while not necessary for an initial acquaintance with mathematics, lays a solid foundation for a potential professional study of it.

Even with this discussion of the role of mathematics in the real world, some readers may still wonder, is mathematics real? Does it exist in any sense other than a meaningless game? Some of the undeniable proofs of

---

[3] Colin C. Adams, *The Knot Book: An Elementary Introduction to the Mathematical Theory of Knots*, American Mathematical Society, 1994.

[4] Jeffrey R. Weeks, *The Shape of Space*, Marcel Dekker, New York, 1985.

[5] David S. Richeson, *Euler's Gem: The Polyhedron Formula and the Birth of Topology*, Princeton University Press, 2008.

[6] John H. Conway, Heidi Burgiel, Chaim Goodman-Strauss, *The Symmetries of Things*, A.K. Peters, 2008.

[7] Mario Livio, *The Equation That Couldn't Be Solved: How Mathematical Genius Discovered the Language of Symmetry*, Simon & Schuster, 2006.

the presence of mathematics in reality are in the patterns that we see in nature: the golden ratio in plants; the elliptic orbits of planets; the fractal nature of blood vessels, coastlines, and trees; and many other examples that we cite in the book, as well as many more that we were not able to include. One extreme viewpoint shared by some mathematicians is that *the world is mathematics* in the sense that the physics of the universe, which preceded all existence and will continue beyond all collapse, is governed by mathematical laws and equations.

Mathematics is also present, though still somewhat isolated, in the world of thoughts. As Gelfand has said, "Any really good work in mathematics always has in it: beauty, simplicity, exactness and crazy ideas."[8] This is a combination that also characterizes poetry, music, philosophy, art, and, in fact, all of the greatest achievements of the human mind. As we hope you have seen, mathematics deserves its place in our collective intellectual heritage. Some writers recite theorems in the same way they recite poems. Perhaps you will too. We read in *Life: A User's Manual*, by the French novelist Georges Perec: *"One of the folders is open at a page partly covered with equations written out in a small, fine hand:*

If $f \in \mathrm{Hom}(\eta, \mu)$ (resp. $g \in \mathrm{Hom}(\xi, \eta)$) is a homogeneous morphism whose degree is the

matrix α (resp. β), $f \circ g$ is homogeneous and its degree is the product matrix $\alpha\beta$ ... "[9]

For a mathematician, it is certainly reassuring to read that a mathematical argument should be a part of a user's manual to life.

---

[8] I.M. Gelfand, *Mathematics as an Adequate Language*, in P. Etingof, V.S. Retakh, I.M. Singer (editors), *The Unity of Mathematics: In Honor of the Ninetieth Birthday of I.M. Gelfand*, Progress in Mathematics, Birkhäuser, 2005.

[9] Georges Perec, *Life: A User's Manual*, translated by David Bellos, David R. Godine, 1988.

Solutions to odd-numbered exercises

Chapter 1. Algebra: The art and craft of computation

Question 1.1. Expand the following expressions.

1. $(a - 2)(b - 2) = ab - 2a - 2b + 4$.

2. $(x + 1)(2x - 3) = 2x^2 - x - 3$.

3. $(y^2 + 1)(y^2 - 1) = y^4 - 1$.

Question 1.3. Let us denote the smaller of the two consecutive integers by n; then the larger is $n + 1$. Now the difference between their squares can be computed using the distributive law:

$$(n + 1)^2 - n^2 = (n^2 + n + n + 1) - n^2 = 2n + 1,$$

and as this difference is one more than a multiple of 2, it is odd.

Question 1.5. Solve the following equations.

1. $x^4 = 25$.

   $(x^2)^2 = 5^2$, so $x^2 = \pm 5$. Now there is no real number whose square is -5, so $x^2 = 5$. Therefore $x = \pm\sqrt{5} \simeq \pm 2.236$.

2. $(x - 1)^2(x - 2)(x + 3) = 0$.

   If a product of real numbers is zero, then at least one of them is zero. Therefore, $x - 1 = 0$, or $x - 2 = 0$, or $x + 3 = 0$. Therefore $x = 1, 2$, or -3.

3. $y^2 + 8y + 15 = 0$.

Using the formula for the quadratic equation, we compute:

$$y = \frac{-8 \pm \sqrt{8^2 - 4 \cdot 1 \cdot 15}}{2 \cdot 1} = \frac{-8 \pm \sqrt{4}}{2} = -4 \pm 1 = -5, -3.$$

4. $A^2 - 3A - 4 = 0$.

Using the formula for the quadratic equation, we compute:

$$A = \frac{3 \pm \sqrt{(-3)^2 + 4 \cdot 1 \cdot 4}}{2 \cdot 1} = \frac{3 \pm \sqrt{25}}{2} = -1, 4.$$

Question 1.7. Solving $y^2 = 6y$ by using the usual quadratic formula requires writing it in "standard form" first: $y^2 - 6y + 0 = 0$. Therefore,

$$y = \frac{6 \pm \sqrt{(-6)^2 - 4 \cdot 1 \cdot 0}}{2 \cdot 1} = \frac{6 \pm 6}{2} = 3 \pm 3 = 0, 6.$$

Another way to solve the equation is to factorize: $y^2 - 6y = 0$ means that $y(y - 6) = 0$ using the distributive law. It follows that either $y = 0$ or $y - 6 = 0$. We once again obtain $y = 0$ or 6.

Question 1.9. Solve the quadratic equations.

1. $2\beta^2 + 22\beta + 60 = 0$.

   $2(\beta^2 + 11\beta + 30) = 2(\beta + 5)(\beta + 6) = 0$. Then $\beta = -6, -5$. You can also use the general formula for the solutions of a quadratic equation.

2. $y^2 - 3y + 3 = 0$.

   Discriminant $= (-3)^2 - 4 \cdot 1 \cdot 3 = -3 < 0$, so the equation has no real solutions.

Question 1.11. Evaluate the roots of each of the following equations, or show using the discriminant that the corresponding quadratic equation has no solution.

1. $x^2 - 7x + 2 = 0$.

   Using the formula for the quadratic equation, we compute:

$$x = \frac{7 \pm \sqrt{(-7)^2 - 4 \cdot 1 \cdot 2}}{2 \cdot 1} = \frac{7 \pm \sqrt{41}}{2}.$$

2. $3y^2 - 6y + 12 = 0$.

   The discriminant of this quadratic equation is $b^2 - 4ac = (-6)^2 - 2 \cdot 3 \cdot 12 = 36 - 72 = -36 < 0$. Therefore the quadratic equation has no real roots.

3. $x^2 + 4x - 1 = -5$.

   First rewrite this quadratic equation in standard form: $x^2 + 4x + 4 = 0$. Now using the formula for the quadratic equation, we compute:

   $$x = \frac{-4 \pm \sqrt{4^2 - 4 \cdot 4}}{2 \cdot 1} = \frac{-4}{2} = -2.$$

   Alternatively, verify that $x^2 + 4x + 4 = (x+2)^2$. Therefore the only (repeated) root of this quadratic equation is $x = -2$.

4. $x^4 - 3x^2 + 2 = 0$. *Hint:* Try to solve the equation for $y = x^2$ first.

   We see that setting $x^2 = y$ makes the equation $y^2 - 3y + 2 = (x^2)^2 - 3x^2 + 2 = 0$. We first solve for $y = x^2$ from this equation, using the usual quadratic formula:

   $$y = x^2 = \frac{3 \pm \sqrt{(-3)^2 - 4 \cdot 1 \cdot 2}}{2} = \frac{3 \pm 1}{2} = 1, 2.$$

   Now taking square roots yields: $x = \pm 1, \pm \sqrt{2}$.

Question 1.13. The following equations may not seem like quadratic equations, but each of them can be reduced to one. Solve them.

1. $D^5 - 16D^4 + 60D^3 = 0$. *Hint:* Take out a common factor.

   The factor D^3 is common to all the terms, so we can rewrite the equation as:
   $$D^3(D^2 - 16D + 60) = 0.$$

   Therefore either $D^3 = 0$ (in which case $D = 0$), or $D^2 - 16D + 60 = 0$. This last quadratic equation can be solved using the usual formula:

   $$D^2 - 16D + 60 = 0$$
   $$\implies \quad D = \frac{16 \pm \sqrt{(-16)^2 - 4 \cdot 1 \cdot 60}}{2 \cdot 1} = \frac{16 \pm \sqrt{16}}{2} = 8 \pm 2 = 6, 10.$$

   Therefore the complete set of solutions to the equation is: $D = 0, 6, 10$.

2. $t - 5\sqrt{t} + 6 = 0$. *Hint:* This is a quadratic polynomial – in which variable?

If we set $x = \sqrt{t}$, then the new equation becomes: $x^2 - 5x + 6 = 0$. Now the formula for the quadratic equation yields:

$$x = \frac{5 \pm \sqrt{(-5)^2 - 4 \cdot 1 \cdot 6}}{2 \cdot 1} = \frac{5 \pm 1}{2} = 2, 3.$$

Therefore $x = \sqrt{t} = 2, 3$, so that $t = x^2 = 4, 9$.

Question 1.15. Suppose the product of two successive integers is 30. Denote the lesser of the two numbers by N. Then the other number is $N + 1$, and we are to solve for N in: $N(N + 1) = 30$. Rewriting this equation yields: $N^2 + N - 30 = 0$, which is a quadratic equation. Solving it, we obtain:

$$N = \frac{-1 \pm \sqrt{1^2 - 4 \cdot 1 \cdot (-30)}}{2 \cdot 1} = \frac{-1 \pm \sqrt{121}}{2} = \frac{-1 \pm 11}{2} = -6, 5.$$

The successive integers are $(-6, -5)$ and $(5, 6)$.

Question 1.17. Denote the length of the shorter of the two perpendicular sides of the right triangle by x (in centimeters). Then the side perpendicular to it has length $x + 5$, and using the formula for the area, we obtain:

$$\frac{1}{2}x(x + 5) = 75.$$

This yields: $x(x + 5) = 150$. Subtracting 150 from both sides yields a quadratic equation: $x^2 + 5x - 150 = 0$. Solving this equation, we obtain:

$$x = \frac{-5 \pm \sqrt{5^2 - 4 \cdot 1 \cdot (-150)}}{2 \cdot 1} = \frac{-5 \pm \sqrt{625}}{2} = \frac{-5 \pm 25}{2} = -15, 10.$$

Because we are dealing with a physical quantity (the side length in a right triangle), x cannot be negative. Therefore we obtain the answer for the shorter side: $x = 10$ cm. Then the longer side is 15 centimeters.

Question 1.19. Here is a "magic trick": We claim that if one squares any odd number and subtracts 1, then the difference is always a multiple of 8.

What makes this "trick" work? Let us write an odd number as 1 plus an even number, e.g., $2n + 1$, where n is some integer. In this case, we obtain:

$$(2n + 1)^2 - 1 = 4n^2 + 4n + 1 - 1 = 4n^2 + 4n = 4n(n + 1) = 8 \cdot \frac{n(n + 1)}{2}.$$

Now we observe that if n is odd, then $n + 1$ is even, and vice versa. Therefore $n(n + 1)$ is always an even number, so that $n(n + 1)/2$ is also an integer. We conclude that $(2n + 1)^2 - 1$ is eight times an integer.

Chapter 2. Velocity: On the road

Question 2.1.

1. Because the villain is 120 miles away from the border, and traveling at a speed of 80 miles per hour, he will take $120/80 = 1.5$ h to reach the border. Having started at noon, he will reach the border at 1:30 pm. James Bond must therefore reach the same border by 1:30 pm. Now Bond will cover 140 miles at the speed of 100 miles per hour in $140/100 = 1.4$ h – which is precisely 84 minutes. Therefore he must start, at the latest, 84 minutes prior to 1:30 pm – namely, at 12:06 pm, six minutes after the villain starts driving.

2. If James Bond leaves Spyburg at 12:30 pm, he has to travel 140 miles by 1:30 pm (by the calculations in the previous part), i.e., in exactly one hour. Therefore his average speed must be at least 140 miles per hour.

Question 2.3.

1. Let us compute the time it takes for cars A and B to meet, and the time it takes for cars B and C to meet. (This automatically answers the third part as well as the first part.)

   Because car A is traveling 15 miles per hour faster than car B, it covers 15 miles more than car B per hour of travel. Because car B is only $5 = 15/3$ mi ahead, car A will cover this gap in $1/3$ hour, i.e., in 20 minutes.

   Next, cars B and C are traveling toward each other at a relative speed of 100 miles per hour. Therefore 100 miles of distance between the cars would be reduced each hour. As there are only $30 = 100 \cdot (3/10)$ mi of distance between the cars, it takes $3/10$ hour to cover this distance – i.e., 18 minutes before cars B and C meet.

   Therefore car B meets car C first.

2. After 18 min $= 3/10$ h of travel, because car A is traveling 15 miles per hour faster than car B, it has covered $15 \cdot 3/10 = 4.5$ mi more than

car B. As the initial gap was 5 miles, car A is now only $5 - 4.5 = 0.5$ mi behind car B.

3. By the calculations in the first part, the time between the meeting of car B with the other two cars is 2 minutes.

Question 2.5. As in Example 2.3, we only need to find the first meeting time. Beyond that time, the bees together have to cover 60 feet before they meet again, exactly as discussed in Example 2.3. Because the bees have a relative speed of 5 feet per second, successive meeting times occur 12 seconds apart.

To compute when the bees meet for the first time, note that at 9:00:05, Yolanda is starting from the western wall while Zoe has already flown for 5 seconds. Therefore, Zoe is 15 feet away from Yolanda at 9:00:05. As the relative speed of the two bees is 5 feet per second, it takes three more seconds for them to meet for the first time. Therefore, the first three meeting times are 9:00:08, 9:00:20, 9:00:32.

Question 2.7. Suppose the traveler is walking at a constant speed of v kilometers per hour (or replace kilometers with a unit of distance of your choice); therefore in one hour he has walked v kilometers. Now suppose the pigeon flies for t hours before catching up with the traveler. In that time, the traveler has walked vt kilometers, while the pigeon has flown a total of $v + v + vt$ kilometers (to Pigeonville and back from there). Moreover, this quantity, which equals $v(2 + t)$, must also equal $(8v/5)t$ kilometers, because the pigeon flies at $8v/5$ kilometers per hour. Equating the two distances, we obtain:

$$v(2 + t) = 8vt/5 \quad \Longrightarrow \quad 2 + t = 8t/5 \quad \Longrightarrow \quad 2 = t\left(\frac{8}{5} - 1\right) = 3t/5.$$

Therefore the pigeon takes $t = 2 \cdot 5/3 = 10/3$ h, i.e., 3 hours and 20 minutes, to catch up with the traveler.

Question 2.9. The same analysis as in Question 2.7 yields:

$$v(2 + t) = kvt \quad \Longrightarrow \quad 2 + t = kt \quad \Longrightarrow \quad t(k - 1) = 2.$$

We conclude that the pigeon catches up with the traveler in $t = 2/(k - 1)$ h, at least when $k > 1$ (i.e., the pigeon is traveling faster than the traveler). Now suppose $k \leq 1$. First, if $k = 1$ then the equation $t(k - 1) = 2$ has no solution in t. If $k < 1$, then $k - 1$ is negative, and we get $t = 2/(k - 1)$ to be a negative number. These results are consistent with the physical reality of the situation: if the pigeon is not traveling strictly faster than the traveler, it can never again catch up with the traveler in any positive (finite) amount of time.

Chapter 3. Acceleration: After the apple falls

Question 3.1. Because the initial height and velocity are $s_0 = 80$ ft and $v_0 = 0$ ft/sec in this question, the equations of motion are: $s(t) = 80 - 16t^2, v(t) = -32t$.

(a) To find the time t at which the stone is 40 feet above the ground, we write $80 - 16t^2 = 40$, so $40 = 16t^2$. This implies $t^2 = 40/16 = 5/2 = 2.5$, so $t = \pm\sqrt{2.5} \simeq \pm 1.58$ sec. Because this is a physical situation, we disregard the negative square root and arrive at $t \simeq 1.58$ sec. At this time, the velocity of the stone is $-32 \cdot t \simeq -50.56$ ft/sec.

(b) To find the time t at which the stone hits the ground, we write $80 - 16t^2 = 0$. Then $t^2 = 5$, so $t = \pm\sqrt{5} \simeq 2.24$ sec (again disregarding the negative square root). At this time, the velocity of the stone is $-32 \cdot t \simeq -71.68$ ft/sec.

(c) Although the distance in part (b) is twice that covered by the stone in part (a), the time taken to cover the distance in part (b) is less than twice the time taken in part (a). This can be explained by the fact that the stone is falling faster and faster because of the acceleration due to gravity, so it takes less time to travel successive distances of 40 feet.

Question 3.3. In order to compute the maximum height, we can use the formula $s_{\max} = v_{\mathrm{gr}}^2/2g$, derived in this chapter, where v_{gr} is the velocity as the ball takes off from the ground. Therefore, we need to compute the velocities when the ball takes off after the first and second impacts; these are four-fifths of the velocities at the first and second impacts.

Let us compute the velocity of the ball at the time of the first impact with the ground (in meters per second). The equations of motion involve initial height and velocity equal to $s_0 = 5$ m, $v_0 = -5$ m/sec. In these units, $g = 9.8$ m/sec^2. We first compute the time t_0 of first impact using: $s(t_0) = 0$. Therefore,

$$0 = s(t_0) = 5 - 5t_0 - 4.9t_0^2 \quad \Longrightarrow \quad 4.9t_0^2 + 5t_0 - 5 = 0.$$

Solving this quadratic equation for t_0 yields:

$$t_0 = \frac{-5 \pm \sqrt{5^2 - 4 \cdot 4.9 \cdot (-5)}}{2 \cdot 4.9} = \frac{-5 \pm \sqrt{123}}{9.8}.$$

Because t_0 is positive in this problem, $t_0 = (\sqrt{123} - 5)/9.8$ m/sec.

We can now compute the velocity at the time of first impact:

$$v(t_0) = v_0 - 9.8t_0 = -5 - 9.8(\sqrt{123} - 5)/9.8 = -5 - \sqrt{123} + 5 = -\sqrt{123}.$$

When the ball takes off from the ground the first time, its velocity is therefore $v_1 = -4v(t_0)/5 = 4\sqrt{123}/5$ m/sec.

The maximum height attained by the ball after the first impact is

$$s_1 = \frac{v_1^2}{2g} = \frac{4^2 \cdot 123}{5^2 \cdot 2 \cdot 9.8} \simeq 4.016 \text{ m}.$$

To compute the second maximum height, note that the downward velocity at the time of second impact is exactly the reverse of v_1, i.e., $-4\sqrt{123}/5$ m/sec. Therefore the velocity as the ball takes off the second time is $v_2 = (4/5) \cdot v_1 = 4^2\sqrt{123}/5^2 = 16\sqrt{123}/25$. Then the second maximum height is

$$s_2 = \frac{v_2^2}{2g} = \frac{16^2 \cdot 123}{25^2 \cdot 2 \cdot 9.8} \simeq 2.57 \text{ m}.$$

Question 3.5. Solve Example 3.9 with the assumption that the balls meet at seven-eighths of the initial height of the yellow ball.

As in the solution to Example 3.9, let us denote by $s_y(t)$ and $s_w(t)$ the position of the yellow and the white balls, respectively, as they change with time. Then we have

$$s_y(t) = s_0 - \frac{1}{2}gt^2, \qquad s_w(t) = v_0t - \frac{1}{2}gt^2,$$

where $s_0 = 5$ m is the initial height of the yellow ball, and v_0 is the unknown initial velocity of the white one. At the moment of the meeting, t_{meet}, the balls have the same vertical position:

$$s_0 - \frac{1}{2}gt_{\text{meet}}^2 = v_0t_{\text{meet}} - \frac{1}{2}gt_{\text{meet}}^2.$$

Then we have

$$s_0 = v_0t_{\text{meet}} \quad \Longrightarrow \quad t_{\text{meet}} = \frac{s_0}{v_0}.$$

(See Example 3.9 for why gravity does not seemingly play a role in the previous computations.) We now compute the position of their meeting s_{meet} with respect to the ground level:

$$s_{\text{meet}} = s_y(t_{\text{meet}}) = s_0 - \frac{1}{2}gt_{\text{meet}}^2 = s_0 - \frac{g}{2} \cdot \left(\frac{s_0}{v_0}\right)^2.$$

We are given that this height equals $7s_0/8$. Then

$$s_{\text{meet}} = \frac{7}{8}s_0 = s_0 - \frac{g}{2} \cdot \left(\frac{s_0}{v_0}\right)^2 \quad \Longrightarrow \quad \frac{g}{2} \cdot \left(\frac{s_0}{v_0}\right)^2 = \frac{1}{8}s_0 \quad \Longrightarrow \quad \frac{s_0}{v_0^2} = \frac{1}{4g}.$$

This leads to

$$v_0^2 = s_0 \cdot 4g \quad \Longrightarrow \quad v_0 = \sqrt{s_0 \cdot 4g} = \sqrt{5 \cdot 39.2} = \sqrt{196} = 14,$$

where we have discarded the negative answer for the velocity. The white ball was launched upward at a velocity of 14 meters per second.

Question 3.7. Let us start measuring time from the moment of impact with the barrier. At this time, the initial height and velocity are s_0 feet and 150 feet per second, respectively. After eight seconds, the equations of motion yield:

$$s(8) = s_0 + 150 \cdot 8 - \frac{1}{2}g \cdot 8^2 = s_0 + 1,200 - \frac{32}{2} \cdot 64 = s_0 + 1,200 - 1,024$$

$$= s_0 + 176.$$

Because the cannonball is given to be 226 feet high at this time, we obtain that $226 = s(8) = s_0 + 176$. Therefore $s_0 = 226 - 176 = 50$ ft.

We now compute the times at which the cannonball is 226 feet high:

$$226 = s(t) = s_0 + v_0 t - \frac{1}{2}gt^2 = 50 + 150t - 16t^2.$$

Solving the quadratic equation $16t^2 - 150t + 176 = 0$, we get:

$$t = \frac{150 \pm \sqrt{(-150)^2 - 4 \cdot 16 \cdot 176}}{2 \cdot 16} = \frac{150 \pm \sqrt{11,236}}{32} = \frac{150 \pm 106}{32}$$

$$= \frac{11}{8}, 8 \text{ sec.}$$

Therefore, the given information corresponds to the cannonball being at 226 feet and falling downward; it was at this height previously on its way up. Finally, the velocities at these times can also be computed using the equations of motion:

$$v(11/8) = v_0 - g \cdot \frac{11}{8} = 150 - \frac{32 \cdot 11}{8} = 150 - 44 = 106 \text{ ft/sec,}$$

$$v(8) = v_0 - g \cdot 8 = 150 - 32 \cdot 8 = 150 - 256 = -106 \text{ ft/sec.}$$

Question 3.9. We may assume that the water at the top of the falls has zero vertical velocity at the moment it starts to fall. Let us compute the time t_0 it takes for the water to fall down (disregarding the effects of air resistance). Because Niagara Falls is about 165 feet high, we obtain:

$$s(t_0) = -165 = -\frac{1}{2}gt_0^2 = -16t_0^2.$$

Therefore $t_0 = \sqrt{165/16} = \sqrt{165}/4$ sec. Then the vertical velocity of the water stream at the bottom is given by

$$v(t_0) = -gt_0 = -32\sqrt{165}/4 = -8\sqrt{165} \simeq -102.76 \text{ ft/sec}.$$

The answer is negative because the water is traveling downward.

Question 3.11. The stopping distance as a function of the deceleration a_{car} of the car and the velocity v_0 at the time of braking is: $SD = v_0^2/(2a_{\mathrm{car}})$. Therefore, we compute for the first car:

$$SD = \frac{90^2}{2 \cdot 10} = 405 \text{ ft,}$$

and for the second car:

$$SD = \frac{80^2}{2 \cdot 7} \simeq 457.14 \text{ ft.}$$

The second car travels farther before it stops.

Chapter 4. Irrational: The golden mean and other roots

Question 4.1. Which of the following numbers are irrational?

1. $\dfrac{3\sqrt{2} - 2}{5}$.

   This number is irrational. Suppose to the contrary that $y = (3\sqrt{2} - 2)/5$ was rational, say, $y = p/q$ for some integers p and q. Then $5y + 2 = 5(p/q) + 2 = (5p + 2q)/q$ is also rational, and hence $(5y + 2)/3 = (5p + 2q)/(3q)$ is also rational. But then the number

   $$\frac{5y + 2}{3} = \frac{(3\sqrt{2} - 2) + 2}{3} = \sqrt{2}$$

should also be rational, but it is known to be irrational. This is a contradiction, so our initial supposition that y is rational must be false. It follows that $y = (3\sqrt{2} - 2)/5$ is irrational.

2. $\dfrac{3\sqrt{\frac{9}{4}} - 2}{5}.$

This number is rational: $\dfrac{3\sqrt{\frac{9}{4}} - 2}{5} = \dfrac{3 \cdot (3/2) - 2}{5} = \dfrac{9/2 - 2}{5} = \dfrac{5/2}{5} = \dfrac{1}{2}.$

Question 4.3. If x and y are two distinct irrational numbers, their product can be a rational number. For instance, let $x = \sqrt{2} \simeq 1.414$ and $y = 1/\sqrt{2} \simeq 0.707$. Then $x \cdot y = 1$ is rational.

Question 4.5.   1. What is the length of the $B2$ size?

As mentioned in the chapter, $l(A1) = l(A0)/\sqrt{2} = \sqrt[4]{2}/\sqrt{2}$ and $l(A2) = l(A0)/2 = \sqrt[4]{2}/2$. Therefore,

$$l(B2) = \sqrt{l(A1) \cdot l(A2)} = \sqrt{\frac{\sqrt[4]{2} \cdot \sqrt[4]{2}}{\sqrt{2} \cdot 2}} = \sqrt{\frac{\sqrt{2}}{2\sqrt{2}}} = \sqrt{1/2} = 1/\sqrt{2}$$

$\simeq 0.707$ m.

2. What is the area of the $B3$ size?

As in the previous part (and the chapter), $l(A2) = \sqrt[4]{2}/2$ and $l(A3) = l(A2)/\sqrt{2} = \sqrt[4]{2}/(2\sqrt{2})$. Now the length and width of $B3$ are

$$l(B3) = \sqrt{l(A2) \cdot l(A3)}, \qquad w(B3) = \frac{l(B3)}{\sqrt{2}} = \frac{\sqrt{l(A2) \cdot l(A3)}}{\sqrt{2}}.$$

Therefore the area of the $B3$ size is

$$l(B3) \cdot w(B3) = \frac{l(A2) \cdot l(A3)}{\sqrt{2}} = \frac{(\sqrt[4]{2}/2) \cdot (\sqrt[4]{2}/(2\sqrt{2}))}{\sqrt{2}}$$

$$= \frac{\sqrt[4]{2} \cdot \sqrt[4]{2}}{2 \cdot 2\sqrt{2} \cdot \sqrt{2}} = \frac{1}{4\sqrt{2}},$$

and this equals (approximately) 0.177 m^2.

3. What is the reduction factor between the lengths of the $A4$ size and the $B6$ size?

From the chapter, $l(A5) = \dfrac{l(A4)}{\sqrt{2}}$ and $l(A6) = \dfrac{l(A5)}{\sqrt{2}} = \dfrac{l(A4)}{2}$.

Therefore,

$$l(B6) = \sqrt{l(A5) \cdot l(A6)} = \sqrt{\frac{l(A4) \cdot l(A4)}{\sqrt{2} \cdot 2}} = \sqrt{\frac{\sqrt{2} \cdot l(A4)^2}{\sqrt{2} \cdot \sqrt{2} \cdot 2}}$$

$$= \frac{l(A4)\sqrt{\sqrt{2}}}{\sqrt{4}} = \frac{l(A4)\sqrt[4]{2}}{2}.$$

We conclude that the reduction factor is $l(B6)/l(A4) = \sqrt[4]{2}/2 \simeq 0.594$.

Question 4.7.

1. We are to show that the new paper sizes $B'n$ satisfy Lichtenberg conditions (1) and (2). Note that (2) is exactly the same as (2b'), so it remains to show that Lichtenberg condition (1) holds. To see why this holds, recall that $l(An)/\sqrt{2} = l(A(n+1))$ for all $n \geq 0$. Because the width of $B'n$ is $1/\sqrt{2}$ times $l(B'n)$, we compute:

$$w(B'n) = \frac{l(B'n)}{\sqrt{2}} = \frac{l(A(n-1)) + l(An)}{2\sqrt{2}} = \frac{\frac{l(A(n-1))}{\sqrt{2}} + \frac{l(An)}{\sqrt{2}}}{2}$$

$$= \frac{l(An) + l(A(n+1))}{2} = l(B'(n+1)).$$

It follows that $l(B'n)/l(B'(n+1)) = \sqrt{2}$, so that if we cut a $B'n$-paper in half across the longer side, each of the two sheets has length $l(B'(n+1)) = l(B'n)/\sqrt{2}$, and width $l(B'n)/2 = \frac{l(B'n)/\sqrt{2}}{\sqrt{2}} = l(B'(n+1))/\sqrt{2}$. But these are precisely the dimensions of a paper of size $B'(n+1)$, as desired.

2. Using the above information, we compute:

$$l(B'4) = \frac{l(A3) + l(A4)}{2} = \frac{\sqrt{2}l(A4) + l(A4)}{2} = l(A4) \cdot \frac{\sqrt{2}+1}{2}$$

$$\simeq 1.207\, l(A4).$$

On the other hand,

$$l(B4) = \sqrt{l(A3) \cdot l(A4)} = \sqrt{\sqrt{2}l(A4) \cdot l(A4)} = \sqrt[4]{2}\sqrt{l(A4)^2}$$
$$= l(A4) \cdot \sqrt[4]{2} \simeq 1.189 \, l(A4).$$

Therefore $l(B'4) > l(B4)$.

3. Given $a, b > 0$, we claim that $\frac{a+b}{2}$ is always at least as large as $\sqrt{ab}$. This is a well-known inequality called the *AM-GM inequality*. Here, "AM" stands for the arithmetic mean, or average $\frac{a+b}{2}$, of two numbers a and b, and "GM" stands for their geometric mean $\sqrt{ab}$.

To see why the AM-GM inequality holds, because $a, b > 0$, we can write $a = A^2, b = B^2$. Then,

$$\frac{a+b}{2} - \sqrt{ab} = \frac{A^2 + B^2}{2} - AB = \frac{1}{2}(A^2 + B^2 - 2AB) = \frac{1}{2}(A - B)^2$$
$$= \frac{1}{2}(\sqrt{a} - \sqrt{b})^2 \geq 0,$$

because every square is nonnegative. It follows that $\dfrac{a+b}{2} \geq \sqrt{ab}$ for all positive numbers a and b. Moreover, the equality is achieved if and only if $\sqrt{a} = \sqrt{b}$. Because $a, b > 0$, this happens if and only if $a = b$.

Question 4.9. To show that $\phi = \sqrt{1 + \sqrt{1 + \sqrt{1 + ...}}}$, let us denote the right-hand side by x. Then we compute:

$$x^2 = 1 + \sqrt{1 + \sqrt{1 + ...}} = 1 + x.$$

In other words, x satisfies the equation $x^2 - x - 1 = 0$, which is precisely equation (4.1). As discussed in the chapter, this quadratic equation has two solutions, and ϕ is the positive solution among them. The same is true for x, because by definition we always consider the square root to be nonnegative. Therefore, x equals the positive root of equation (4.1), which is precisely ϕ.

Question 4.11. We are to show that

$$\left(\frac{1}{\phi}\right)^n = \left(\frac{1}{\phi}\right)^{n-2} - \left(\frac{1}{\phi}\right)^{n-1}.$$

First note by equation (4.1) that $1 = \phi^2 - \phi$. Now multiply this by $1/\phi^2$ to obtain $(\frac{1}{\phi})^2 = 1 - \frac{1}{\phi}$. Finally, multiply this equation by $(1/\phi)^{n-2}$ for any n to obtain the desired equation.

Question* 4.13.   1. From Question 4.11, we know that

$$\left(\frac{1}{\phi}\right)^2 = \left(1 - \left(\frac{1}{\phi}\right)\right),$$

and so $\left(\frac{1}{\phi}\right)^3 = \left(\frac{1}{\phi}\right) - \left(\frac{1}{\phi}\right)^2 = \left(\frac{1}{\phi}\right) - \left(1 - \left(\frac{1}{\phi}\right)\right) = -1 + 2\left(\frac{1}{\phi}\right)$.
This proves the formula

$$\left(-\frac{1}{\phi}\right)^n = f_{n-1} - f_n\left(\frac{1}{\phi}\right)$$

for $n = 2, 3$. Now suppose the formula holds for some integer n. Multiply both sides by $-\frac{1}{\phi}$:

$$\left(-\frac{1}{\phi}\right)^{n+1} = f_{n-1}\left(-\frac{1}{\phi}\right) + f_n\left(\frac{1}{\phi}\right)^2 = f_{n-1}\left(-\frac{1}{\phi}\right) + f_n\left(1 - \frac{1}{\phi}\right)$$

$$= f_n - \frac{1}{\phi}(f_{n-1} + f_n) = f_n - f_{n+1}\left(\frac{1}{\phi}\right).$$

In the last equality we used the Fibonacci numbers relation, $f_{n+1} = f_n + f_{n-1}$. The equality we obtained is exactly the required formula written for $(n + 1)$ instead of n. Therefore, if we assume that the formula holds for n, it must also hold for the next consecutive integer $(n + 1)$. Because we know that the formula holds for $n = 3$, this implies that it holds for $n = 4, 5, \ldots$, and thus for all natural numbers. The argument we just used is called *mathematical induction*, and it comes in handy when proving statements that involve an arbitrary integer. We will see other applications of the same method in subsequent chapters.

2. Subtracting f_{n-1} from both sides of the identity in part (1), and dividing by f_n, we get

$$\frac{1}{\phi} = \frac{f_{n-1}}{f_n} - \frac{1}{f_n} \cdot \left(-\frac{1}{\phi}\right)^n = \frac{f_{n-1}}{f_n} - \frac{(-1)^n}{f_n} \cdot \frac{1}{\phi^n}.$$

Because $\phi \simeq 1.618$ is greater than 1, the fraction $\frac{1}{\phi^n}$ becomes very small as n grows. Also, because the Fibonacci sequence grows, $\frac{(-1)^n}{f_n}$ becomes close to zero, and the last term of the equation $\frac{(-1)^n}{f_n} \cdot \frac{1}{\phi^n}$ becomes insignificant as n grows. Then we have

$$\frac{f_{n-1}}{f_n} \longrightarrow \frac{1}{\phi} \quad \text{as } n \text{ grows.}$$

Chapter 5. Exponents: How much would you pay for the island of Manhattan?

Question 5.1. Simplify the following expressions. All variables are assumed to take positive values only.

1. $\dfrac{\sqrt[4]{A}}{\sqrt[5]{A}} = \dfrac{A^{1/4}}{A^{1/5}} = A^{(1/4)-(1/5)} = A^{1/20} = \sqrt[20]{A}.$

2.
$$\sqrt[3]{56} \cdot 7^{-\frac{1}{3}} = 56^{1/3} \cdot 7^{-1/3} = (8 \cdot 7)^{1/3} 7^{-1/3} = 8^{1/3} \cdot 7^{1/3} \cdot 7^{-1/3}$$
$$= \sqrt[3]{8} \cdot 7^{1/3-1/3} = 2 \cdot 7^0 = 2.$$

3. $2^{45}/32^9 = 2^{45}/(2^5)^9 = 2^{45}/2^{9 \cdot 5} = 2^{45-45} = 1.$

4. We compute:
$$3abc - \sqrt{ab}\sqrt{bc}\sqrt{ca} = 3abc - (ab)^{1/2}(bc)^{1/2}(ca)^{1/2}$$
$$= 3abc - (a^2b^2c^2)^{1/2},$$

and this equals $2abc$.

Question 5.3. Suppose \$1,000 was borrowed from a bank for five years with an annual compound interest rate of 8%.
(a) If the bank compounds monthly, then the total amount after five years is:
$$1,000 \cdot \left(1 + \frac{8}{1,200}\right)^{12 \cdot 5} \simeq 1,000 \cdot 1.4898457 = 1,489.85.$$

Therefore the interest accumulated is \$489.85.

(b) If the bank compounds annually, then the total amount after five years is:

$$1,000 \cdot \left(1 + \frac{8}{100}\right)^5 \simeq 1,000 \cdot 1.469328 = 1,469.33.$$

Therefore the interest accumulated is $469.33.

Question 5.5. Suppose a deposit of $A = \$20,000$ in a bank that compounds quarterly accumulates \$3,912.36 of interest in three years.

1. Let us denote the annual rate of interest by $r\%$. Then

$$23,912.36 = 20,000\left(1 + \frac{r}{400}\right)^{4 \cdot 3} = 20,000\left(1 + \frac{r}{400}\right)^{12}.$$

In order to solve for r, we first divide both sides by $20,000$ and then take the twelfth root, to get:

$$1 + \frac{r}{400} = (23,912.36/20,000)^{1/12} = 1.195618^{1/12} = 1.015.$$

Solving for r yields: $r/400 = 0.015$, so that $r = 6\%$.

2. At the end of six years, we have:

$$A(6) = A(0)\left(1 + \frac{r}{400}\right)^{4 \cdot 6} = 20,000 \cdot 1.015^{24} = 28,590.056.$$

Therefore the amount in the bank after six years is approximately \$28,590.06.

3. We compute: $A(3)^2/A = 28,590.048$, which is essentially the same as the previous answer (up to rounding errors). We claim that in fact this is mathematically correct. This is because we can compute:

$$\frac{A(6)}{A(0)} = \left(1 + \frac{r}{400}\right)^{4 \cdot 6} = 1.015^{24} = (1.015^{12})^2 = \left(\frac{A(3)}{A(0)}\right)^2.$$

Now multiplying both sides by $A = 20,000$ yields: $A(6) = A(3)^2/A$.

Question 5.7. If \$12,000 is invested for twenty years at an annual rate of 7% compounded monthly, the amount equals

$$12,000\left(1 + \frac{7}{1,200}\right)^{12 \cdot 20} = \$48,464.87.$$

If on the other hand the investment is at an annual rate of 7.15% compounded annually, the amount at the end of 20 years is

$$12,000 \left(1 + \frac{7.15}{100}\right)^{20} = \$47,755.65.$$

The first investment yields a greater return.

Question 5.9. Recall the defining formula for the effective compounding rate:

$$1 + \frac{r_{\text{eff}}}{100} = \left(1 + \frac{r}{n \cdot 100}\right)^{n},$$

where r is the annual rate of interest and n is the number of compoundings per year. Therefore, if one invests A amount of money for t years under the scheme of annual rate r and n compoundings per year, then the amount at the end of t years can be computed as follows, using the properties of exponents:

$$A \left(1 + \frac{r}{n \cdot 100}\right)^{n \cdot t} = A \left(\left(1 + \frac{r}{n \cdot 100}\right)^{n}\right)^{t} = A \left(1 + \frac{r_{\text{eff}}}{100}\right)^{t}.$$

Therefore if we have two different compounding schemes to compare, with effective rates $r_{1,\text{eff}}, r_{2,\text{eff}}$, then investing \$15,000 over three years yields the amounts:

$$15,000 \left(1 + \frac{r_{1,\text{eff}}}{100}\right)^{3}, \qquad 15,000 \left(1 + \frac{r_{2,\text{eff}}}{100}\right)^{3}.$$

It is easy to verify that if $r_{1,\text{eff}} > r_{2,\text{eff}}$, then the first amount above is larger than the second. Therefore, in order to compute which bank yields a greater rate of interest on \$15,000 over 3 years (or on *any* amount over *any* number of years), it is enough to compare the two effective rates. The effective rate for bank A is computed as follows:

$$1 + \frac{r_{\text{A,eff}}}{100} = \left(1 + \frac{6.4}{400}\right)^{4} \simeq 1.065552 \qquad \Longrightarrow \qquad r_{\text{A,eff}} = 6.5552\%.$$

Similarly, the effective rate for bank B is computed as follows:

$$1 + \frac{r_{\text{B,eff}}}{100} = \left(1 + \frac{6}{1200}\right)^{12} \simeq 1.061678 \qquad \Longrightarrow \qquad r_{\text{B,eff}} = 6.1678\%.$$

It follows that bank A has a greater effective annual rate.

Chapter 6. Logarithms I: Money grows on trees, but it takes time

Question 6.1. In what follows, $\log = \log_{10}$. Compute or solve the following:

1. We compute:

$$\log\left(\frac{10^{11}}{100^{3.5}}\right) = \log(10^{11}) - \log(100^{3.5}) = 11 - 3.5\log(100)$$
$$= 11 - 3.5 \cdot 2 = 11 - 7 = 4.$$

2. $\log \sqrt[7]{1,000} = \log(1,000^{1/7}) = \frac{1}{7}\log(1,000) = \frac{3}{7}.$

3. Solve for x: $10^{(7x+1)} = \sqrt[3]{100}$. Taking the log of both sides, we compute:

$$(7x + 1)\log(10) = \log(100^{1/3}) = \frac{1}{3}\log(100)$$
$$\implies \quad 7x + 1 = \frac{2}{3} \quad \implies \quad 7x = \frac{2}{3} - 1 = \frac{-1}{3},$$

which implies that $x = -1/21$.

Question 6.3. Solve the equations:

1. $e^{(2t+1)} = 18$. Taking the natural logarithm of both sides, we compute:

$$\ln 18 = (2t + 1)\ln(e) = 2t + 1$$
$$\implies t = \frac{(\ln 18 - 1)}{2} = \frac{\ln(18) - \ln(e)}{2} = \frac{1}{2}\ln(18/e) = \ln\sqrt{18/e}.$$

2. $(0.5)^{(10x-8)} = 3$. Taking the natural logarithm of both sides, we compute:

$$\ln 3 = (10x - 8)\ln(1/2) \quad \implies \quad 10x - 8 = \frac{\ln 3}{\ln(1/2)}$$
$$\implies \quad 10x = 8 + \frac{\ln 3}{\ln 1/2} \quad \implies \quad x = \frac{8}{10} + \frac{\ln 3}{10\ln 1/2} = \frac{4}{5} - \frac{\ln 3}{10\ln 2}.$$

Question 6.5. In what follows, $\log = \log_{10}$. Solve the equations:

1. $\log(x^2 - 16) - \log(x - 4) = 2$. We compute

$$2 = \log(x^2 - 16) - \log(x - 4) = \log \frac{x^2 - 16}{x - 4} = \log(x + 4).$$

This implies that $x + 4 = 10^2 = 100$, whence $x = 96$. Then we have to verify that both $x^2 - 16$ and $x - 4$ are indeed positive, for the equation to make sense.

2. $\log(x^2 + 5x + 7) = 0$. The given equation can be reformulated as: $x^2 + 5x + 7 = 1$, or equivalently, $x^2 + 5x + 6 = 0$. The quadratic formula now yields:

$$x = \frac{-5 \pm \sqrt{5^2 - 4 \cdot 1 \cdot 6}}{2 \cdot 1} = \frac{-5 \pm 1}{2} = -3, -2.$$

Question 6.7. How long will it take for an investment to triple at an interest rate of 5% compounded:
(a) monthly?
Suppose it takes t months to triple the investment. Assuming A denotes the initial investment, we have:

$$3A = A \left(1 + \frac{5}{1,200} \right)^t = A \left(\frac{1,205}{1,200} \right)^t.$$

Dividing by A and taking the natural logarithm yields: $t \ln \frac{1,205}{1,200} = \ln 3$.
Therefore,

$$t = \ln 3 / \ln \frac{1,205}{1,200} \simeq \frac{\ln 3}{\ln 1.004167} \simeq 264.195 \simeq 264.$$

Therefore it takes 264 months, or 22 years, to triple the investment.

(b) continuously (to the nearest tenth of a year)?
In this case, we have $Ae^{rt/100} = 3A$. Canceling A and taking the natural logarithm yields:

$$\frac{rt}{100} = \ln(3) \quad \Longrightarrow \quad t = \frac{100 \ln(3)}{r} = \frac{100 \cdot 1.0986}{5} \simeq 21.972.$$

To the nearest tenth of a year, this is also equal to 22 years.

Question 6.9. Suppose a bank compounds continuously at an annual interest rate of 3.5%.

1. The time required for the assets to grow from $20,000 to $30,000 is given by:

$$30,000 = 20,000e^{3.5t/100} = 20,000e^{7t/200}.$$

Now divide both sides of the equation by 20,000 and take the natural logarithm:

$$\ln(3/2) = 7t/200 \quad \Longrightarrow \quad t = \frac{200}{7}\ln(3/2) \simeq 11.5847 \text{ yr}.$$

This translates to approximately 139 months, i.e., 11 years and 7 months.

2. The time required for the assets to grow from $30,000 to $40,000 is obtained similarly:

$$t = \frac{200}{7}\ln(4/3) \simeq 8.2195 \text{ yr}.$$

This translates to approximately 99 months, i.e., 8 years and 3 months.

3. Similarly to the analysis in the previous parts, for the time required for the investment to grow from $50,000 to $60,000, we compute:

$$t = \frac{200}{7}\ln(6/5) \simeq 5.2092 \text{ yr}.$$

This translates to approximately 63 months, i.e., 5 years and 3 months.

Question 6.11. Let us count how many seconds there are in three years. This is at most:

$$(365 + 365 + 366) \cdot 24 \cdot 60 \cdot 60 = 94,694,400 \text{ sec}.$$

We would like the adversary to take at least this long to crack the encryption key. Let us round up the number of seconds and say that the adversary should take at least 10^8 seconds to crack the code. The adversary's processor cluster can make 10^{15} tries per second, and hence, at most $10^{15} \cdot 10^8 = 10^{23}$ tries in 3 years. Suppose that the sequence is k bits long. Because the adversary has to check only half of them, we have the equation for k:

$$10^{23} = \frac{1}{2} 2^k = 2^{k-1}.$$

Taking the natural logarithm of both sides, we have

$$23 \ln(10) = (k-1)\ln(2) \quad \Longrightarrow \quad k-1 = \frac{23\ln(10)}{\ln(2)} \simeq 76.404$$

$$\Longrightarrow \quad k \simeq 77.404.$$

If the encryption key is 78 or more bits long, then the adversary will require at least 3 years to crack it by brute force.

Chapter 7. Logarithms II: Rescaling the world

Question 7.1. Use $\log = \log_{10}$ or $\ln = \log_e$, the base change formula, and your calculator to solve the following equations or compute the following expressions.

1. Solve for t: $10^{3t+1} = 80$. Using the properties of exponents, $10^{3t+1} = 10 \cdot 10^{3t} = 80$. Dividing by 10, we get $10^{3t} = 8$. Now taking logarithms,

$$\log 8 = \log 10^{3t} = 3t \quad \Longrightarrow \quad t = \frac{1}{3}\log 8 = \log 8^{1/3} = \log 2.$$

2. $\log_{13.5} 7,300$. By the base change formula, $\log_{13.5} 7,300 = \dfrac{\log 7,300}{\log 13.5}$
   $\simeq 3.41786$.

3. $\log_{\sqrt{e}} 80$. Again by the base change formula, $\log_{\sqrt{e}} 80 = \dfrac{\ln 80}{\ln e^{1/2}} \simeq$
   $\dfrac{4.382}{1/2} = 8.764$.

Question 7.3. Let L_{vc}, P_{vc}, respectively, denote the decibel levels and the sound intensity of the vacuum cleaner, and L_{nv}, P_{nv} the same for a normal voice. Then we have:

$$\frac{P_{vc}}{P_{nv}} = 31.6, \qquad L_{nv} = 10\log(P_{nv}/P_0) = 60.$$

We need to find

$$L_{\text{vc}} = 10\log(P_{\text{vc}}/P_0) = 10\log\left(\frac{P_{\text{vc}}}{P_{\text{nv}}} \cdot \frac{P_{\text{nv}}}{P_0}\right).$$

Using the additive property of logarithms, we obtain

$$L_{\text{vc}} = 10\log\left(\frac{P_{\text{vc}}}{P_{\text{nv}}}\right) + 10\log\left(\frac{P_{\text{nv}}}{P_0}\right) = 10\log 31.6 + L_{\text{nv}}$$

$$\simeq 10 \cdot 1.5 + 60 = 75 \text{ dB}.$$

Question 7.5. Let F_S, m_S, respectively, denote the flux (brightness) and the apparent magnitude of the sun, and F_M, m_M the same for the moon. By the definition of the apparent magnitude,

$$m_S = -\frac{5}{2}\log(F_S/F_0) = -26.74, \qquad m_M = -\frac{5}{2}\log(F_M/F_0) = -12.74.$$

Solving for the ratios of brightness, we have

$$\frac{F_S}{F_0} = 10^{-26.74\cdot(-2/5)} = 10^{10.696}, \qquad \frac{F_M}{F_0} = 10^{-12.74\cdot(-2/5)} = 10^{5.096}.$$

Finally, dividing the first equation by the second yields the desired ratio of brightness:

$$\frac{F_S}{F_M} = 10^{10.696}/10^{5.096} = 10^{10.696-5.096} = 10^{5.6} \simeq 398,107.$$

Therefore the Sun appears to be approximately $398,000$ times as bright as the Moon.

Question 7.7. By the given formula, $M = m - 5\log\frac{d}{10}$, where M and m denote the absolute and apparent magnitudes of the light source, and d the distance from Earth. Therefore, for Sirius we compute:

$$M_{\text{Sirius}} = -1.46 - 5\log 0.26 \simeq 1.465,$$

while for Rigel the absolute magnitude is

$$M_{\text{Rigel}} = 0.12 - 5\log 43 \simeq -8.047,$$

and for Canopus,

$$M_{\text{Canopus}} = -0.72 - 5\log 2.27 \simeq -2.5.$$

Therefore the stars are ranked in increasing order of absolute magnitude (or decreasing order of brightness) as Rigel, then Canopus, and finally Sirius. Sirius appears brighter to us because it is far closer than most stars.

Question 7.9. We use the same equation for the absolute magnitude, substituting $m = 6$ for the apparent magnitude and solving for the distance d. Therefore,

$$-2.7 = 6 - 5 \log \frac{d}{10}.$$

This yields $5 \log(d/10) = 5(-1 + \log d) = 8.7$, and therefore $\log d = 2.74$. It follows that $d = 10^{2.74} \simeq 549.54$ pc.

Question 7.11. Suppose a denotes the base of the desired logarithmic timescale. In order to compute this base, just as for the "egocentric scale" in the chapter, we also need the reference time point x (in days, say) in the past, from which we measure ahead to all given times. Therefore, a given time point X days in the future would be represented in our logarithmic scale by: $\log_a \frac{X + x}{x}$.

For instance, for the present moment one has $X = 0$, so the logarithmic scale value for it is $\log_a(0 + x)/x = \log_a(1) = 0$, as required. For $X = 3$ days and 1 year ahead (which we take to mean 365 days), we compute:

$$1 = \log_a \frac{3 + x}{x}, \qquad 2 = \log_a \frac{365 + x}{x}.$$

Therefore,

$$\log_a \frac{365 + x}{x} - \log_a \frac{3 + x}{x} = \log_a \frac{365 + x}{3 + x} = 2 - 1 = 1.$$

From here we find:

$$\frac{3 + x}{x} = a^1 = a = \frac{365 + x}{3 + x}.$$

Solving this equation for x, we obtain a quadratic equation, which simplifies to a linear equation:

$$(3+x)^2 = x(365+x) \quad \implies \quad x^2 + 6x + 9 = x^2 + 365x \quad \implies \quad 359x = 9.$$

Therefore $x = 9/359$ days in the past, i.e., about 36 minutes and 6 seconds into the past. From this value we now solve for a:

$$a = \frac{3 + x}{x} = \frac{3 + (9/359)}{9/359} = \frac{1,086/359}{9/359} = \frac{1,086}{9} = 120.(6).$$

Question 7.13.

1. Compare the approximations of a perfect fourth interval (ratio of frequencies $\frac{4}{3}$) provided by the 12-tone, 15-tone, and 19-tone equal temperament scales. Let us find the relative error of approximation of a perfect fourth in the 12-tone equal temperament scale. First we need to find the number of the pitch that produces the closest interval. This number is approximated by k such that

$$2^{\frac{k}{12}} = \frac{4}{3}.$$

Taking $\log_2$ of both sides, we have $\frac{k}{12} = \log_2 \frac{4}{3} \simeq 4.98$. Therefore, we should take the sixth pitch (5 steps up from the first pitch). The ratio of frequencies between the first and the sixth pitch of the equal temperament scale is $2^{\frac{5}{12}}$, which we should compare with $\frac{4}{3}$. We find that the relative error of approximation is

$$\left| \frac{2^{\frac{5}{12}} - \frac{4}{3}}{\frac{4}{3}} \right| \simeq \frac{0.0015}{\frac{4}{3}} \simeq 0.15\%.$$

The computations for the 15- and 19-tone scales are similar. For the 15-tone scale, we compute

$$2^{\frac{k}{15}} = \frac{4}{3} \quad\Longrightarrow\quad k \simeq 6.22.$$

Taking the closest integer, $k = 6$, we find the relative error of approximation:

$$\left| \frac{2^{\frac{6}{15}} - \frac{4}{3}}{\frac{4}{3}} \right| \simeq \frac{0.0138}{\frac{4}{3}} \simeq 1.04\%.$$

For the 19-tone scale, we compute

$$2^{\frac{k}{19}} = \frac{4}{3} \quad\Longrightarrow\quad k \simeq 7.89.$$

Taking the closest integer, $k = 8$, we find the relative error of approximation:

$$\left| \frac{2^{\frac{8}{19}} - \frac{4}{3}}{\frac{4}{3}} \right| \simeq \frac{0.0056}{\frac{4}{3}} \simeq 0.42\%.$$

The best approximation of the perfect fourth is provided by the
12-tone scale, the next best is given by the 19-tone scale, and the
approximation given by the 15-tone scale is the least precise.

2. Compare the approximations of a minor third interval (ratio of frequencies $\frac{6}{5}$) provided by the 12-tone, 17-tone, and 19-tone equal
   temperament scales. Similarly to (1), we compute for the 12-tone
   scale

$$2^{\frac{k}{12}} = \frac{6}{5} \quad \Longrightarrow \quad k \simeq 3.16.$$

Taking the closest integer, $k = 3$, we find the relative error of approximation:

$$\left| \frac{2^{\frac{3}{12}} - \frac{6}{5}}{\frac{6}{5}} \right| \simeq \frac{0.0108}{\frac{6}{5}} \simeq 0.90\%.$$

For the 17-tone scale, we compute

$$2^{\frac{k}{17}} = \frac{6}{5} \quad \Longrightarrow \quad k \simeq 4.47.$$

This is right in the middle between two integers, but our best choice
will be to take $k = 4$. We find the relative error of approximation:

$$\left| \frac{2^{\frac{4}{17}} - \frac{6}{5}}{\frac{6}{5}} \right| \simeq \frac{0.0229}{\frac{6}{5}} \simeq 1.90\%.$$

For the 19-tone scale, we compute

$$2^{\frac{k}{19}} = \frac{6}{5} \quad \Longrightarrow \quad k \simeq 4.998.$$

Taking the closest integer, $k = 5$, we find the relative error of approximation:

$$\left| \frac{2^{\frac{5}{19}} - \frac{6}{5}}{\frac{6}{5}} \right| \simeq \frac{0.0001}{\frac{6}{5}} < 0.01\%.$$

Here the 19-tone scale provides by far the best approximation, followed by the 12-tone and the 17-tone scales.

Chapter 8. e: The queen of growth and decay

Question 8.1. Suppose that $a > 1$ and m is a natural number. We claim that $\sqrt[m]{a} > 1$. Suppose not; i.e., $\sqrt[m]{a} \leq 1$. Now if $x, y \leq 1$, then $x \cdot y \leq x \cdot 1 \leq 1 \cdot 1 = 1$, so $xy \leq 1$ as well. Applying this equation with $x = y = \sqrt[m]{a} = a^{1/m}$ yields: $\sqrt[m]{a^2} = a^{2/m} \leq 1$. Applying this equation again with $x = a^{1/m}, y = a^{2/m}$ yields: $a^{3/m} \leq 1$. Continuing in this fashion, we eventually obtain that $a^{m/m} = a \leq 1$. This contradicts the given hypotheses, so our assumption that $\sqrt[m]{a} \leq 1$ must have been false. Therefore $\sqrt[m]{a} > 1$.

Question 8.3. Suppose a quantity decreases by a factor of 7 every 20 minutes. If $A(t)$ denotes the quantity as it varies over time in minutes (starting from some fixed initial moment), then we know that $A(20) = \frac{1}{7} A(0)$. Therefore,

$$\frac{1}{7} A(0) = A(20) = A(0)e^{20k_1},$$

where k_1 is the decay rate per minute. Dividing both sides by $A(0)$ and taking the natural logarithm yields:

$$\ln\left(\frac{1}{7}\right) = 20k_1 \quad \implies \quad k_1 = \frac{1}{20}\ln\left(\frac{1}{7}\right) \simeq -0.0973/\text{min}.$$

For the second part, we do a similar computation, measuring time in hours. Let k_{60} denote the relative rate of decay per hour. Because 20 minutes equals one-third of an hour, we have

$$\frac{1}{7} A(0) = A\left(\frac{1}{3}\right) = A(0)e^{\frac{1}{3}k_{60}}.$$

Then

$$\ln\left(\frac{1}{7}\right) = \frac{1}{3}k_{60} \quad \implies \quad k_{60} = 3\ln\left(\frac{1}{7}\right) \simeq -5.83773/\text{h}.$$

Even though the numerical values of the two rates of decay are different, they both represent the same process, so they must be equal up to a change of the units of time. Indeed, we can check:

$$60 \cdot k_1 \simeq 60 \cdot (-0.0973) \simeq -5.838 \simeq k_{60}.$$

Question 8.5. Suppose k is the growth rate of the bacteria (per minute). Then
$$A(0) = 1, \qquad A(3) = A(0)e^{3k} = e^{3k} = 2.$$
We would like to compute the number $A(60) = A(1)e^{60k} = e^{60k}$. One way is to compute k by taking the natural logarithm, and then computing e^{60k}. But an easier way is to use the properties of exponents and compute:
$$e^{60k} = (e^k)^{60} = ((e^k)^3)^{20} = (e^{3k})^{20} = A(3)^{20} = 2^{20}.$$

Question 8.7. Suppose the time needed for the population to reach $100,000$ is t years. We compute:
$$100,000 = 45,000e^{0.00693t}.$$

Divide both sides by $45,000$ and take the natural logarithm to solve for t:
$$0.00693t = \ln(100,000/45,000) \simeq 0.7985 \quad \Longrightarrow \quad t = \frac{0.7985}{0.00693} \simeq 115.22 \text{ yr.}$$

Therefore the population reaches $100,000$ in about 115 years and 3 months.
To find the doubling time T_d, we compute
$$T_d = \frac{\ln 2}{k} = \frac{0.693}{0.00693} = 100 \text{ yr.}$$

Question 8.9. We first compute the decay rate $k = k_{242}$ for curium-242 (per day):
$$A(100) = A(0)\frac{1}{100} = A(0)e^{k\cdot 1,063}.$$
Divide by $A(0)$ and take the natural logarithm:
$$\ln(1/100) = k\cdot 1,063 \quad \Longrightarrow \quad k = \frac{1}{1,063}\ln(1/100) \simeq -0.00433/\text{day}.$$

Now the half-life is:
$$T_{1/2} = \frac{\ln(1/2)}{k} \simeq 160 \text{ days.}$$

Question 8.11.

1. If the half-life of carbon-11 is one quarter of the half-life of barium-139, then the relative rate of decay of carbon-11 is four times the relative rate of decay of barium-139. Let k_{139} and k_{11} denote the

decay constants for barium-139 and carbon-11, respectively. Then we have $k_{11} = 4k_{139}$. If a sample of barium-139 decays to 92 percent in 10 minutes, we have the equation

$$0.92 = e^{10k_{139}}.$$

Then to find how much a sample of carbon-11 decays in 10 minutes, we need to compute $x = e^{10k_{11}}$. We have

$$e^{10k_{11}} = e^{10 \cdot 4k_{139}} = (e^{10k_{139}})^4 = 0.92^4 \simeq 0.7164 \simeq 72\%.$$

It will decay to approximately 72% of its initial mass.

2. Let us compute the half-lives and decay constants for both isotopes. If we start with 100 grams of barium-139, then it reduces to 92 grams in 10 minutes. Therefore,

$$92 = 100e^{10k_{139}} \quad \Longrightarrow \quad 0.92 = e^{10k_{139}}$$
$$\Longrightarrow \quad 10k_{139} = \ln(0.92) \simeq -0.08338.$$

Therefore $k_{139} = -0.008338/\text{min}$. Now the half-life for barium-139 equals $\ln(1/2)/k_{139} \simeq 83.1295$ min. Therefore the half-life for carbon-11 is $83.1295/4 = 20.782375$ min.

Question 8.13.

1. Let $k_A, T_{1/2,A}$ denote, respectively, the decay rate and half-life for substance A, and $k_B, T_{1/2,B}$ similarly for substance B. Because $k_A = 5k_B$, we compute:

$$T_{1/2,B} = \frac{\ln(1/2)}{k_B} = \frac{\ln(1/2)}{k_A/5} = 5 \cdot \frac{\ln(1/2)}{k_A} = 5T_{1/2,A}.$$

The half-life of substance B is 5 times that of substance A.

2. Suppose we start with m grams each of substances A and B. Let $A(t), B(t)$ denote the amounts remaining after time t (in years). Then

$$A(1) = 0.98m = me^{k_A \cdot 1} = me^{k_A} = me^{5k_B},$$
$$B(1) = me^{k_B \cdot 1} = m(e^{k_A})^{1/5}.$$

Then $B(1)/B(0) = B(1)/m = 0.98^{1/5} \simeq 0.996$. Therefore, substance B will reduce to 99.6% of its initial mass in one year.

Chapter 9. Finite series: Summing up your mortgage, geometrically

Question 9.1. We use the finite geometric series formula to find the sums:

1. $1 + 5 + 5^2 + 5^3 + \ldots + 5^{16} = \frac{1-5^{17}}{1-5} = \frac{5^{17}-1}{4} = 190,734,863,281.$

2. $\sum_{k=0}^{10}(-\frac{1}{2})^k = \frac{1-(-1/2)^{11}}{1-(-1/2)} = \frac{1-(-1/2^{11})}{3/2} = \frac{2}{3}(1+(1/2^{11})) = \frac{2}{3} \cdot$
   $\frac{2^{11}+1}{2^{11}} = \frac{2 \cdot 2049}{3 \cdot 2048} = \frac{2,049}{3,072}.$

3. To find the sum $1 + 1.25^4 + 1.25^8 + 1.25^{12} + \ldots + 1.25^{40}$, we use the properties of exponents to rewrite it as $1 + (1.25^4) + (1.25^4)^2 + \cdots + (1.25^4)^{10}$. This is a geometric series with common ratio $(1.25)^4 = (5/4)^4 = 625/256$, so we compute its sum to be

$$\frac{1-(625/256)^{11}}{1-(625/256)} = \frac{(625/256)^{11}-1}{369/256} \simeq 12,741.79.$$

Question 9.3. Under the scheme of quarterly compounding at an annual interest rate of $r = 12\%$, depositing \$1 in the bank for N quarters makes the value appreciate by $\left(1 + \frac{12}{400}\right)^N = 1.03^N$. The 2013 value of a dollar in 2023 is forty compoundings earlier, and hence equal to

$$1.03^{-40} \simeq 0.30656.$$

The 2013 value of a dollar in 2003 is forty compoundings later, and hence equal to
$$1.03^{40} \simeq 3.262.$$

The 2013 value of a dollar in 2035 is $22 \cdot 4 = 88$ compoundings earlier, and hence equal to
$$1.03^{-88} \simeq 0.0741.$$

We conclude that the 2013 values of a dollar in 2023, 2003, and 2035 are approximately 31 cents, \$3.26, and 7 cents, respectively.

Question 9.5. We apply the general mortgage formula with the mortgage amount $M = \$150,000$, the compound annual rate $r = 4\%$, $n = 12$ compounding periods per year, and $t = 20$ yr. Now if we denote $p =$

$(1 + r/(n \cdot 100)) = 1 + (4/1,200) = 1.00(3)$, then the amount to be paid each month is

$$A = M \cdot \frac{p^{nt}(1-p)}{1-p^{nt}} = 150,000 \cdot \frac{1.00(3)^{240} \cdot 0.00(3)}{1.00(3)^{240} - 1} \simeq \$908.94.$$

Question 9.7. The compounding factor is $1 + \frac{r}{n \cdot 100} = 1 + \frac{10}{1,200} = 1.008(3)$. Now suppose the monthly installment amount is $\$A$. The first installment is paid today, hence compounded by a factor of $1.008(3)^{72}$, the second installment is paid a month from now and compounded by a factor of $1.008(3)^{71}$, and so on. The total value of the monthly installments 6 years from now is:

$$A\left(1.008(3)^{72} + 1.008(3)^{71} + \cdots + 1.008(3)^1\right)$$
$$= 1.008(3)A\left(1 + \cdots + 1.008(3)^{71}\right)$$
$$= 1.008(3) \cdot A \cdot \frac{1 - 1.008(3)^{72}}{1 - 1.008(3)} \simeq A \cdot 98.9289.$$

Equating this value to $\$100,000$, we obtain:

$$A = \frac{100,000}{98.9289} \simeq \$1,010.83.$$

Question* 9.9. Suppose a mortgage or loan M is to be paid back in equal installments of A' each, over t years with n installments per year. If r is the annual rate of interest, the compounding factor is $p = 1 + \frac{r}{n \cdot 100}$.

1. If the first installment is paid at the time of taking the loan, then this installment is compounded nt times in t years. The second installment is compounded $nt - 1$ times in t years, and so on until the final installment, which is compounded exactly once. Therefore the total value of all of the installments after t years have passed is:

$$A'(p^{nt} + p^{nt-1} + \cdots + p^1) = A'p(1 + p + p^2 + \cdots + p^{nt-1}) = A'p \cdot \frac{1 - p^{nt}}{1 - p}.$$

This amount must equal the loan after compounding nt times. Thus,

$$A'p \cdot \frac{1 - p^{nt}}{1 - p} = M \cdot p^{nt}.$$

Solving for A' in terms of M, we obtain:

$$A' = M \cdot \frac{p^{nt}(1-p)}{p(1 - p^{nt})}.$$

2. We have a formula for the installment amount A in the scheme where the first installment is paid one compounding period after borrowing the loan: $A = M \cdot \dfrac{p^{nt}(1-p)}{1-p^{nt}}$. Dividing A by A', we obtain:

$$\frac{A}{A'} = M \cdot \frac{p^{nt}(1-p)}{(1-p^{nt})} \cdot \frac{1}{M} \cdot \frac{p(1-p^{nt})}{p^{nt}(1-p)} = p.$$

The ratio is p because each payment A' spends one compounding period longer in the bank, and therefore results in p times more money accumulated in the bank by the closing date.

Question 9.11. The total amount of all prizes is the sum

$$1,000 + 1,000 \cdot \frac{4}{5} + 1,000 \cdot \left(\frac{4}{5}\right)^2 + \cdots + 1,000 \cdot \left(\frac{4}{5}\right)^{19}$$

$$= 1,000 \left(1 + \frac{4}{5} + \left(\frac{4}{5}\right)^2 + \cdots + \left(\frac{4}{5}\right)^{19}\right)$$

$$= 1,000 \cdot \frac{1 - \left(\frac{4}{5}\right)^{20}}{1 - \frac{4}{5}} \simeq 1,000 \cdot \frac{1 - 0.8^{20}}{\frac{1}{5}} = 5,000(1 - 0.8^{20}) \simeq 4,942.35.$$

Therefore, the organizers will need to have a total of $4,942.35 for the prize fund.

Question 9.13. If the decoration is made with n pieces for any integer n, then the first piece is 1 inch long, the second is $1 \cdot (7/8)$ inches long, and so on, until the last piece is $1 \cdot (7/8)^{n-1}$ inches long. Therefore the total length of the golden wire needed is

$$1 + 1 \cdot \frac{7}{8} + 1 \cdot \left(\frac{7}{8}\right)^2 + \cdots + 1 \cdot \left(\frac{7}{8}\right)^{n-1} = \frac{1 - (7/8)^n}{1 - (7/8)} = 8(1 - (7/8)^n).$$

Now to answer the given questions: for $n = 10$, the length of the golden wire needed is

$$8(1 - (7/8)^{10}) \simeq 5.895 \text{ in},$$

while for $n = 30$ pieces, the required length is

$$8(1 - (7/8)^{30}) \simeq 7.854 \text{ in}.$$

Chapter 10. Infinite series: Fractals and the myth of forever

Question 10.1. Let us compute the distance Achilles has to cover before he overtakes the tortoise.

After Achilles travels 100 feet, he is 10 feet behind the tortoise.

After Achilles travels $100 + 10 = 110$ ft, he is 1 foot behind the tortoise, and so on to infinity.

This yields an infinite geometric series:

$$100 + 10 + 1 + 0.1 + \cdots = 100 \left(1 + \frac{1}{10} + \frac{1}{100} + \frac{1}{1,000} + \cdots \right)$$

$$= 100 \cdot \frac{1}{1 - (1/10)} = \frac{100}{9/10} = \frac{1,000}{9}.$$

Therefore, Achilles will catch up with the tortoise when they have both covered $1,000/9 = 111.(1)$ ft. This computation shows that even though we have to sum up infinitely many distances, these distances are decreasing so fast that the sum is still finite. Therefore, Achilles will overtake the tortoise in finite time. For instance, if v feet per second is the speed of Achilles, then Achilles will overtake the tortoise after $\frac{1,000}{9v}$ seconds.[1]

We can also deduce the same result using the methods in Chapter 2, where we studied constant speed motion along a straight line. Say the speed of Achilles is v feet per second, and that of the tortoise $\frac{1}{10}v$ feet per second, and the tortoise is 100 feet ahead. Then we have the following equation for the time point t when Achilles will overtake the tortoise:

$$vt = 100 + \frac{1}{10}vt \quad \Longrightarrow \quad \frac{9}{10}vt = 100 \quad \Longrightarrow \quad t = \frac{1000}{9v} \text{ sec.}$$

The distance covered by Achilles by that time is $vt = \frac{1000}{9} = 111.(1)$ ft, just as above.

Question 10.3.

1. In the first minute, 1 gallon of water is discharged; in the second minute, $1 \cdot (4/5)$ gallons are discharged; in the third minute, $1 \cdot (4/5)^2$

---

[1] The Zeno paradox has another aspect: what does it mean for a real process to be subdivided into infinitely many parts? How does a mental experiment apply in reality? These questions belong to the realm of philosophy and are not addressed in this book.

gallons, and so on. Therefore the total amount of water discharged is well approximated by the sum of the infinite geometric series

$$1 + \left(\frac{4}{5}\right) + \left(\frac{4}{5}\right)^2 + \cdots = \frac{1}{1 - (4/5)} = 5 \text{ gal.}$$

2. To compute the time elapsed until the discharge rate falls below 0.0093, we compute: $(4/5)^t = 0.0093$. Taking the natural logarithm of both sides yields:

$$t \ln(4/5) = t \ln(0.8) = \ln 0.0093$$

$$\implies \quad t = \frac{\ln(0.0093)}{\ln(0.8)} \simeq 20.96 \simeq 21 \text{ min.}$$

3. The amount of water (in gallons) discharged from the roof in the first 21 minutes can be computed by adding up the finite geometric series:

$$1 + \left(\frac{4}{5}\right) + \left(\frac{4}{5}\right)^2 + \cdots + \left(\frac{4}{5}\right)^{20} = \frac{1 - (4/5)^{21}}{1 - (4/5)} = 5(1 - (4/5)^{21})$$

$$\simeq 4.95388 \text{ gal.}$$

Question 10.5. Find the exact value of each of the following repeating decimals:

1. $3.4343434\ldots$. We can rewrite the number using a geometric series:

$$3.(43) = 10 \cdot 0.(34) = 10 \left(\frac{34}{10^2} + \frac{34}{10^4} + \frac{34}{10^6} + \cdots\right)$$

$$= 10 \cdot \frac{34}{10^2} \left(1 + \frac{1}{10^2} + \frac{1}{10^4} + \cdots\right).$$

This is an infinite geometric series with initial term $10 \cdot 0.34 = 3.4$, and ratio $1/10^2 = 1/100$ between successive terms. Therefore, it sums to $\frac{3.4}{1-(1/100)} = \frac{3.4}{99/100} = \frac{340}{99}$.

Here is another method to solve the same question: let $x = 3.(43)$ be the number. If we multiply x by 100, we obtain 343.43, which has the same repeating part beyond the decimal point, so that it can be expressed again in terms of x:

$$100x = 343.(43) = 340 + 3.4343\cdots = 340 + x \quad \implies \quad 99x = 340.$$

Solving this equation for x, we obtain $x = 340/99$ as before.

2. $12.213\ 213\ 213\ldots$.

The non-repeating part of the decimal is 12, and the repeating part can be written as

$$0.(213) = \frac{213}{10^3} + \frac{213}{10^6} + \frac{213}{10^9} + \cdots = \frac{213}{10^3}\left(1 + \frac{1}{10^3} + \frac{1}{10^6} + \cdots\right)$$
$$= \frac{213}{1,000} \cdot \frac{1}{1-(1/1,000)} = \frac{213}{1,000(1-(1/1,000))} = \frac{213}{999}.$$

Therefore, $12.(213) = 12 + \dfrac{213}{999} = \dfrac{11,988+213}{999} = \dfrac{12,201}{999}.$

Alternatively, if $12.(213) = x$, then

$$1,000x = 12,213.(213) = 12,201 + x,$$

so that $x = 12,201/999$.

3. $148.78787\ldots$.

Let us write the non-repeating part of the decimal as 148, and the repeating part as

$$0.(78) = \frac{78}{10^2} + \frac{78}{10^4} + \frac{78}{10^6} + \cdots = \frac{78}{10^2}\left(1 + \frac{1}{10^2} + \frac{1}{10^4} + \cdots\right)$$
$$= \frac{78}{100} \cdot \frac{1}{1-(1/100)} = \frac{78}{100(1-(1/100))} = \frac{78}{99}.$$

Therefore, $148.(78) = 148 + \dfrac{78}{99} = \dfrac{14,652+78}{99} = \dfrac{14,730}{99}.$

Alternatively, if $148.(78) = x$, then $100x = 14,878.(78) = 14,730 + 148.(78) = 14,730 + x$, so that $x = 14,730/99$.

Question 10.7. At the end of the first hour, the amount of agar in the petri dish is 0.05 gram, because 0.95 gram is absorbed. At the end of the second hour, the amount is $0.05(1 + 0.05) = 0.05 + 0.05^2$, because first an extra gram was added to the petri dish, and then 95% of the 1.05 grams of agar in the dish was absorbed. In general, after the nth hour, the minimum amount of agar in the plate is

$$0.05(1 + 0.05 + \cdots + 0.05^{n-1}) = 0.05\frac{1-0.05^n}{1-0.05} = \frac{5}{100}\frac{1-0.05^n}{0.95}$$
$$= \frac{5}{95}(1-0.05^n) = \frac{1-0.05^n}{19}.$$

Alternatively, use the given formula in the chapter with $D = 1, f = 0.95$ to obtain the same answer. Therefore after 2 hours, the minimum amount of agar in the plate is

$$\frac{1}{19}(1 - 0.05^2) = 0.0525 \text{ g} = 52.5 \text{ mg}.$$

After one day, i.e., 24 hours, the minimum amount is

$$\frac{1}{19}(1 - 0.05^{24}) \simeq 52.63 \text{ mg}$$

to two decimal places (in milligrams, or five decimal places in grams). In the long term, the minimum amount is $1/19$ of a gram, i.e., 52.63 milligrams to two decimal places.

Question 10.9. We will compute the fractal dimension of any solid triangle. To do so, connect the midpoints of each side of the original triangle. The four obtained triangles are congruent. They also have the same angles as the original triangle, and their side lengths measure half of the side lengths of the original triangle.

Equilateral and general triangles divided into four congruent parts.

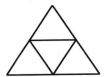 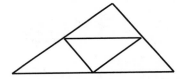

Therefore, the triangle is a self-similar object which gets divided into four smaller objects under a magnification factor of 2. Using the definition, we conclude that the fractal dimension of the triangle is $\dfrac{\log 4}{\log 2} = \dfrac{2 \log 2}{\log 2} = 2.$

Question 10.11. The fractal boundary is known as the *quadratic Koch curve* of type 1. The boundary is made up of four identical curves, each of which is self-similar, so we compute the fractal dimension of each of these four parts. It is clear that one side of the original square is replaced in the next iteration by five sides, each exactly similar in structure to the original side. Moreover, the magnification factor is 3 because each of the smaller sides is one-third the length of the original side. Therefore the fractal dimension of the boundary is $\dfrac{\log 5}{\log 3} \simeq 1.465.$

Question* 10.13. In the first iteration, we begin with one interval $[0, 1]$ of length 1, and remove the middle third of this interval, of length $1/3$. In the second iteration, we remove two intervals each of length $(1/3)^2$. The total length removed in the first two stages is therefore

$$1 \cdot \frac{1}{3} + 2 \cdot \frac{1}{3^2}.$$

In the third iteration, we remove four intervals, each of length $(1/3)^3$, and so on. Therefore the total length removed after infinitely many steps is the infinite geometric series

$$1 \cdot \frac{1}{3} + 2 \cdot \frac{1}{9} + 4 \cdot \frac{1}{3^3} + 8 \cdot \frac{1}{3^4} + \cdots = \frac{1}{3} \left(1 + \frac{2}{3} + \frac{2^2}{3^2} + \frac{2^3}{3^3} + \cdots \right)$$

$$= \frac{1}{3} \cdot \frac{1}{1 - (2/3)} = \frac{1}{3} \cdot \frac{1}{1/3} = 1.$$

Therefore the total length of the part removed from the interval is 1, which shows that the Cantor set has zero "length."

Question 10.15. The initial volume of the balloon equals the sum of the amounts of helium lost every second. It equals

$$1 + 1 \cdot \frac{1}{3} + 1 \cdot \frac{1}{3^2} + 1 \cdot \frac{1}{3^3} + \cdots \frac{1}{1 - (1/3)} = \frac{1}{2/3} = \frac{3}{2} = 1.5 \, \text{ft}^3.$$

Chapter 11. Estimation: What is your first guess?

Question 11.1. We will use first-order approximation to compute the following values:

1. The cube and cube root of 1.015. Because 0.015 is small compared to 1, we compute using the first-order approximation formula for exponents:

$$1.015^3 \approx 1 + 3 \cdot 0.015 = 1.045,$$

$$\sqrt[3]{1.015} = 1.015^{1/3} \approx 1 + \frac{1}{3} \cdot 0.015 = 1.005.$$

2. The cube of 2.999. We note that $2.999 = 3 + (-0.001)$, and that we need to transform the expression to the form $(1 + x)^t$ before we can use the first-order approximation formula. We compute using the properties of exponents:

$$2.999^3 = \left(3 - \frac{1}{1,000}\right)^3 = [3(1 - 1/3,000)]^3 = 3^3 \left(1 + \frac{-1}{3,000}\right)^3$$

$$\approx 3^3 \left(1 + 3 \cdot \frac{-1}{3,000}\right) = 27(1 - 1/1,000) = 27 - \frac{27}{1,000} = 26.973.$$

3. The number $e^{0.012}$. Because 0.012 is small compared to 1, using the first-order approximation formula for e^x, we compute: $e^{0.012} \approx 1 + 0.012 = 1.012$.

Question 11.3. The volume of a (right circular) cylinder is $\pi r^2 h$, where r and h denote the radius and height of the cylinder, respectively.

1. Suppose a cylinder is 4.001 inches high with a radius of 3.995 inches. Then the volume of the cylinder is $\pi r^2 h = \pi \cdot 4.001^2 \cdot 3.995$. We can apply the formula for the first-order approximation for a product of three factors:

$$(a_1 + \Delta a_1)(a_2 + \Delta a_2)(a_3 + \Delta a_3) \approx a_1 a_2 a_3 \left(1 + \frac{\Delta a_1}{a_1} + \frac{\Delta a_2}{a_2} + \frac{\Delta a_3}{a_3}\right).$$

Now use this formula in the special case $a_1 = a_2 = a_3 = 4$, $\Delta a_1 = \Delta a_2 = 0.001$, and $\Delta a_3 = -0.005$. Therefore, the volume of the cylinder is approximately equal to

$$\pi \cdot 4.001^2 \cdot 3.995 \approx \pi \cdot 4^3 \left(1 + \frac{0.001}{4} + \frac{0.001}{4} + \frac{-0.005}{4}\right)$$

$$= 64\pi \left(1 - \frac{0.003}{4}\right) = \pi(64 - 16 \cdot 0.003)$$

$$= \pi(64 - 0.048) = 63.952\pi \text{ in}^3.$$

2. If the height of the cylinder is 4.002 inches and the radius 1.001 inches, then using the same formula as above, we find the volume of the cylinder to be approximately equal to

$$\pi \cdot 1.001^2 \cdot 4.002 \approx \pi \cdot 1^2 \cdot 4 \left(1 + \frac{0.001}{1} + \frac{0.001}{1} + \frac{0.002}{4}\right)$$

$$= 4\pi \left(1 + 0.002 + 0.0005\right) = 4\pi(1.0025) = 4.01\pi \text{ in}^3.$$

Question 11.5. Let us find the maximum possible deviation. The volume of the space under the table is given by $l^2 h$, where l is the side length of the square tabletop and h the height of the leg. If l_0, h_0 are the specified dimensions for the table, then the manufactured units can have lengths up to $1.04 l_0$ and heights up to $1.01 h_0$. Therefore the maximum possible volume equals

$$(1.04 l_0)^2 \cdot (1.01 h_0) = l_0^2 h_0 (1.04^2 \cdot 1.01).$$

We approximate the above product using the first-order approximation formula for a product of three terms:

$$1.04^2 \cdot 1.01 \approx 1^3 \left(1 + \frac{0.04}{1} + \frac{0.04}{1} + \frac{0.01}{1}\right) = 1(1 + 0.09) = 1.09.$$

Therefore the maximum percentage error in the volume is $\dfrac{(1.09 - 1) l_0^2 h_0}{l_0^2 h_0}$
$= 0.09 = 9\%$.

Question 11.7. Recall that the formula for continuous compounding with initial amount $A(0)$ and annual interest rate $r\%$ is $A(t) = A(0)e^{rt/100}$. If $rt/100$ is small, then the first-order approximation for the exponential function gives

$$A(t) \approx A(0)(1 + \frac{rt}{100}) = A(0) + A(0)\frac{rt}{100}.$$

The simple interest accumulated over time t is equal to $A(0)rt/100$. Therefore the above first-order approximation equals the initial amount plus the simple interest accumulated over the same time period at the same annual interest rate.

Question 11.9. A hundred years after one gram of radium-226 was placed in the container, the amount left is

$$A(100) = A(0)e^{k \cdot 100} = 1 \cdot \exp(-4.28 \cdot 10^{-4} \cdot 100) = \exp(-4.28 \cdot 10^{-2})$$
$$= \exp(-0.0428) \approx 1 + (-0.0428) = 0.9572 \text{ g}.$$

Question* 11.11. Consider the expressions for $(1 + x)^n$ for $n = 1, 2$:

$$(1 + x)^1 = 1 + x, \quad (1 + x)^2 = 1 + 2x + x^2.$$

The first-order parts of these expressions are $1 + x$ and $1 + 2x$. Let us suppose that in the expression $(1 + x)^n$, the first-order part is $1 + nx$.

Assuming this, what is the first-order part of the expression $(1 + x)^{n+1}$? We compute:

$$(1 + x)^{n+1} = (1 + x) \cdot (1 + x)^n = (1 + x)(1 + nx + \text{higher-order terms})$$
$$= 1 + x + nx + nx^2 + \text{higher-order terms}.$$

Taking only the first-order part of the obtained expression, we have

$$(1 + x)^{n+1} = 1 + (n + 1)x + \text{higher-order terms},$$

because x^2 is a second-order term and is listed together with the higher-order terms. Now we have: the statement that the first-order term of $(1 + x)^n$ equals $1 + nx$ holds for $n = 1$. Moreover, if it holds for n, then it also holds for $n + 1$. Therefore, it holds for $n = 2, 3, 4, \ldots$, and so for all natural n. This is an application of the method of mathematical induction that we already used before in the solution to Question 4.13.

Chapter 12. Modular arithmetic I: Around the clock and the calendar

Question 12.1.   1. Note that $3 \cdot 3 \equiv 9 \equiv 1 \pmod 8$, so using the properties of congruences, we have $3 \cdot 3 \cdot 3 \equiv 1 \cdot 3 \equiv 3 \pmod 8$. Similarly, $3 \cdot 3 \equiv 9 \equiv 0 \pmod 9$, so $3 \cdot 3 \cdot 3 \equiv 0 \cdot 3 \equiv 0 \pmod 9$.

2. $5 \cdot 5 \cdot 5 \equiv 125 \equiv 25 \pmod{100}$.

3. Note that $5 \cdot 5 \equiv 25 \equiv 4 \pmod 7$, so $5 \cdot 5 \cdot 5 \equiv 4 \cdot 5 \equiv 20 \equiv 6 \pmod 7$.

4. First note that $5 \cdot 5 \equiv 25 \equiv 4 \pmod 7$. Therefore by the properties of congruences, $5^4 \equiv (5^2)^2 \equiv (4)^2 \equiv 16 \equiv 2 \pmod 7$. Finally, $5^8 \equiv (5^4)^2 \equiv 2^2 \equiv 4 \pmod 7$.

Question 12.3. First note that the first of March is $31 + 3 = 34$ days behind April 4. Therefore, the last day of February is one more day, i.e., 35 days, behind April 4. As this is exactly five weeks (i.e., $35 \equiv 0 \pmod 7$), the last day of February falls on the same day as April 4 in any calendar year.

Next, May 2 is precisely $26 + 2 = 28$ days ahead of April 4 – i.e., exactly four weeks ahead. Similarly, July 4 is $26 + 31 + 30 + 4 = 91$ days ahead of April 4, which is exactly 13 weeks. Therefore both of these dates also fall on the same day as April 4. Therefore, using Example 12.11, 3/0 (last day of February), 4/4, 5/2, 6/6, 7/4, 8/8, 10/10, and 12/12 all fall on the same day of the week in any year.

Question 12.5. To solve these questions, we need a reference point. We will choose the day on which this solution is being written: April 19, 2014, which is a Saturday. Therefore April 18, 2014, was a Friday, so two weeks back, April 4, 2014, was also a Friday.

1. To compute the day of Christmas 1990, we first note that 12/12/2014 also falls on a Friday by the Doomsday rule. Therefore two weeks later, 26 December is also on a Friday, and Christmas 2014 therefore falls on a Thursday. Now computing backward, we must subtract one day for each year and one additional day for each leap year. There are 24 years back to 1990, six of which are leap years $(1992, 1996, \ldots, 2012)$. Therefore we have to subtract $24+6 = 30$ days back from Thursday. Because $30 \equiv 2 \pmod 7$, we subtract only 2 days, which yields a Tuesday.

2. To compute the day of July 20, 1969, first note by a previous question that July 4, 2014, also falls on a Friday (because April 4, 2014, does). Therefore July 20, 2014, falls on a Sunday. Now we subtract 45 days for the 45 years back to 1969, and an additional 11 days for the leap years in between $(1972, \ldots, 2012)$. Because $45 + 11 \equiv 56 \equiv 0 \pmod 7$, we subtract 0 days back, and get the answer: Sunday.

3. To compute the day the Declaration of Independence was adopted, first note as in the previous part that July 4, 2014, falls on a Friday. Therefore we need to subtract $2014 - 1776 = 238$ days for the 238 years elapsed since 1776, and an additional number of days for the leap years in between. The leap years closest to, and in between, July 1776 and July 2014 are 1780 and 2012. Dividing $2012 - 1780 = 232$ by 4, we obtain 58. Therefore there were 59 years divisible by 4 in between 1776 and 2014. Of these, 1800 and 1900 were not leap years, which leaves 57 leap years. Therefore we subtract $238 + 57 = 295$ days back from July 4, 2014, which fell on a Friday. But $295 = 280 + 14 + 1 \equiv 1 \pmod 7$, so we subtract 1 day back, and obtain our answer: Thursday.

Question 12.7.   1. To find the day of the week of October 12, 1492, according to the proleptic Gregorian calendar, recall from Example 12.18 that October 15, 1582, was a Friday. Then October 12, 1582 should have been a Tuesday. (In fact this date didn't exist because of the calendar adjustment that occurred just then.) Between 1492 and 1582 there are 90 years, of which 1496, 1504, and all subsequent fourth years to 1580 are leap years. Totally, there are 21 leap years. We need to subtract one day for each year and one additional day for each leap year, totaling $90 + 21 = 111$ days. We compute $111 \equiv 105 + 6 \equiv 6 \pmod 7$. Counting 6 days back from a Tuesday, we obtain a Wednesday.

2. According to the Julian calendar, October 4, 1582, was a Thursday. Then October 12, 1582, should have been a Friday. Each fourth year is a leap year in the Julian calendar, therefore there were 22 leap years between October 12, 1582, and October 12, 1492. The total number of days to subtract is $90 + 22 = 112$, which is divisible by 7. Therefore, October 12, 1492, was a Friday according to the Julian calendar. This was most probably the calendar used in the expedition.

Another way to find the answer is to recall that the difference between the Julian and the Gregorian calendars in the sixteenth century was 10 days. In the fifteenth century it was one day less, because year 1500 was a leap year according to Julian, but not to the Gregorian calendar. Counting 9 days from a Wednesday, we obtain a Friday.

Question 12.9.   1. To find the time $5,000$ hours after 7 pm, we first compute $5,000 \pmod{24}$, using the properties of congruences. Because $50 \equiv 2 \pmod{24}$, we have:

$$5,000 \equiv 50 \cdot 100 \equiv 50 \cdot 50 \cdot 2 \equiv 2 \cdot 2 \cdot 2 \equiv 8 \pmod{24}.$$

So $5,000$ hours after 7 pm is the same time (but on a different day) as 8 hours after 7 pm, which yields: 3 am.

2. To find the month $10,000$ months after January, we compute modulo 12:

$$10,000 \equiv 100^2 \equiv 50^2 \cdot 2^2 \equiv 2^2 \cdot 2^2 \equiv 16 \equiv 4 \pmod{12}.$$

Therefore, four months ahead of January is May.

Question 12.11. In this question, we have to compute modulo 8 because there are 8 possible positions where the hamster could be at the end of each minute.

1. After 33 minutes, the hamster has covered $25 \cdot 33$ feet. Using the properties of congruences, we compute modulo 8:

$$25 \cdot 33 \equiv 1 \cdot 1 \equiv 1 \pmod{8}.$$

Therefore if the hamster starts on the east side, he is at the northeast side after 33 minutes.

2. After 51 minutes, the hamster has covered $25 \cdot 51$ feet. Using the properties of congruences, we compute modulo 8:

$$25 \cdot 51 \equiv 1 \cdot 3 \equiv 3 \pmod{8}.$$

Therefore if the hamster starts on the east side, he is at the northwest side after 51 minutes.

Question 12.13. As mentioned in the chapter, the divisibility test for 7 is a "backward" test. Therefore, suppose we want to test whether or not a number n, written in decimal notation as

$$n = a_k a_{k-1} \ldots a_2 a_1 a_0,$$

is divisible by 7. Here, $a_0, a_1, \ldots, a_k$ denote the digits of n, so $n = a_0 + a_1 10^1 + \cdots + a_k 10^k$. Now recall from the chapter that the remainders of the first seven nonnegative powers of 10 modulo 7 have been computed:

$$10^0 \equiv 1, \ 10^1 \equiv 3, \ 10^2 \equiv 2, \ 10^3 \equiv 6, \ 10^4 \equiv 4, \ 10^5 \equiv 5, \ 10^6 \equiv 1 \pmod{7}.$$

By the properties of congruences, it follows that the next six powers of 10 leave the same remainders when divided by 7:

$$10^7 \equiv 10^6 \cdot 10^1 \equiv 1 \cdot 3 \equiv 3, \qquad 10^8 \equiv 10^6 \cdot 10^2 \equiv 1 \cdot 2 \equiv 2,$$

and similarly one computes:

$$10^9 \equiv 6, \ 10^{10} \equiv 4, \ 10^{11} \equiv 5, \ 10^{12} \equiv 1 \pmod{7}.$$

Again, this allows us to compute the remainders of the powers 10^{13} through 10^{18}, and so on. It follows that in order to check whether or not

$$n = a_k a_{k-1} \ldots a_2 a_1 a_0 = a_0 + 10 a_1 + \cdots + 10^k a_k$$

is divisible by 7, we should check if the number s is divisible by 7, where s is

$$s = a_0 + 3a_1 + 2a_2 + \cdots + (10^k \pmod{7}) a_k.$$

Because of the periodicity of the residues modulo 7, this amounts to checking whether or not the sum of digits, starting from the last, multiplied by the periodically repeated sequence of numbers $1, 3, 2, 6, 4, 5$, is divisible by 7.

Question 12.15.

1. Note that $8^{12} = 68,719,476,736 = 687,194,767 \cdot 100 + 36$ using a calculator. Now, 100 is divisible by 25, so we obtain: $8^{12} \equiv 36 \equiv 11$ (mod 25).

   Alternatively, one can avoid using a calculator and solve this problem using congruences alone. First compute: $8^2 = 64 \equiv 14$ (mod 25). Therefore $8^3 \equiv 14 \cdot 8 \equiv 112 \equiv 12$ (mod 25). Now we compute using the properties of exponents and congruences:
   $$8^6 \equiv (8^3)^2 \equiv 12^2 \equiv 144 \equiv (-6) \pmod{25},$$
   so $8^{12} \equiv (8^6)^2 \equiv (-6)^2 \equiv 36 \equiv 11$ (mod 25) as above.

2. Note that $12,121,212 = 6 \cdot 02,020,202 = 6 \cdot 2,020,202$, so $12,121,212 \equiv 0$ (mod 6). It follows that
   $$98,989,898 \cdot 74,747,474 \cdot 12,121,212 \equiv 98,989,898 \cdot 74,747,474 \cdot 0$$
   $$\equiv 0 \pmod{6}.$$

Chapter 13. Modular arithmetic II: How to keep (and break) secrets

Question 13.1. Because T is the twentieth letter in the alphabet, and B the second, to encipher the message, the translation is by $2 - 20 \equiv -18 \equiv 8$ (mod 26). Therefore to decipher the message we need to move each letter back by eight letters. This yields: "T O D A Y A T N I N E," i.e., "today at nine."

Question 13.3.

1. Upon checking the pattern for powers of 5 modulo 7, we obtain:
   $$5^1 \equiv 5 \pmod 7, \quad 5^2 \equiv 25 \equiv 4 \pmod 7,$$
   $$5^3 \equiv 4 \cdot 5 \equiv 20 \equiv 6 \pmod 7, \quad 5^4 \equiv 6 \cdot 5 \equiv 30 \equiv 2 \pmod 7,$$
   $$5^5 \equiv 2 \cdot 5 \equiv 10 \equiv 3 \pmod 7, \quad 5^6 \equiv 3 \cdot 5 \equiv 15 \equiv 1 \pmod 7,$$
   $$5^7 \equiv 1 \cdot 5 \equiv 5 \pmod 7, \quad \dots.$$

Therefore the powers of 5 will repeat every six times. This last statement also follows directly from Fermat's Little Theorem: because 7 is prime and 5 is not divisible by 7, it follows that $5^6 \equiv 5^{7-1} \equiv 1$ (mod 7) and $5^7 \equiv 5$ (mod 7). The above computations verify that the sequence of powers does not repeat more frequently.

2. Using the previous part and the properties of exponents,

$$5^{27} \equiv 5^{24+3} \equiv (5^6)^4 \cdot 5^3 \equiv 1^4 \cdot 6 \equiv 6 \pmod 7,$$
$$5^{601} \equiv 5^{600+1} \equiv (5^6)^{100} \cdot 5 \equiv 1^{100} \cdot 5 \equiv 5 \pmod 7.$$

Question 13.5.

1. Because 13 is prime and 9 is not divisible by 13, Fermat's Little Theorem implies that $9^{12} \equiv 1$ (mod 13). But then $9^{36} \equiv (9^{12})^3 \equiv 1$ (mod 13). Therefore $x = 1$.

2. Because 11 is prime and 5 is not divisible by 11, Fermat's Little Theorem implies that $5^{10} \equiv 1$ (mod 11). Now using the properties of exponents,

$$5^{132} \equiv 5^{10 \cdot 13 + 2} \equiv (5^{10})^{13} \cdot 5^2 \equiv 1^{13} \cdot 25 \equiv 25 \equiv 3 \pmod{11}.$$

Therefore $x = 3$.

Question 13.7. As discussed in the chapter, the last digit of a power of a number N equals the last digit of the same power of the unit's digit of N. This helps simplify the problem considerably.

1. The last digit of 323^{516} equals the last digit of 3^{516} (which is evaluated modulo 10). Now note that $3^2 \equiv 9$ (mod 10) and $3^4 \equiv 9^2 \equiv 81 \equiv 1$ (mod 10). (The second equality also follows by using Euler's theorem.) Therefore,

$$3^{516} \equiv 3^{4 \cdot 129} \equiv (3^4)^{129} \equiv 1^{129} \equiv 1 \pmod{10}.$$

It follows that the last digit of 323^{516} is the last digit of 3^{516}, which is 1.

2. The last digit of 78^{117} is the last digit of 8^{117}. Note that we can no longer apply Euler's theorem because 8 and 10 are not coprime. However, we directly compute the powers:

$$8^2 \equiv 64 \equiv 4 \pmod{10}, \quad 8^4 \equiv (8^2)^2 \equiv 4^2 \equiv 16 \equiv 6 \pmod{10},$$
$$8^5 \equiv 8^4 \cdot 8 \equiv 6 \cdot 8 \equiv 48 \equiv 8 \pmod{10}, \quad \dots.$$

It follows that the remainders of the powers of 8 repeat every four times: $8, 4, 2, 6, 8, 4, \ldots$. Because $117 = 4 \cdot 29 + 1$, it follows that $8^{117} \equiv 8^1 \equiv 8 \pmod{10}$. Therefore the last digit of 78^{117} is also 8.

Question 13.9.

1. Because 11 is prime and 3 is coprime to 11, Fermat's Little Theorem implies that $3^{10} \equiv 1 \pmod{11}$. We now compute using the properties of exponents:

$$3^{111} \equiv 3^{10 \cdot 11 + 1} = (3^{10})^{11} \cdot 3^1 \equiv 1^{11} \cdot 3 \equiv 3 \pmod{11}.$$

   Therefore $x = 3$.

2. Although 9 is not prime, 7 is coprime to 9. Therefore we can use Euler's theorem instead. We compute: $\varphi(9) = 6$, because $1, 2, 4, 5, 7, 8$ are coprime to 9. Therefore $7^6 \equiv 1 \pmod 9$ by Euler's theorem, and we compute using the properties of exponents:

$$7^{111} \equiv 7^{6 \cdot 18 + 3} \equiv (7^6)^{18} \cdot 7^3 \equiv 1^{18} \cdot (-2)^3 \equiv -8 \equiv 1 \pmod 9.$$

   Therefore $x = 1$.

Question 13.11. In all of the computations, we can use Fermat's Little Theorem, because $p = 17$ is prime and 3 is not divisible by 17. Therefore $3^{16} \equiv 1 \pmod{17}$. Now using the properties of exponents, we compute:

$$x \equiv g^a \equiv 3^{50} \equiv 3^{16 \cdot 3 + 2} \equiv (3^{16})^3 \cdot 3^2 \equiv 1^3 \cdot 9 \equiv 9 \pmod{17},$$
$$y \equiv g^b \equiv 3^{67} \equiv 3^{16 \cdot 4 + 3} \equiv (3^{16})^4 \cdot 3^3 \equiv 1^4 \cdot 27 \equiv 10 \pmod{17},$$
$$K \equiv y^a \equiv 10^{50} \equiv 10^{16 \cdot 3 + 2} \equiv (10^{16})^3 \cdot 10^2 \equiv 1^3 \cdot 100 \equiv 15 \pmod{17}.$$

Therefore $x = 9$, $y = 10$, and $K = 15$.

Chapter 14. Probability: Dice, coins, cards, and winning streaks

Question 14.1. If we toss seven fair coins, then the total number of possible outcomes is $2^7 = 128$. Now we are to compute the probability of the following events:

1. At least two coins are tails.

   Let us compute the probability of the complementary event E^c: i.e., E^c is the event that only one coin is a tail, or no coins are tails. Because the coins are fair, there is only one outcome when no coins are tails: HHHHHHH. Similarly, if there is exactly one tail, then that tail occurs in one of seven places:

   $$\text{THHHHHH, HTHHHHH, \ldots, HHHHHHT.}$$

   There are seven possible outcomes. Together, E^c has eight outcomes. Because the total number of outcomes is 128, the event E^c has probability $P(E^c) = \dfrac{8}{128} = \dfrac{1}{16}$. Therefore by the laws of probability, $P(E) = 1 - P(E^c) = \dfrac{15}{16}$.

2. At least one of the coins yields a head.

   Once again it is easier to compute the probability of the complementary event E^c. In this case this consists of only one possible outcome: TTTTTTT. Therefore, $P(E^c) = 1/128$, whence $P(E) = 1 - P(E^c) = 127/128$.

3. The third and fifth coins yield tails, while the second coin yields a head.

   If we fix the outcomes of the second, third, and fifth coins, then the favorable outcomes include all possible outputs for the other four coins. (Recall that outputs are of individual coins, while outcomes are the sets of seven outputs.) Therefore the number of favorable outcomes for our desired event E is:

   $$2 \cdot 1 \cdot 1 \cdot 2 \cdot 1 \cdot 2 \cdot 2 = 16.$$

   Then $P(E) = 16/128 = 1/8$.

Question 14.3. The total number of balls of yarn inside the bag is $3 + 4 + 5 = 12$. Therefore if a ball of yarn is selected at random from the bag, then $P(\text{blue}) = 5/12$. Similarly, the ball is not red if and only if it is either green or blue, so $P(\text{not red}) = (3 + 5)/12 = 2/3$.

Question 14.5. Find the probabilities of the following events:

1. Tossing a fair coin five times and obtaining exactly two heads.

   The total number of outcomes equals $2^5 = 32$. Now observe that there are exactly ten ways in which to obtain two heads and three tails in five coin tosses:

$$\text{HHTTT} \quad \text{HTHTT} \quad \text{HTTHT} \quad \text{HTTTH} \quad \text{THHTT}$$
$$\text{THTHT} \quad \text{THTTH} \quad \text{TTHHT} \quad \text{TTHTH} \quad \text{TTTHH}$$

   Therefore there are ten favorable outcomes, so the desired probability is $10/32 = 5/16$.

2. Rolling three dice and obtaining the sum of 3 or 4.

   The total number of outcomes equals $6 \cdot 6 \cdot 6 = 216$. The only way to obtain a sum of three numbers (each taking values in $\{1, \ldots, 6\}$) to equal 3 or 4 is: $1+1+1, 1+1+2, 1+2+1, 2+1+1$. Therefore, there are four favorable outcomes, and the desired probability equals $4/216 = 1/54$.

Question 14.7. If four fair coins were tossed, then there are $2^4 = 16$ possible outcomes. Now we compute the probabilities of the following events by simply listing the favorable outcomes in each case, and dividing by 16.

1. Exactly two of the outputs are heads.

   There are six possible outcomes with two heads and two tails:

$$\text{HHTT} \quad \text{HTHT} \quad \text{HTTH} \quad \text{THHT} \quad \text{THTH} \quad \text{TTHH}$$

   so that the desired probability equals $6/16 = 3/8$.

2. At least two of the outputs are heads.

   The complementary event E^c consists of zero or one head, with the remaining outputs being tails. There are exactly five such outcomes:

$$\text{TTTT} \quad \text{HTTT} \quad \text{THTT} \quad \text{TTHT} \quad \text{TTTH},$$

   and hence the desired probability equals

$$P(E) = 1 - P(E^c) = 1 - \frac{5}{16} = \frac{11}{16}.$$

3. An even number of tails occurs.

   The number of tails that can occur are zero, two, or four. There is exactly one way in which zero tails can occur, and one way for four tails to occur: HHHH and TTTT, respectively. By the first part, there are six outcomes when precisely two tails occur. Therefore, the number of favorable outcomes is $1 + 1 + 6 = 8$, yielding a probability of $8/16 = 1/2$.

4. An odd number of heads occurs.

   This event is precisely the complement of the event in the previous part. Therefore by the previous part and the laws of probability, its probability is $1 - \frac{1}{2} = \frac{1}{2}$.

Question 14.9. If three (fair) dice are rolled then the number of possible outcomes is $6 \cdot 6 \cdot 6 = 216$. There are precisely four outcomes where the sum of the three outputs is less than five:

$$(1, 1, 1), \quad (2, 1, 1), \quad (1, 2, 1), \quad (1, 1, 2).$$

Therefore there are $216 - 4 = 212$ outcomes whose sum is five or more. This answers the second question: the probability that the sum of the outputs is at least five, equals $212/216 = 53/54$.

For the other question, we now compute the number of outcomes whose sum of outputs is exactly 5. These outcomes are:

$$(3, 1, 1), \quad (1, 3, 1), \quad (1, 1, 3), \quad (2, 2, 1), \quad (2, 1, 2), \quad (1, 2, 2).$$

Therefore there are six such outcomes. This means there are $4 + 6 = 10$ outcomes whose sum is 5 or less. Therefore the probability that the sum of the outputs is at most five, equals $10/216 = 5/108$.

Question 14.11. The total number of outcomes in rolling a die and tossing a coin are $2 \cdot 6 = 12$. Let us denote the outcomes as pairs, such as $(H, 4), (T, 6)$, etc.

1. There are two possibilities for getting 6: $(H, 5)$ and $(T, 6)$. Because there are exactly two favorable outcomes, the desired probability equals $2/12 = 1/6$.

2. The favorable outcomes are precisely $(H, 1), (T, 1), (T, 2)$, so the desired probability is $3/12 = 1/4$.

3. The favorable outcomes are $(H, 2), (H, 4), (H, 6), (T, 1), (T, 3), (T, 5)$. Therefore the desired probability is $6/12 = 1/2$.

Chapter 15. Permutations and combinations: Counting your choices

Question 15.1. These are straightforward computations:

$$_8C_3 = {}_8C_5 = \frac{8!}{5!3!} = \frac{8 \cdot 7 \cdot 6 \cdot 5!}{5! \cdot 6} = 8 \cdot 7 = 56.$$

Next, we first compute $_{14}P_8$:

$$_{14}P_8 = \frac{14!}{(14-8)!} = \frac{14!}{6!} = 14 \cdot 13 \ldots 8 \cdot 7 = 121,080,960.$$

Then by definition of $_{14}C_8$, we have $_{14}C_8 = {}_{14}P_8/{}_8P_8 = {}_{14}P_8/8! = 3,003$.

Question 15.3.   1. Because there is one T, four E's, two N's, and two S's, the number of permutations is

$$\frac{9!}{1! \cdot 4! \cdot 2! \cdot 2!} = \frac{9 \cdot 8 \cdot 7 \cdot 6 \cdot 5 \cdot 4!}{4 \cdot 4!} = 9 \cdot 2 \cdot 7 \cdot 6 \cdot 5 = 3,780.$$

2. There are only four different letters in the word: T, E, N, S. The number of different collections of three letters out of four is the number of combinations of 3 out of 4 items, which equals $_4C_3 = 4$. Each collection can be arranged in $_3P_3 = 3!$ number of ways. Then the number of all possible permutations of three different letters of this word is $_4C_3 \cdot {}_3P_3 = 4 \cdot 3! = 24$.

Question 15.5. Because the same five songs played in two different orders would constitute two different programs, we need to use permutations here and not combinations. Therefore, the number of possible programs of five songs equals

$$_{15}P_5 = 15 \cdot 14 \cdot 13 \cdot 12 \cdot 11 = 360,360.$$

Question 15.7. Because the order of the colors matters in the flag, we use permutations here and not combinations. Because there are five colors and three bands, the number of flags with colors arranged horizontally is $_5P_3 = 5 \cdot 4 \cdot 3 = 60$. The same calculation yields 60 flags with colors arranged vertically. Therefore the total number of possible flags equals $60 + 60 = 120$.

Question 15.9. The total number of ways to choose three out of nine transistors is $_9C_3 = (9 \cdot 8 \cdot 7)/(1 \cdot 2 \cdot 3) = 84$.

1. The number of ways of selecting all three defective transistors is $_3C_3 = 1$. Therefore, the desired probability is $1/84$.

2. The number of ways of selecting all three functioning transistors is $_6C_3 = (6 \cdot 5 \cdot 4)/(1 \cdot 2 \cdot 3) = 20$. Therefore, the desired probability is $20/84 = 5/21$.

Question 15.11. The total number of possible outcomes from 15 consecutive fair coin tosses is 2^{15}.

1. To find the probability of getting exactly 11 tails, we count the number of possible ways in which one obtains exactly 11 tails. This equals $_{15}C_{11} = {}_{15}C_4$, so the desired probability is

$$\frac{_{15}C_4}{2^{15}} = \frac{1}{32,768} \cdot \frac{15 \cdot 14 \cdot 13 \cdot 12}{1 \cdot 2 \cdot 3 \cdot 4} = \frac{1,365}{32,768}.$$

2. To find the probability of getting at least 11 tails, observe that obtaining at least 11 tails is the same as obtaining $11, 12, 13, 14,$ or 15 tails. Similar to the previous problem, this means that the number of favorable outcomes equals

$$\begin{aligned}
&{}_{15}C_{11} + {}_{15}C_{12} + {}_{15}C_{13} + {}_{15}C_{14} + {}_{15}C_{15} \\
&= {}_{15}C_4 + {}_{15}C_3 + {}_{15}C_2 + {}_{15}C_1 + {}_{15}C_0 \\
&= \frac{15 \cdot 14 \cdot 13 \cdot 12}{1 \cdot 2 \cdot 3 \cdot 4} + \frac{15 \cdot 14 \cdot 13}{1 \cdot 2 \cdot 3} + \frac{15 \cdot 14}{1 \cdot 2} + \frac{15}{1} + 1 \\
&= 1,365 + 455 + 105 + 15 + 1 = 1,941.
\end{aligned}$$

Therefore the desired probability equals $1,941/32,768$.

3. To get 11 tails in the first 11 tosses, we have to obtain exactly one possible output (a tail) in each of the first 11 tosses, and two possible outputs (a head and a tail) in the remaining four coin tosses. Therefore, the total number of favorable outcomes is $1^{11}2^4$, so that the desired probability is

$$\frac{2^4}{2^{15}} = 2^{-11} = \frac{1}{2,048}.$$

Question 15.13. A box holds 25 muffins, including 10 blueberry muffins, 6 carrot muffins, 5 chocolate chip muffins, and 4 apple muffins. The total number of ways to choose 3 muffins out of 25 is $_{25}C_3 = \frac{25 \cdot 24 \cdot 23}{3 \cdot 2 \cdot 1} = 25 \cdot 4 \cdot 23 = 2,300$.

1. To find the probability that all three muffins are chocolate chip, we count the number of ways to choose 3 chocolate chip muffins out of 5 available. It equals $_5C_3 = \frac{5 \cdot 4}{2} = 10$. Then the desired probability is

$$\frac{_5C_3}{_{25}C_3} = \frac{10}{2,300} = \frac{1}{230}.$$

2. Now we find the probability that no muffins are blueberry, and exactly one of them is apple. The number of ways to choose one apple muffin out of four is $_4C_1 = 4$. These have to be multiplied by the number of ways to choose the remaining 2 muffins, which must be either carrot or chocolate chip. There are $_{11}C_2 = \frac{11 \cdot 10}{2} = 55$ ways to choose these two muffins. Finally, the desired probability is

$$\frac{_4C_1 \cdot _{11}C_2}{_{25}C_3} = \frac{4 \cdot 55}{2,300} = \frac{11}{115}.$$

Question 15.15. We first write down the two expressions:

$$_{n-1}C_{k-1} = \frac{(n-1)!}{(n-1-(k-1))!(k-1)!} = \frac{(n-1)!}{(n-k)!(k-1)!},$$

$$_{n-1}C_k = \frac{(n-1)!}{(n-1-k)!k!}.$$

If we multiply both the numerator and the denominator in the first expression by k, and similarly for the second expression by $(n-k)$, then we obtain:

$$_{n-1}C_{k-1} = \frac{(n-1)! \cdot k}{(n-k)!k!}, \qquad _{n-1}C_k = \frac{(n-1)! \cdot (n-k)}{(n-k)!k!}.$$

Note that both of these expressions now have the same denominator. Therefore, adding them up, we obtain:

$$_{n-1}C_{k-1} + _{n-1}C_k = \frac{(n-1)! \cdot k}{(n-k)!k!} + \frac{(n-1)! \cdot (n-k)}{(n-k)!k!}$$

$$= \frac{(n-1)! \cdot (k+n-k)}{(n-k)!k!} = \frac{(n-1)! \cdot n}{(n-k)!k!} = _nC_k,$$

which concludes the proof.

Chapter 16. Bayes' law: How to win a car... or a goat

Question 16.1. Are the following pairs of events A and B independent? Justify your answer by computing the probabilities $P(A)$, $P(B)$, and $P(A \cap B)$.

1. Toss three fair coins. A: the first is a head. B: there are exactly two heads.

   The total number of outcomes is $2^3 = 8$. Then $P(A) = 4/8 = 1/2$, $P(B) = {}_3C_2/8 = 3/8$ (corresponding to HHT, HTH, THH), and $P(A \cap B) = 2/8$ (corresponding to HHT, HTH). Therefore $P(A \cap B) \neq P(A)P(B)$, so A and B are not independent.

2. Draw two cards from a deck without replacement. A: the first card is a number from 2 to 10; B: the second card is a king or an ace.

   The total number of outcomes is $52 \cdot 51$. Then $P(A) = 36/52 = 9/13$. The probability $P(B)$ is harder to compute. There are two types of favorable outcomes: first, the first card is any of the 44 cards which is not a king or an ace, and the second card is a king or an ace. There are precisely $44 \cdot 8$ such outcomes. Otherwise, the first and second cards are both a king or an ace. There are precisely $8 \cdot 7$ such outcomes. Therefore,

   $$P(B) = \frac{44 \cdot 8 + 8 \cdot 7}{52 \cdot 51} = \frac{8 \cdot 51}{52 \cdot 51} = \frac{8}{52} = \frac{2}{13}.$$

   (d) $P(A \cap B) = \dfrac{36 \cdot 8}{52 \cdot 51}$ (corresponding to 36 choices for the first card, and 8 choices for the second card given any of these 36 choices).

   Therefore $P(A \cap B) = \dfrac{36}{52} \cdot \dfrac{8}{51} = \dfrac{9}{13} \cdot \dfrac{8}{51} \neq \dfrac{9}{13} \cdot \dfrac{8}{52} = P(A)P(B)$, so A, B are not independent.

3. Roll two fair dice. A: the sum of the outputs is 7. B: the first output is odd.

   The total number of outcomes is $6 \cdot 6 = 36$. Then $P(A) = 6/36 = 1/6$ (corresponding to $(1, 6), (2, 5), \ldots, (6, 1)$), and $P(B) = 3 \cdot 6/36 = 1/2$ (corresponding to $1, 3, 5$ for the first output, and $1, \ldots, 6$ for the second). The probability $P(A \cap B) = 3/36 = 1/12$ (corresponding to $(1, 6), (3, 4), (5, 2)$).

Therefore $P(A \cap B) = P(A)P(B)$, so A, B are independent.

Question 16.3.

1. Let B denote the event that the sum is odd. Then either the first output is odd and the second is even, or vice versa. There are $3 \cdot 3$ outcomes of each type, which shows that B has 18 outcomes out of a total of $6 \cdot 6 = 36$ outcomes.

   Next, $A \cap B$ is the event that the sum of the outputs is 11 and that it is odd (which is now redundant, because 11 is odd). There are precisely two ways in which this happens: $(5, 6)$ and $(6, 5)$. Therefore,

   $$P(A|B) = \frac{P(A \cap B)}{P(B)} = \frac{2/36}{18/36} = \frac{2}{18} = \frac{1}{9}.$$

2. For either draw, the total number of possible outcomes is $_{175}C_2$, because there are 175 marbles in the bag. Let B be the event that both marbles drawn randomly (without replacement) are not green, and A the event that they are both red. Then $A \cap B = A$. Now because there are 125 marbles which are not green, and 100 which are red, the number of favorable outcomes for B and A are $_{125}C_2$ and $_{100}C_2$, respectively. Therefore,

   $$P(A|B) = \frac{P(A \cap B)}{P(B)} = \frac{P(A)}{P(B)} = \frac{_{100}C_2/_{175}C_2}{_{125}C_2/_{175}C_2} = \frac{_{100}C_2}{_{125}C_2}$$
   $$= \frac{100 \cdot 99/2!}{125 \cdot 124/2!} = \frac{4 \cdot 99}{5 \cdot 124} = \frac{99}{155}.$$

Question 16.5. The number of ways to choose 2 books out of the 5 in French (respectively, the 10 not in Spanish) is $_5C_2$ (respectively, $_{10}C_2$). Similarly, the number of all possible outcomes is $_{15}C_2$. Therefore,

$$P(A) = \frac{_5C_2}{_{15}C_2} = \frac{5 \cdot 4/2!}{15 \cdot 14/2!} = \frac{2}{21},$$
$$P(B) = \frac{_{10}C_2}{_{15}C_2} = \frac{10 \cdot 9/2!}{15 \cdot 14/2!} = \frac{3}{7}.$$

Moreover, $A \cap B = A$, so we also compute:

$$P(A|B) = \frac{P(A \cap B)}{P(B)} = \frac{P(A)}{P(B)} = \frac{2/21}{3/7} = \frac{2 \cdot 7}{3 \cdot 21} = \frac{2}{9},$$
$$P(B|A) = \frac{P(A \cap B)}{P(A)} = \frac{P(A)}{P(A)} = 1.$$

Because $P(B|A) = 1 \neq P(B)$, the events A, B are not independent. (One can also compute that $P(A \cap B) \neq P(A)P(B)$.)

Question 16.7.

1. If two cards are drawn from a deck without replacement, the number of possible outcomes are $52 \cdot 51$. Out of these, $13 \cdot 12$ outcomes involve both cards being clubs. On the other hand, getting four heads in four fair coin tosses involves one favorable outcome out of $2^4 = 16$ possible outcomes. Therefore, we need to compare these two probabilities:

$$\frac{13 \cdot 12}{52 \cdot 51} = \frac{12}{4 \cdot 51} = \frac{3}{51} = \frac{1}{17} < \frac{1}{16}.$$

   Therefore getting four heads in four fair coin tosses is the more likely event.

2. Note that $P(A|B) = \dfrac{P(A \cap B)}{P(B)}$. Now B is the event that the sum of the outputs is 3 – the only way that this happens is that the outcome is $(1, 1, 1)$. In this case the second die cannot yield 2, so $A \cap B$ has zero outcomes, and hence $P(A|B) = 0$.

Question 16.9. We will use the grid method to solve the problem. Suppose 300 students go on the picnic. Start by filling out the last row of the following grid. Now using the column totals and the given information, fill out the entries in the first two columns. Finally, add the entries to write down the row totals. This yields the following grid.

	Boys	Girls	Total
Jeans	150	100	250
Skirts	0	50	50
Total	150	150	300

Therefore the probability that a randomly picked student wearing jeans is a girl is $100/250 = 2/5$.

Alternatively, the same question can be solved using Bayes' law. Let G denote the event of a student being a girl, and J the event of a student wearing jeans. We need to find the conditional probability $P(G|J)$. The reverse probability $P(J|G)$ is given to be $\frac{1}{3}$. We also know that $P(G) = \frac{1}{2}$. To find $P(J)$, we sum up the probabilities of independent events that the student is a girl wearing jeans $\frac{1}{2} \cdot \frac{2}{3}$ or a boy wearing jeans $\frac{1}{2} \cdot 1$. Finally, by Bayes' law we compute

$$P(\text{G}|\text{J}) = \frac{P(\text{J}|\text{G}) \cdot P(\text{G})}{P(\text{J})} = \frac{\frac{2}{3} \cdot \frac{1}{2}}{\frac{1}{2} \cdot \frac{2}{3} + \frac{1}{2} \cdot 1} = \frac{\frac{1}{3}}{\frac{1}{3} + \frac{1}{2}} = \frac{2}{5}.$$

Another method that can be used in this and many other cases is a tree diagram. Because all three methods are equivalent, we will leave it to readers to choose their favorite way. We will solve the remaining questions in this chapter by the grid method.

Question 16.11. The computations carried out by the grid method in Example 16.8 immediately yield the answer: $P(D|-) = 10/865 \simeq 1.156\%$.

Question 16.13. We will use the grid method to solve the problem. Suppose $1,000$ patients are being tested, and we know that 25 out of the $1,000$ are diseased. Start by filling out the last row of the following grid. Now using the column totals and the given information, fill out the entries in the first two columns. Finally, add the entries to write down the row totals. This yields the following grid.

	Diseased (D)	Not Diseased (N)	Total
Positive (+)	24.75	48.75	73.5
Negative (−)	0.25	926.25	926.5
Total	25	975	1,000

Recall that obtaining decimals is not a problem, because eventually we are interested in probabilities. Now the desired probability is $P(D|+) = 24.75/73.5 \simeq 33.673\%$.

Question 16.15. We once again use the grid method to solve this problem. The columns contain information about your incoming spam and non-spam emails, while the rows contain information about the numbers of such emails containing the word "prince." For convenience, we assume that your inbox software has received $10,000$ emails over a long period of time, say. Once again we follow the algorithm in the previous solution, for filling out the grid, to obtain:

	Spam (S)	Not Spam (N)	Total
Contains "prince" (+)	7	0.99	7.99
Doesn't contain "prince" (−)	93	9,899.01	9,992.01
Total	100	9,900	10,000

Therefore the desired probability is $P(S|+) = 7/7.99 \simeq 87.61\%$.

Question 16.17.

1. Regardless of which door the player chooses, initially it is assumed to be equally likely for the car to be behind any door. Therefore the answer is: $P(A) = P(B) = \cdots = P(E) = 1/5$.

2. Using the various parts of the question, we obtain the following table. Note that the nonzero entries in any fixed column must be the same, because it is equally likely that the host will open any of the three possible pairs of doors not containing A and the door for that column. Because there are six experiments per column, this means that three entries in each column are 2, and the remaining entries are 0. We therefore obtain the following table:

Host opens door	Car behind door A	Car behind door B	Car behind door C	Car behind door D	Car behind door E	Total
BC	1	0	0	2	2	5
BD	1	0	2	0	2	5
BE	1	0	2	2	0	5
CD	1	2	0	0	2	5
CE	1	2	0	2	0	5
DE	1	2	2	0	0	5
Total	6	6	6	6	6	30

The above grid also answers parts (3) and (4).

5. Now suppose the host has opened doors B and E after the participant chose door A. Then sticking to the original door A yields a probability of $P(A|B, E) = 1/5$ of success. On the other hand, switching to either door C or D yields a probability of $P(C|B, E) = 2/5 = P(D|B, E)$ of success. Hence, the participant should switch.

Moreover, these numbers do not depend on which pair of doors the host opens, because the corresponding nonzero entries in the above table are always 1 in the first column and 2 in all other columns, and because the row sums are always 5.

Chapter 17. Statistics: Babe Ruth and Barry Bonds

Question 17.1. Denote the scores by x_1 through x_7. It is then easy to compute that the mean score is $\overline{x} = 71.2857$ (to four decimal places) and $x_1^2 + \cdots + x_7^2 = 36,351$. Therefore the variance of the scores is $\sigma_X^2 = 111.3469$, and the standard deviation is $\sigma_X = 10.5521$. Moreover, arranging the scores in increasing order, we conclude that the median is 72.

In order to compute which of the scores are statistically significant, we first compute the one-sigma limits $\overline{x} \pm \sigma_X$. These equal

$$\overline{x} + \sigma_X = 71.2857 + 10.5521 = 81.8378,$$
$$\overline{x} - \sigma_X = 71.2857 - 10.5521 = 60.7336.$$

Therefore the students with scores 85 and 50 have, respectively, over- and underperformed compared to the class. Their statistical significances are computed to be:

$$ss(x_4) = ss(85) = \frac{85 - \overline{x}}{\sigma_X} = \frac{85 - 71.2857}{10.5521} = 1.3,$$
$$ss(x_7) = ss(50) = \frac{50 - \overline{x}}{\sigma_X} = \frac{50 - 71.2857}{10.5521} = -2.0172.$$

We conclude that the score of 50 is statistically more significant than the other outlier of 85.

Question 17.3.

1. Denote the scores by x_1 through x_{10}. It is then easy to compute that the mean score is $\overline{x} = 52.7$ and $x_1^2 + \cdots + x_{10}^2 = 43,059$. Therefore the variance of the scores is $\sigma_X^2 = 1,528.61$, and the standard deviation is $\sigma_X = 39.0974$. Moreover, arranging the scores in increasing order, we conclude that the median is 29.5.

2. In order to compute which of the scores are statistically significant, we compute the one-sigma limits $\overline{x} \pm \sigma_X$. These equal

$$\overline{x} + \sigma_X = 52.7 + 39.0974 = 91.7974,$$
$$\overline{x} - \sigma_X = 52.7 - 39.0974 = 13.6026.$$

Therefore the scores $7, 12, 92$ and the two scores of 111 are statistically significant.

Question 17.5. Let f_i, s_i denote the scores for the first (F) and second (S) midterms, respectively. As in Example 17.8, we then fill in the following table:

Index (i)	f_i	s_i	f_i^2	s_i^2	$f_i s_i$
1	45	33	2,025	1,089	1,485
2	35	36	1,225	1,296	1,260
3	34	40	1,156	1,600	1,360
4	41	43	1,681	1,849	1,763
5	50	48	2,500	2,304	2,400
Total	205	200	8,587	8,138	8,268

1. The mean and median scores are both equal to 41 for the first midterm, and both are equal to 40 for the second midterm. We now compute to four decimal places (with $n = 5$):

$$\sigma_F^2 = \frac{1}{5} \sum_{i=1}^{5} f_i^2 - \overline{f}^2 = \frac{8,587}{5} - 41^2 = 36.4,$$

$$\sigma_S^2 = \frac{1}{5} \sum_{i=1}^{5} s_i^2 - \overline{s}^2 = \frac{8,138}{5} - 40^2 = 27.6.$$

Therefore the standard deviations for the midterm scores are: $\sigma_F = \sqrt{36.4} \simeq 6.0332$ and $\sigma_S = \sqrt{27.6} \simeq 5.2536$, respectively.

2. Using the above table, we further compute:

$$\text{Cov}(F, S) = \frac{1}{5} \sum_{i=1}^{5} f_i s_i - \overline{f} \cdot \overline{s} = \frac{8,268}{5} - 41 \cdot 40 = 13.6,$$

$$\rho_{F,S} = \frac{\text{Cov}(F, S)}{\sigma_F \sigma_S} = \frac{13.6}{\sqrt{36.4 \cdot 27.6}} = 0.4291.$$

Thus the correlation between the scores from both midterms is about 0.4291.

Chapter 18. Regression: Chasing connections in big data

Question 18.1. Let u_i, p_i denote the numbers (in millions) of total users and total photos at a given time. Then we compute the least squares coefficients using the following table.

Index (i)	u_i	p_i	u_i^2	p_i^2	$u_i p_i$
1	8	150	64	22,500	1,200
2	10	170	100	28,900	1,700
3	11	183	121	33,489	2,013
4	12	195	144	38,025	2,340
5	15	231	225	53,361	3,465
Total	56	929	654	176,275	10,718

Now using the formulas in Chapters 17 and 18, we compute (with $n = 5$):

$$\overline{u} = \frac{1}{5} \sum_{i=1}^{5} u_i = \frac{56}{5} = 11.2,$$

$$\overline{p} = \frac{1}{5} \sum_{i=1}^{5} p_i = \frac{929}{5} = 185.8,$$

$$\sigma_U^2 = \frac{1}{5}\sum_{i=1}^{5} u_i^2 - \overline{u}^2 = \frac{654}{5} - 11.2^2 = 5.36,$$

$$\sigma_P^2 = \frac{1}{5}\sum_{i=1}^{5} p_i^2 - \overline{p}^2 = \frac{176,275}{5} - 185.8^2 = 733.36,$$

$$\mathrm{Cov}(U, P) = \frac{1}{5}\sum_{i=1}^{5} u_i p_i - \overline{u}\cdot\overline{p} = \frac{10,718}{5} - 11.2\cdot 185.8 = 62.64,$$

$$\beta_{LS} = \frac{\mathrm{Cov}(U, P)}{\sigma_U^2} = \frac{62.64}{5.36} = 11.6866,$$

$$\alpha_{LS} = \overline{p} - \beta_{LS}\overline{u} = 185.8 - 11.6866\cdot 11.2 = 54.91,$$

$$SSE_{LS} = n(\sigma_P^2 - \beta_{LS}^2\sigma_U^2) = 5(733.36 - 11.6866^2\cdot 5.36) = 6.5466.$$

Thus the least squares model is $P = 54.91 + 11.6866U$. The (relatively) low value of the residual sum of squares SSE_{LS} indicates that the model fits the data well.

Finally, to predict the number of (millions of) photographs for 20 million users, we compute:

$$\widehat{p} = \alpha_{LS} + \beta_{LS}\cdot 20 = 54.91 + 11.6866\cdot 20 = 288.642 \text{ million photographs.}$$

Question 18.3. Let t_i denote the times of measurement, and $l_i = \ln A(t_i)$ be the natural logarithm of the amount of radioactive substance measured at time t_i. Also define $\alpha = \ln A(0)$ and $\beta = \ln(1/2)/T_{1/2}$. Then we have to estimate α and β from the simple linear regression model $L = \alpha + \beta T$. To do so, we use the following table.

Index (i)	t_i	$A(t_i)$	l_i	t_i^2	l_i^2	$t_i l_i$
1	10	1	0	100	0	0
2	15	0.4	-0.9163	225	0.8396	-13.7445
Total	25		-0.9163	325	0.8396	-13.7445

1. To estimate α and β, we use the formulas in Chapters 17 and 18 and compute (with $n = 2$):

$$\overline{t} = \frac{1}{2}\sum_{i=1}^{2} t_i = \frac{25}{2} = 12.5,$$

$$\bar{l} = \frac{1}{2} \sum_{i=1}^{2} l_i = \frac{-0.9163}{2} = -0.45815,$$

$$\sigma_T^2 = \frac{1}{2} \sum_{i=1}^{2} t_i^2 - \bar{t}^2 = \frac{325}{2} - 12.5^2 = 6.25$$

$$\sigma_L^2 = \frac{1}{2} \sum_{i=1}^{2} l_i^2 - \bar{l}^2 = \frac{0.8396}{2} - (-0.45815)^2 = 0.2099,$$

$$\text{Cov}(T, L) = \frac{1}{2} \sum_{i=1}^{2} t_i l_i - \bar{t} \cdot \bar{l} = \frac{-13.7445}{2} - 12.5 \cdot (0.45815)$$

$$= -1.145375,$$

$$\beta_{LS} = \frac{\text{Cov}(T, L)}{\sigma_T^2} = \frac{-1.145375}{6.25} = -0.18326,$$

$$\alpha_{LS} = \bar{l} - \beta_{LS}\bar{t} = -0.45815 - (-0.18326) \cdot 12.5 = 1.8326.$$

Note that we should expect β_{LS} to be negative because during radioactive decay, the amount of material decreases with time.

Having obtained the least squares estimates α_{LS} and β_{LS}, we solve for the initial amount $A(0)$, because $\alpha_{LS} = \ln A(0)$:

$$A(0) = e^{\alpha_{LS}} = e^{1.8326} = 6.25 \text{ g}.$$

Now $\beta_{LS} = \ln(1/2)/T_{1/2}$, so the half-life is estimated to be

$$T_{1/2} = \frac{\ln(1/2)}{\beta_{LS}} = \frac{-0.693147}{-0.18326} = 3.7823 \text{ yr}.$$

2. In order to predict the amount of radioactive substance 25 years from the initial time point, we compute:

$$A(25) = A(0)e^{\ln(1/2) \cdot 25/T_{1/2}} = 6.25 \cdot e^{-0.693147 \cdot 25/3.7823} = 0.064 \text{ g}.$$

3. We compute the residual sum of squares of errors:

$$SSE_{LS} = n(\sigma_L^2 - \beta_{LS}^2 \sigma_T^2) = 2(0.2099 - (-0.18326)^2 \cdot 6.25) = 0.$$

This is not a coincidence: if there are only two data points in a linear regression problem, the line of best fit is in fact the unique line passing through them. We explain this phenomenon in general in Question 18.9.

We could have obtained the values $A(0)$ and $A(25)$ using the radioactive decay model directly. Following the method of Chapter 8, we know that the radioactive substance decays by equal factors in equal intervals of time. In this case, the substance decays by a factor of 0.4 every five years. In other words, if we go backward in time by five years, the amount of the substance was greater by a factor of $1/0.4 = 2.5$ times. Thus if it is 0.4 gram fifteen years from the initial time point, it would be $0.4 \cdot 2.5 = 1$ g five years prior to that (as given). Going back five years at a time, the amount would have been $0.4 \cdot 2.5 \cdot 2.5 \cdot 2.5 = 6.25$ g at the initial time point.

This explains part (1) of the problem – i.e., why $A(0) = 6.25$ g. Similarly, ten years after it is 0.4 grams, the amount should reduce to $0.4 \cdot 0.4^2 = 0.064$ g, which is the answer in part (3). Thus, the exact fit of the model to the data is reflected in our estimates of the initial and future amounts.

Question 18.5. Let v_i denote the volume of the ith measured cube, so that $v_i = l_i^3$ for $i = 1, 2, \ldots, 4$. Then the model is $W = \alpha + \beta V$, which is a simple linear regression model with W and V representing the weight and volume of the cube of clay (plus metal frame), respectively. Therefore we carry out the usual calculations in order to estimate α and β.

Index (i)	l_i	w_i	$v_i = l_i^3$	v_i^2	w_i^2	$v_i w_i$
1	5	180	125	15,625	32,400	22,500
2	7	503	343	117,649	253,009	172,529
3	8	750	512	262,144	562,500	384,000
4	10	1,410	1,000	1,000,000	1,988,100	1,410,000
Total		2,843	1,980	1,395,418	2,836,009	1,989,029

To estimate α and β, we use the formulas in Chapters 17 and 18 and compute (with $n = 4$):

$$\bar{v} = \frac{1}{4} \sum_{i=1}^{4} v_i = \frac{1,980}{4} = 495,$$

$$\bar{w} = \frac{1}{4} \sum_{i=1}^{4} w_i = \frac{2,843}{4} = 710.75,$$

$$\sigma_V^2 = \frac{1}{4} \sum_{i=1}^{4} v_i^2 - \bar{v}^2 = \frac{1,395,418}{4} - 495^2 = 103,829.5$$

$$\sigma_W^2 = \frac{1}{4}\sum_{i=1}^{4} w_i^2 - \overline{w}^2 = \frac{2,836,009}{4} - 710.75^2 = 203,836.6875,$$

$$\mathrm{Cov}(V,W) = \frac{1}{4}\sum_{i=1}^{4} v_i w_i - \overline{v}\cdot\overline{w} = \frac{1,989,029}{4} - 495 \cdot 710.75 = 145,436,$$

$$\beta_{LS} = \frac{\mathrm{Cov}(V,W)}{\sigma_V^2} = \frac{145,436}{103,829.5} = 1.4007,$$

$$\alpha_{LS} = \overline{w} - \beta_{LS}\overline{v} = 710.75 - 1.4007 \cdot 495 = 17.4035.$$

Thus the density of the clay is estimated to be $\beta_{LS} = 1.4007$ g/cm^3, and the weight of the adjustable metal frame is estimated to be $\alpha_{LS} = 17.4035$ g. Finally, we can predict the weight of the original lump of the clay as follows: because the edge length is 100 centimeters, the weight is approximately

$$\widehat{w} = \alpha_{LS} + \beta_{LS}(100^3) = 17.4035 + 1.4007 \cdot 10^6 = 1,400,717.4035 \text{ g}.$$

Therefore the weight of the clay is approximately 1.4 tonnes.

Question* 18.7. Once we convert the problem to the simple linear regression model $W = \alpha + \beta V$, the computations for decomposing the sum of squares of errors are exactly the same as carried out toward the end of Chapter 18 (except that X and Y should be replaced by V and W, respectively, and similarly for x_i and y_i). Now the formula for α_{LS} was proved in the chapter, and we obtained that

$$SSE = n(\alpha - (\overline{w} - \beta v))^2 + A_\beta,$$

where $A_\beta = \sum_{i=1}^{n}(w_i - \beta v_i)^2 - n(\overline{w} - \beta\overline{v})^2$ does not depend on α. To solve for β_{LS}, we now need to minimize the expression A_β with respect to β. To do this, we write A_β as a combination of powers of β:

$$A_\beta = \sum_{i=1}^{n}(w_i - \beta v_i)^2 - n(\overline{w} - \beta\overline{v})^2$$

$$= \sum_{i=1}^{n} w_i^2 - \sum_{i=1}^{n} 2\beta w_i v_i + \sum_{i=1}^{n} \beta^2 v_i^2 - n\overline{w}^2 + 2n\beta\overline{w}\overline{v} - n\beta^2\overline{v}^2$$

$$= \left(\beta^2\sum_{i=1}^{n} v_i^2 - n\beta^2\overline{v}^2\right) - \left(2\beta\sum_{i=1}^{n} w_i v_i - 2n\beta\overline{w}\overline{v}\right) + \left(\sum_{i=1}^{n} w_i^2 - n\overline{w}^2\right).$$

We have thus separated A_β into three expressions corresponding to its quadratic, linear, and constant terms in β. Now verify that taking $n\beta^2$ outside the first expression yields $n\beta^2 \cdot \sigma_V^2$. Similarly, taking $2n\beta$ outside the second expression yields: $2n\beta\text{Cov}(V, W)$; and the last expression is precisely $n\sigma_W^2$. Therefore,

$$A_\beta = n\beta^2\sigma_V^2 - 2n\beta\text{Cov}(V, W) + n\sigma_W^2.$$

Now complete the square in β, formed by the first two terms. Thus,

$$A_\beta = n(\beta^2\sigma_V^2 - 2\beta\text{Cov}(V, W)) + n\frac{\text{Cov}(V, W)^2}{\sigma_V^2} - n\frac{\text{Cov}(V, W)^2}{\sigma_V^2} + n\sigma_W^2$$

$$= n\left(\beta\sigma_V - \frac{\text{Cov}(V, W)}{\sigma_V}\right)^2 + n(\sigma_W^2 - \text{Cov}(V, W)^2/\sigma_V^2).$$

Again, this is minimized over all possible values of β precisely when the first term is minimized. As the first term is a perfect square, it is smallest when it is zero, which happens precisely when $\beta\sigma_V = \text{Cov}(V, W)/\sigma_V$. This yields: $\beta_{LS} = \text{Cov}(V, W)/\sigma_V^2$, as claimed.

Therefore we have proved that the least squares coefficients are as claimed, for the general model $W = \alpha + \beta V$. In fact we have done even more: we showed the following *decomposition of the SSE*:

$$SSE = n\left(\alpha - (\overline{w} - \beta\overline{v})\right)^2 + n\left(\beta\sigma_V - \frac{\text{Cov}(V, W)}{\sigma_V}\right)^2$$
$$+ n(\sigma_W^2 - \text{Cov}(V, W)^2/\sigma_V^2).$$

Now if we set $\beta_{LS} = \text{Cov}(V, W)/\sigma_V^2$ and $\alpha_{LS} = \overline{w} - \beta_{LS}\overline{v}$ as above, then

$$SSE_{LS} = 0 + 0 + n\left(\sigma_W^2 - \frac{\text{Cov}(V, W)^2}{\sigma_V^2}\right) = n\left(\sigma_W^2 - \frac{\text{Cov}(V, W)^2}{\sigma_V^4} \cdot \sigma_V^2\right)$$
$$= n(\sigma_W^2 - \beta_{LS}^2\sigma_V^2),$$

as claimed.

Question* 18.9. We will use the second method outlined in the hint. Suppose there are only two data points (x_1, y_1) and (x_2, y_2). Applying our change of variables, we have (v_1, w_1) and (v_2, w_2), to which we are to fit the linear regression model $W = \alpha + \beta V$. We are given that $v_1 \neq v_2$.

Now we first compute the mean and variance of v_1, v_2 to be:

$$\overline{v} = \frac{1}{2}(v_1 + v_2),$$

$$\sigma_V^2 = \frac{1}{2}\left(\left(v_1 - \frac{v_1 + v_2}{2}\right)^2 + \left(v_2 - \frac{v_1 + v_2}{2}\right)^2\right)$$

$$= \frac{1}{2}\left(((v_1 - v_2)/2)^2 + ((v_2 - v_1)/2)^2\right)$$

$$= \frac{1}{2} \cdot 2 \cdot ((v_1 - v_2)/2)^2 = \frac{1}{4}(v_1 - v_2)^2.$$

Similarly, the variance of w_1, w_2 is $(w_1 - w_2)^2/4$. Moreover, the covariance is

$$\text{Cov}(V, W) = \frac{1}{2}\left((v_1 - \overline{v})(w_1 - \overline{w}) + (v_2 - \overline{v})(w_2 - \overline{w})\right)$$

$$= \frac{1}{2}\left(\frac{v_1 - v_2}{2} \cdot \frac{w_1 - w_2}{2} + \frac{v_2 - v_1}{2} \cdot \frac{w_2 - w_1}{2}\right)$$

$$= \frac{1}{2} \cdot 2 \cdot \frac{(v_1 - v_2)(w_1 - w_2)}{4} = \frac{(v_1 - v_2)(w_1 - w_2)}{4}.$$

Now using the above analysis, together with the formula for SSE_{LS} from the chapter, we compute:

$$SSE_{LS} = n\left(\sigma_W^2 - \frac{\text{Cov}(V, W)^2}{\sigma_V^2}\right)$$

$$= 2\left(\frac{(w_1 - w_2)^2}{4} - \frac{((v_1 - v_2)(w_1 - w_2)/4)^2}{(v_1 - v_2)^2/4}\right)$$

$$= 2 \cdot \frac{(w_1 - w_2)^2}{4} - 2 \cdot \frac{(w_1 - w_2)^2}{4} = 0.$$

Practice exams

Practice exams for Chapters 1-9

These exams have a total of 100 points. You have 75 minutes. Try to show your work, so that if you make a mistake you will at least get partial credit. You may use any result arrived at in class. The points attached to each problem are indicated beside the problem. You are not allowed books, notes, computers, or cell phones.

Here is a list of formulas that can be useful:

- For any positive real A and B and any real c and d, the following properties hold:

$$A^c \cdot A^d = A^{c+d}, \qquad A^c/A^d = A^{c-d}, \qquad (A^c)^d = A^{cd},$$
$$A^c \cdot B^c = (AB)^c, \qquad A^c/B^c = (A/B)^c.$$

- For positive numbers a and B, the logarithm $\log_a B$ is defined as the solution to the equation:

$$a^x = B \qquad \Longleftrightarrow \qquad x = \log_a B.$$

The number a is the base of $\log_a$. The standard notations are

$$\log = \log_{10}, \quad \ln = \log_e.$$

For positive numbers a, B, C and any real number r, the following formulas hold:

$$\log_a(B \cdot C) = \log_a B + \log_a C; \qquad \log_a(B^r) = r \cdot \log_a B.$$

423

- In the case of a compounding scheme with compound interest rate $r\%$ compounded n times per year, the amount accumulated (or owed) after t years on the principal of A dollars is

$$A(t) = A\left(1 + \frac{r}{n \cdot 100}\right)^{nt}.$$

- If $r\%$ is the compound annual rate, and it is compounded n times per year, then the effective annual rate r_{eff} is computed according to the formula

$$\frac{r_{\text{eff}}}{100} = \left(1 + \frac{r}{n \cdot 100}\right)^{n} - 1.$$

- To model a process of exponential growth or decay use the following formula:

$$A(t) = A(0)e^{kt},$$

where $A(t)$ is the quantity at the moment t, $A(0)$ is the initial quantity, and k is the relative growth rate (if $k > 0$) or the relative rate of decay (if $k < 0$). If this is a process of growth, then the relative growth rate k is related to the doubling time T_d by the formula:

$$k = \frac{\ln 2}{T_d}, \quad T_d = \frac{\ln 2}{k} \qquad (k > 0).$$

If this is a process of decay, then the relative rate of decay k is related to the half-life $T_{1/2}$ by the formula:

$$k = -\frac{\ln 2}{T_d}, \quad T_d = -\frac{\ln 2}{k} \qquad (k < 0).$$

- The sum of a geometric series $a(1 + r + r^2 + \cdots + r^n)$ with finitely many terms is

$$\frac{a(1 - r^{n+1})}{1 - r}, \quad \text{if} \quad r \neq 1.$$

- Suppose a person takes a loan or mortgage of M dollars from a bank that compounds n times per year at $r\%$ annual interest rate. The mortgage has to be paid back in t years by paying equal installments of A dollars n times per year, and the first installment is paid one compounding period from today. Then to find A, we have to equate the value of the mortgage in t years

$$M \cdot p^{nt},$$

with the total value of all installments:

$$M \cdot p^{nt} = A \left(1 + p + p^2 + \cdots + p^{nt-1}\right) = \frac{A(1 - p^{nt})}{1 - p},$$

where
$$p = \left(1 + \frac{r}{n \cdot 100}\right)$$

is the compounding factor.

EXAM 1

---

1. (30) Solve the following equations:

   (a)
   $$5 \cdot 3^y = \frac{5}{9}.$$

   (b)
   $$\sqrt[3]{x^{30}} \cdot x^2 \cdot x^{-5} \cdot (x^{-3})^2 = 7.$$

   Hint: Use properties of exponents.

2. (20) Suppose you make a deposit in a bank that compounds quarterly at an annual interest rate of 8%. How long does it take for the investment to double? Give your answer to the nearest quarter of a year.

3. (25) A population of yellow bugs P_y doubles every half-month: $P_y(\frac{1}{2}) = 2P_y(0)$, and a population of blue bugs P_b triples every three-fourths of a month: $P_b(\frac{3}{4}) = 3P_b(0)$.

   (a) Find the relative growth rate per month for each population.

   (b) If $P_y(0) = P_b(0)$, find the ratio P_y/P_b after 3 months.

4. (25) A person borrows $20,000 from a bank that compounds quarterly at an annual interest rate of 8%. The loan has to be repaid in 16 quarterly installments, $X each, over 4 years. The first installment has to be paid a quarter from now. Find the quarterly payment X. Give your answer to the nearest cent.

$$\boxed{\text{EXAM 2}}$$

1. (25) Solve the following equations.

   (a)
   $$\frac{8\,x^{-\frac{6}{5}}x^7}{\sqrt[5]{x^4}} = x^2.$$

   (b) Here, log stands for $\log_{10}$.
   $$\log\left(\frac{36y^3}{5}\right) - 2\log 6 + \log 5 = -6.$$

2. (30) A bank uses a compound interest model with 6% annual interest rate compounded monthly.

   (a) How much should one deposit now to accumulate $80,000 in 20 years? Give your answer up to a cent.

   (b) Find the effective annual rate. Give your answer up to two decimal points of a percent.

   (c) What is the present value of a dollar in 2007 (at the same time of the year) under this compounding model? Give your answer to the nearest cent.

3. (20) At 10 am, snails Bob and Carl are 4 feet east from snail Alex on a straight path. Alex and Bob move east at a constant speed of 5 and 2 feet per hour, respectively. Carl moves west at a constant speed of 1 foot per hour. When Alex catches up with Bob, he turns around and starts moving west at the same constant speed of 5 feet per hour. When will Alex catch up with Carl? Give your answer in hours and minutes.

4. (25) (a) A fragment of a bone found in an archaeological site contains 42% of the atmospheric level of carbon-14. Assuming that the atmospheric level of carbon-14 is constant and its half-life is 5,730 years, estimate the age of the bone. Give your answer to the nearest hundred years.

   (b) A population of grasshoppers grows by a factor of 1.5 every 1.5 weeks. A population of silver-spotted skipper butterflies grows by a factor of 3 every 3 weeks. Which population grows faster? Find the doubling time in each case.

Practice exams for Chapters 9-16

These exams have a total of 100 points. You have 75 minutes. Try to show your work, so that if you make a mistake you will at least get partial credit. You may use any result arrived at in class. The points attached to each problem are indicated beside the problem. You are not allowed books, notes, computers, or cell phones.

Here is a list of formulas that can be useful:

- The sum of a geometric series $A(1 + r + r^2 + \cdots + r^n)$ with finitely many terms is

$$\frac{A(1 - r^{n+1})}{1 - r}, \quad \text{if} \quad r \neq 1.$$

- The sum of a geometric series $A(1 + r + r^2 + \cdots)$ with infinitely many terms is

$$\frac{A}{1 - r}, \quad \text{if} \quad -1 < r < 1.$$

- Suppose a person takes a loan or mortgage of M dollars from a bank that compounds n times per year at $r\%$ annual interest rate. The mortgage has to be paid back in t years by paying equal installments of A dollars n times per year, and the first installment is paid one compounding period from today. Then to find A, we have to equate the value of the mortgage in t years

$$M \cdot p^{nt},$$

with the total value of all installments:

$$M \cdot p^{nt} = A \left(1 + p + p^2 + \cdots + p^{nt-1}\right) = \frac{A(1 - p^{nt})}{1 - p},$$

where

$$p = \left(1 + \frac{r}{n \cdot 100}\right)$$

is the compounding factor.

- The number of permutations (the order of collection matters) of k items out of n items is

$$_nP_k = n \cdot (n - 1) \cdot (n - 2) \cdots (n - k + 1) = \frac{n!}{(n - k)!}.$$

- The number of combinations (the order of collection doesn't matter) of k items out of n items is

$$_nC_k = \frac{_nP_k}{_kP_k} = \frac{n!}{(n-k)!k!} = \frac{n \cdot (n-1) \cdot (n-2) \cdots (n-k+1)}{k \cdot (k-1) \cdots 1}.$$

- The probability of any event E is the quotient of the number of outcomes that constitute E divided by the number of all possible outcomes of the experiment:

$$P(E) = \frac{\#E}{\#\Omega}.$$

- Two events are *independent* if the occurrence of one has no influence on the probability of the other. Two events A and B are independent if and only if the probability of both of them happening (probability of their intersection) equals the product of their probabilities:

$$P(A \cap B) = P(A) \cdot P(B).$$

This formula holds if and only if A and B are independent.

- The *conditional probability* of A given B is the probability of event A occurring *given* that event B has occurred. It is denoted by $P(A|B)$. By definition, it equals the probability of both events happening $P(A \cap B)$ divided by the probability of B:

$$P(A|B) = \frac{P(A \cap B)}{P(B)}.$$

For any two events A and B we have:

$$P(A \cap B) = P(B) \cdot P(A|B), \quad \text{and}$$

$$P(A \cap B) = P(A) \cdot P(B|A).$$

EXAM 3

---

1. (30) (a) Find the exact value of the repeating decimal

$$4.84848484\ldots$$

(b) An exam has infinitely many questions. The first question is worth 30 points. Each next question is worth $\frac{7}{10}$ of the previous question. What is the total number of points for all questions? To get an A, a student needs to earn at least 90 points. Jessica solved the first 7 questions. Will she get an A?

2. (20) A person takes out a loan of $200,000 from a bank that compounds monthly with a compound annual interest rate of 8%. The loan is to be paid back in 15 years. The first payment is made a month from now. Find the monthly payment.

3. (15) (a) In how many ways can 8 operas be chosen and scheduled to be staged in the next season if the company is ready to perform 12 different operas and insists on starting the season with *Norma* and finishing with *Pagliacci*? Each of the 8 operas can be performed only once.

(b) In how many ways can 6 shirts and 3 pairs of pants, out of 10 shirts and 5 pairs of pants, be chosen to take on a trip?

4. (35) (a) What is greater: the probability of getting more than one head in four coin tosses or the probability of drawing two number cards (1 to 10) in a row from a deck? Compute both probabilities.

(b) Suppose you roll two dice. Event A: the sum of the outputs is greater than 8. Event B: the first result is a 4. Are the events independent? Find the probabilities $P(A)$, $P(B)$, and $P(A \cap B)$.

(c) In question (b), what is the conditional probability $P(A|B)$ that the sum is greater than 8, given that the first die yields 4? What is the conditional probability $P(B|A)$ that the first die yields 4 if the sum of the two is greater than 8?

EXAM 4

1. (30) (a) Find the exact value of the repeating decimal

$$3.2424242424\ldots$$

(b) Once activated, a coffee machine starts pouring out drops of coffee. The first drop weighs 1 ounce, and each next drop is $\frac{11}{12}$ of the previous drop. How much coffee will you get if you let the machine run forever? Will the first 11 drops be enough to get 8 ounces of coffee?

2. (15) A person takes a loan of \$400,000 from a bank that compounds monthly with a compound annual interest rate of 5%. The loan is to be paid back in 20 years. The first payment is made a month from now. Find the monthly payment.

3. (15) Two journalists, one philosopher, and three astronomers need to be chosen to participate in a panel discussion.

(a) In how many ways can the group be formed out of 4 journalists, 5 philosophers, and 10 astronomers?

(b) Once the participants are chosen, in how many ways can they be scheduled to speak? (Assume that all participants are distinct; for example the schedule with journalist A followed by journalist B is different from the schedule with journalist B followed by journalist A.)

4. (40) (a) Find the probability of getting at least 4 tails in 6 coin tosses.

(b) Suppose you draw two cards from a deck (without replacement). Event A: the first card is a spade. Event B: the second card is a heart. Compute the probabilities $P(A)$, $P(B)$, and $P(A \cap B)$. Are events independent?

(c) Suppose you roll two dice. Event A: the sum of the outputs is 5. Event B: the second output is greater than or equal to 3. Find the conditional probabilities $P(A|B)$ and $P(B|A)$. Are events A and B independent?

Practice final exams

These exams have a total of 200 points. You have 2 hours 30 minutes. Try to show your work, so that if you make a mistake you will at least get partial credit. You may use any result arrived at in class. The points attached to each problem are indicated beside the problem. You are not allowed books, notes, computers, or cell phones.

Here is a list of formulas that can be useful:

- For any positive real A and B and any real c and d, the following properties hold:

$$A^c \cdot A^d = A^{c+d}, \qquad A^c/A^d = A^{c-d}, \qquad (A^c)^d = A^{cd},$$
$$A^c \cdot B^c = (AB)^c, \qquad A^c/B^c = (A/B)^c.$$

- For positive numbers a and B, the logarithm $\log_a B$ is defined as the solution to the equation:

$$a^x = B \qquad \Longleftrightarrow \qquad x = \log_a B.$$

The number a is the base of $\log_a$. The standard notations are

$$\log = \log_{10}, \qquad \ln = \log_e.$$

For positive numbers a, B, C and any real number r, the following formulas hold:

$$\log_a(B \cdot C) = \log_a B + \log_a C; \qquad \log_a(B^r) = r \cdot \log_a B.$$

- In the case of a compound interest model with annual interest rate $r\%$ compounded n times per year, the amount accumulated (or owed) after t years on the principal of A dollars is

$$A(t) = A\left(1 + \frac{r}{n \cdot 100}\right)^{nt}.$$

- If $r\%$ is the compound annual rate, and it is compounded n times per year, then the effective annual rate r_{eff} is computed according to the formula
$$\frac{r_{\text{eff}}}{100} = \left(1 + \frac{r}{n \cdot 100}\right)^n - 1.$$

- To model the process of exponential growth or decay use the following formula:
$$A(t) = A(0)e^{kt},$$

where $A(t)$ is the quantity at the moment t, $A(0)$ is the initial quantity, and k is the relative growth rate (if $k > 0$) or the relative rate of decay (if $k < 0$). If this is a process of growth, then the relative growth rate k is related to the doubling time T_d by the formula:

$$k = \frac{\ln 2}{T_d}, \qquad T_d = \frac{\ln 2}{k} \qquad (k > 0).$$

If this is a process of decay, then the relative rate of decay k is related to the half-life $T_{1/2}$ by the formula:

$$k = -\frac{\ln 2}{T_{1/2}}, \quad T_{1/2} = -\frac{\ln 2}{k} \qquad (k < 0).$$

- The sum of a geometric series $a(1 + r + r^2 + \cdots + r^n)$ with finitely many terms is

$$\frac{a(1 - r^{n+1})}{1 - r}, \quad \text{if} \quad r \neq 1.$$

- The sum of a geometric series $a(1 + r + r^2 + \cdots)$ with infinitely many terms is

$$\frac{a}{1 - r}, \quad \text{if} \quad -1 < r < 1.$$

- Suppose a person takes a loan or mortgage of M dollars from a bank that compounds n times per year at $r\%$ annual interest rate. The mortgage has to be paid back in t years by paying equal installments of A dollars n times per year, and the first installment is paid one compounding period from today. Then to find A, we have to equate the value of the mortgage in t years

$$M \cdot p^{nt},$$

with the total value of all installments:

$$M \cdot p^{nt} = A\left(1 + p + p^2 + \cdots + p^{nt-1}\right) = \frac{A(1 - p^{nt})}{1 - p},$$

where

$$p = \left(1 + \frac{r}{n \cdot 100}\right)$$

is the compounding factor.

- The number of permutations (the order of collection matters) of k items out of n items is

$$_nP_k = n \cdot (n - 1) \cdot (n - 2) \cdots (n - k + 1) = \frac{n!}{(n - k)!}.$$

- The number of combinations (the order of collection doesn't matter) of k items out of n items is

$$_nC_k = \frac{_nP_k}{_kP_k} = \frac{n!}{(n - k)!k!} = \frac{n \cdot (n - 1) \cdot (n - 2) \cdots (n - k + 1)}{k \cdot (k - 1) \cdots 1}.$$

- The probability of any event E is the quotient of the number of outcomes that constitute E divided by the number of all possible outcomes of the experiment:

$$P(E) = \frac{\#E}{\#\Omega}.$$

- Two events are *independent* if the occurrence of one has no influence on the probability of the other. Two events A and B are independent if and only if the probability of both of them happening (probability of their intersection) equals the product of their probabilities:

$$P(A \cap B) = P(A) \cdot P(B).$$

This formula holds if and only if A and B are independent.

- The *conditional probability* of A given B is the probability of event A occurring *given* that event B has occurred. It is denoted by $P(A|B)$.

$$P(A|B) = \frac{P(A \cap B)}{P(B)}.$$

For any two events A and B we have:

$$P(A \cap B) = P(B) \cdot P(A|B), \quad \text{and}$$

$$P(A \cap B) = P(A) \cdot P(B|A).$$

- *Bayes' law* of conditional probabilities: for any two events A and B with positive probabilities, the conditional probability of the event B given the event A is computed according to the following formula:

$$P(B|A) = \frac{P(A|B) \cdot P(B)}{P(A)}.$$

- Let a and x be integers and n be a positive integer. We say that a is *congruent* to x *modulo* n if $(a - x)$ is divisible by n. The notation is

$$a \equiv x \mod n$$

In other words, $a \equiv x \mod n$ if a and x differ by a multiple of n.

- Suppose
$$a \equiv x \mod n \quad \text{and} \quad b \equiv y \mod n.$$

  Attention: n is the same in both formulas. Then

$$a \cdot b \equiv x \cdot y \mod n.$$

- If s is a positive integer, and $a \equiv x \mod n$, then the following holds as well:
$$a^s \equiv x^s \mod n.$$

EXAM 5

---

1. (25) Solve the equation:
$$24 \cdot 2^{5x+1} = 18.$$

2. (25) A bank compounds monthly at an annual compound interest rate of $r\%$. With this compounding scheme, the value of \$1 in 15 years will be \$2.

   (a) Find the annual compound interest rate r up to two decimal digits of a percent.

   (b) With the same compounding scheme, what will be the value of \$1 in 8 years? Give your answer to the nearest cent.

3. (30) A population of blue dragonflies increased by 50% in 3 weeks.

   (a) Find the relative growth rate (per week) of this population. Give your answer up to five decimals.

   (b) How many weeks does it take for this population to grow from $10,000$ to $60,000$? Give your answer up to a day.

4. (30) In a marathon competition with infinitely many competitors, the prize fund of \$6,561 is divided as follows: the winner gets $\frac{1}{3}$ of the prize, the runner arriving second $\frac{1}{9}$ of the prize, the runner arriving third $\frac{1}{27}$ of the prize, and so on to infinity. The rest of the prize is left to the organizers.

   (a) What part of the prize did all runners get? *Hint:* You have to sum up all parts of the prize paid to the runners.

(b) How much money did the organizers get? Give your answer to the nearest cent.

(c) How much money did the first five runners get?

5. (35) A bag contains 10 white, 5 red, and 15 blue marbles. Drawing 3 marbles randomly, what is the probability

(a) of getting 3 white marbles?

(b) that none of the marbles is red?

(c) that exactly one of the marbles is red?

6. (30) In London, a weather forecaster predicts rain on $\frac{4}{9}$ of the days, a cloudy sky on $\frac{1}{3}$ of the days, and a sunny sky on $\frac{2}{9}$ of the days. If the forecaster predicts rain, Mr. Pickwick takes his big umbrella with probability $\frac{3}{4}$ and his small umbrella with probability $\frac{1}{4}$. If the forecaster predicts a cloudy sky, there is an equal probability $\left(\frac{1}{3}\right)$ that Mr. Pickwick will take his big umbrella, small umbrella, or no umbrella. If the forecaster predicts a sunny sky, Mr. Pickwick will take his small umbrella with probability $\frac{1}{6}$, and no umbrella otherwise. Find the probability that the forecaster predicted a cloudy sky, given that Mr. Pickwick is carrying his small umbrella. *Hint:* Use Bayes' law.

7. (25) Find the last digit of 447^{447}.

EXAM 6

---

1. (25) Solve the equation
$$6^{2x} \cdot 3^{(x+2)} = \frac{\sqrt[7]{36}}{2^{(x+2)}}.$$

2. (30) A sample of a radioactive substance decays to $\frac{3}{4}$ of its original mass in $\frac{3}{4}$ of a year.

(a) Find the half-life of the substance. Give your answer to the nearest month.

(b) How much will a sample of this substance decay in 1 year? Give your answer to the nearest tenth of a percent.

3. (30) A person makes regular payments of \$250 per month to a bank that compounds monthly with an annual interest rate of 4%.

    (a) How much money will accumulate after 5 years? Assume that the last payment does not get compounded. Give your answer to the nearest cent.

    (b) How many payments will it take to accumulate the total of \$25,000?

    (c) How long will part (b) take? Give your answer in years and months.

4. (30) A trotting horse covers 1 mile in the first 10 minutes, $\frac{10}{11}$ of a mile in the next 10 minutes, and each subsequent 10 minutes $\frac{10}{11}$ of the distance covered in the previous 10 minutes.

    (a) If the horse runs forever, what is the total distance it will cover?

    (b) What distance will the horse cover in the first hour? Give your answer to the nearest one hundredth of a mile.

5. (35) A box contains 4 chocolate-covered cherries, 8 coffee-flavored chocolates, and 12 mint-flavored chocolates. Drawing 4 chocolates randomly, what is the probability

    (a) of getting 4 cherries?

    (b) of getting *exactly* 2 coffee-flavored chocolates?

    (c) of getting *at least* 2 coffee-flavored chocolates?

6. (25) One of the distinctive characteristics of the rare *Cycnoches Pentadactylon* orchid is that its flowers have the scent of marshmallows[1]. About 70% of its plants have this characteristic. However, about 1 in 500 similar-looking common orchids is reported to have the scent of marshmallows as well. Statistically, only about 1 in 1,000 similar-looking orchids is actually *Cycnoches Pentadactylon*. If you find a marshmallow-scented orchid, what is the probability that it is a *Cycnoches Pentadactylon*? *Hint:* Use Bayes' law.

7. (25)   (a) Find the last digit of 9^{312}.

      (b) Compute 4^{77} (mod 11).

---

[1] This question is not based on a confirmed botanical fact.

Solutions to practice exams

EXAM 1

1. Solve equations:

   (a)
   $$5 \cdot 3^y = \frac{5}{9}.$$

   Dividing both sides by 5, we get

   $$3^y = \frac{1}{9} \quad \Longrightarrow \quad 3^y = 3^{-2} \quad \Longrightarrow \quad y = 2.$$

   (b)
   $$\sqrt[3]{x^{30}} \cdot x^2 \cdot x^{-5} \cdot (x^{-3})^2 = 7.$$

   Using properties of exponents, we get

   $$x^{30/3+2-5+(-3)\cdot 2} = 7 \quad \Longrightarrow \quad x^{10-3-6} = 7 \quad \Longrightarrow \quad x = 7.$$

2. Suppose the amount of the deposit is A. We have $r = 8\%$, $n = 4$ (quarters). Then the formula for the accumulated amount is

   $$A\left(1 + \frac{8}{4 \cdot 100}\right)^{4t}.$$

   This should be equated to $2A$ since we want the initial investment to double. We have the equation

   $$2A = A\left(1 + \frac{8}{4 \cdot 100}\right)^{4t}.$$

   The deposit amount A cancels from both sides and $\frac{8}{400} = 0.02$, so we have
   $$2 = (1.02)^{4t}.$$

   Taking the natural logarithm of both sides, we find

   $$\ln 2 = \ln (1.02)^{4t} = 4t \cdot \ln 1.02.$$

Then solving for t, we get

$$t = \frac{1}{4} \cdot \frac{\ln 2}{\ln 1.02} \simeq 8.7507 \simeq 8\frac{3}{4}.$$

It will take about $8\frac{3}{4}$ years.

3. (a) Let k_y be the relative growth rate per month for yellow bugs, and k_b for blue bugs. By the population growth formula, we have

$$2P_y(0) = P_y\left(\frac{1}{2}\right) = P_y(0)e^{k_y \cdot \frac{1}{2}}.$$

$P_y(0)$ cancels on both sides, and we have

$$2 = e^{k_y \cdot \frac{1}{2}}.$$

Taking the natural logarithm, we get

$$\ln 2 = \ln e^{k_y \cdot \frac{1}{2}} = k_y \cdot \frac{1}{2}.$$

Solving for k_y gives

$$k_y = 2\ln 2 \simeq 1.3863/\text{month}.$$

Similarly,

$$3P_b(0) = P_b\left(\frac{3}{4}\right) = P_b(0)e^{k_b \cdot \frac{3}{4}}.$$

Then
$$3 = e^{k_b \cdot \frac{3}{4}}, \quad \text{therefore} \quad \ln 3 = \ln e^{k_b \cdot \frac{3}{4}} = k_b \cdot \frac{3}{4}.$$

Solving for k_b gives

$$k_b = \frac{4}{3} \cdot \ln 3 \simeq 1.4648/\text{month}.$$

(b) Using the population growth model with $k_y = 2\ln 2$ and $k_b = \frac{4}{3}\ln 3$, we have for both populations after 3 months:

$$P_y(3) = P_y(0)e^{k_y \cdot 3} = P_y(0)e^{3\cdot 2\cdot \ln 2} = P_y(0)e^{6\cdot \ln 2}.$$

$$P_b(3) = P_b(0)e^{k_b \cdot 3} = P_b(0)e^{3\cdot \frac{4}{3}\cdot \ln 3} = P_b(0)e^{4\cdot \ln 3}.$$

Now, we have $P_y(0) = P_b(0)$. Therefore,

$$\frac{P_y(3)}{P_b(3)} = \frac{e^{6 \cdot \ln 2}}{e^{4 \cdot \ln 3}}.$$

The numerical value can be computed with the help of a calculator. However, you can do this by hand using the properties of the logarithms:

$$e^{6 \cdot \ln 2} = e^{\ln 2^6} = 2^6 = 64, \quad \text{and} \quad e^{4 \cdot \ln 3} = e^{\ln 3^4} = 3^4 = 81.$$

So the ratio of populations is

$$\frac{P_y(3)}{P_b(3)} = \frac{64}{81}.$$

The same answer can be obtained by the following argument: the population of yellow bugs P_y doubles every $1/2$ of a month; therefore in 3 months it will double 6 times and will amount to $2^6 P_y(0)$. The population of blue bugs P_b triples every $3/4$ of a month; therefore in 3 months it will triple 4 times and will amount to $3^4 P_b(0)$. Since $P_y(0) = P_b(0)$, the ratio after 3 months is

$$\frac{P_y(3)}{P_b(3)} = \frac{2^6 \cdot P_y(0)}{3^4 \cdot P_b(0)} = \frac{64}{81}.$$

4. We have $r = 8\%$, $n = 4$, $A = \$20,000$. First find the value of the loan in 4 years.

$$A(4) = A\left(1 + \frac{8}{400}\right)^{16} = 20,000(1.02)^{16} \simeq 27,455.71.$$

Now find the value of the quarterly payments. The first payment is made 1 quarter from now; therefore it is compounded 15 times: $X\left(1 + \frac{8}{400}\right)^{15}$. The next payment is compounded 14 times, and so on to the last payment of $\$X$, which is not compounded at all since it is made on the day when the loan closed. The total value of all payments is

$$X\left(1 + \frac{8}{400}\right)^{15} + X\left(1 + \frac{8}{400}\right)^{14} + \cdots + X\left(1 + \frac{8}{400}\right) + X =$$

$$= X(1.02)^{15} + X(1.02)^{14} + \cdots + X(1.02) + X.$$

Using the formula for the sum of the finite geometric series with $r = 1.02$ and $n = 15$, we have

$$X(1.02)^{15} + X(1.02)^{14} + \cdots + X(1.02) + X$$
$$= X\frac{(1.02)^{16} - 1}{1.02 - 1} \simeq X \cdot 18.63929.$$

This should be equal to the value of the loan:

$$\$27,455.71 = 18.63929 \cdot X.$$

Therefore,

$$X \simeq \$1,473.00.$$

EXAM 2

1. Solve the equations:

(a)
$$\frac{8\,x^{-\frac{6}{5}}x^7}{\sqrt[5]{x^4}} = x^2.$$

$$8x^{-\frac{6}{5}}x^{-\frac{4}{5}}x^7 = x^2$$

$$8x^{-\frac{10}{5}+7} = x^2$$

$$8x^5 = x^2, \quad \implies \quad x = 0, \quad \text{or}$$

$$8x^3 = 1 \quad \implies \quad x^3 = \frac{1}{8} \quad \implies \quad x = \frac{1}{2}.$$

Finally,

$$x = 0, x = \frac{1}{2}.$$

(b) Here log stands for $\log_{10}$.

$$\log\left(\frac{36y^3}{5}\right) - 2\log 6 + \log 5 = -6.$$

$$\log{(36y^3)} - \log 5 - 2\log 6 + \log 5 = -6.$$

$$2\log 6 + \log y^3 - 2\log 6 = -6.$$

$$\log y^3 = -6.$$

$$10^{-6} = y^3.$$

Then we get

$$y = 10^{-2} = \frac{1}{100} = 0.01.$$

2. (a)

$$80{,}000 = A(1 + \frac{6}{1{,}200})^{12 \cdot 20} = A(1.005)^{240}.$$

$$A = \frac{80{,}000}{(1.005)^{240}} \simeq \$24{,}167.69.$$

(b)

$$1 + \frac{r_{\text{eff}}}{100} = (1 + \frac{6}{1{,}200})^{12} = (1.005)^{12} \simeq 1.0617.$$

$$\frac{r_{\text{eff}}}{100} \simeq 0.0617, \qquad \Longrightarrow \qquad r_{\text{eff}} \simeq 6.17\%.$$

(c)

You had \$1 in 2007 and you want to know its value 5 years later.

$$A(5) = 1(1 + \frac{6}{1{,}200})^{5 \cdot 12} = (1.005)^{60} \simeq \$1.35.$$

3. First we will find the time when Alex meets Bob. Let the speed of Alex be $v_A = 5$ ft/h and the speed of Bob be $v_B = 2$ ft/h. They are moving in the same direction (east), and the initial distance between them is $D_1 = 4$ ft. Therefore, they will meet after

$$t_1 = \frac{D_1}{v_A - v_B} = \frac{4}{5 - 2} = \frac{4}{3} \text{ h} = 1 \text{ h } 20 \text{ min.}$$

Now let us find the time it takes Alex to catch up with Carl, counting from when Alex and Bob meet. Let Alex's speed again be $v_A = 5$ ft/h, and Carl's speed be $v_C = 1$ ft/h. They move in the same direction (west). To find the time, we need to find the distance between them at the moment when Alex meets Bob. At this moment, Bob is

$$2 \cdot \frac{4}{3} = \frac{8}{3} = 2\frac{2}{3} \text{ ft}$$

east from his starting point, because he moved at the speed $v_B = 2$ ft/h during $t_1 = \frac{4}{3}$ h. At the same moment, Carl is

$$1 \cdot \frac{4}{3} = \frac{4}{3} \text{ ft}$$

west from the same point, because he moved all this time at a speed of 1 ft/h. Therefore, the distance D_2 between Alex and Carl after $t_1 = \frac{4}{3}$ h is

$$D_2 = 2\frac{2}{3} + 1\frac{1}{3} = 4 \text{ ft.}$$

Now, we can find the time t_2 it takes Alex to catch up with Carl, counting from when Alex and Bob meet:

$$t_2 = \frac{D_2}{v_A - v_C} = \frac{4}{5 - 1} = 1 \text{ h.}$$

Totally, the time elapsed from 10 am to when Alex and Carl meet, is

$$t = t_1 + t_2 = 1\frac{1}{3} + 1 = 2\frac{1}{3} = 2 \text{ h } 20 \text{ min.}$$

Alex and Carl will meet at 12:20 pm.

4. (a) Let k be the relative rate of decay of carbon-14 and $T_{1/2} = 5,730$ its half-life. Then

$$k = -\frac{\ln 2}{T_{1/2}} = -\frac{\ln 2}{5,730} \simeq -0.00012097/\text{yr.}$$

The age t of the object is given by the equation:

$$0.42 = 1 \cdot e^{kt} = e^{-0.00012097t}$$

$$\implies \ln 0.42 = -0.00012097 \cdot t$$

$$\implies t = \frac{\ln 0.42}{-0.00012097} \simeq 7,171 \simeq 7,200 \text{ yr.}$$

The bone is approximately $7,200$ years old.

(b) Let k_g be the relative growth rate of the grasshoppers (per week), and k_b for the butterflies. Let A_g and A_b be their initial populations. Then we have the following equations:

$$1.5A_g = A_g e^{1.5k_g} \qquad\qquad 3A_b = A_b e^{3k_b}$$
$$\ln 1.5 = 1.5 \cdot k_g \qquad\qquad \ln 3 = 3 \cdot k_b$$
$$k_g = \tfrac{\ln 1.5}{1.5} \qquad\qquad\qquad k_b = \tfrac{\ln 3}{3}$$
$$k_g \simeq 0.2703 \qquad\qquad\qquad k_b \simeq 0.3662.$$

Clearly, the butterfly population grows faster.

For the doubling times, we have

$$T_{d,g} = \tfrac{\ln 2}{k_g} \qquad\qquad\qquad T_{d,b} = \tfrac{\ln 2}{k_b}$$
$$T_{d,g} = \simeq 2.6 \text{ weeks} \qquad\qquad T_{d,b} \simeq 1.9 \text{ weeks.}$$

The butterfly population grows faster. The doubling time for the butterflies is approximately 1.9 weeks, and for the grasshoppers 2.6 weeks.

EXAM 3

1. (a) We compute:

$$4.84848484\ldots = \frac{48}{10} + \frac{48}{1,000} + \cdots$$
$$= \frac{48}{10}\left(1 + \frac{1}{100} + \frac{1}{100^2} + \cdots\right)$$
$$= \frac{48}{10} \cdot \frac{1}{1 - \frac{1}{100}} = \frac{24}{5} \cdot \frac{100}{99} = \frac{160}{33}.$$

(b) The total number of points on the exam is

$$30\left(1 + \frac{7}{10} + \frac{7^2}{10^2} + \cdots\right) = 30 \cdot \frac{1}{1 - \frac{7}{10}} = 30 \cdot \frac{10}{3} = 100.$$

Solving the first 7 questions yields

$$30\left(1 + \frac{7}{10} + \frac{7^2}{10^2} + \cdots + \frac{7^6}{10^6}\right) = 30 \cdot \frac{1 - \frac{7^7}{10^7}}{1 - \frac{7}{10}} \simeq 91.8.$$

Therefore, Jessica will get an A.

2. First we find the number $p = (1 + \frac{8}{1,200}) \simeq 1.00666667$. Now, the amount to be paid is

$$Mp^{nt} = 200,000 \cdot (1.00666667)^{12 \cdot 15} \simeq 661,384.30.$$

Suppose the monthly payment is A. Then the sum of all payments is

$$A\left(1 + p + p^2 + \cdots + p^{179}\right) = A \cdot \frac{1 - p^{180}}{1 - p} \simeq A \cdot \frac{1 - 3.30692}{-0.006667}$$
$$\simeq 346.038A.$$

Then the monthly payment is

$$A = \frac{661,384.30}{346.038} \simeq \$\,1,911.30.$$

3. (a) Because the first and the last operas are already chosen, it remains to choose and schedule 6 out of 10 operas. The order of the performances matters. Therefore, the number of ways is given by the number of permutations of 6 out of 10:

$$_{10}P_6 = \frac{10!}{4!} = 10 \cdot 9 \cdot 8 \cdot 7 \cdot 6 \cdot 5 = 151,200.$$

(b) In this case, the order does not matter, so the answer is given by the product of two numbers of combinations:

$$_{10}C_6 \cdot {}_5C_3 = \frac{10 \cdot 9 \cdot 8 \cdot 7}{4 \cdot 3 \cdot 2 \cdot 1} \cdot \frac{5 \cdot 4}{2 \cdot 1} = 2,100.$$

4. (a) The complementary event to having more than one head in 4 coin tosses is having 0 or 1 head. There is only one outcome with 0 heads, TTTT, and four outcomes with exactly one head, HTTT, THTT, TTHT, TTTH. Alternatively, the number of outcomes with 0 heads is given by $_4C_0 = 1$ and with 1 head by $_4C_1 = 4$. Then the probability of the complementary event is $\frac{1+4}{2^4} = \frac{5}{16}$. Therefore the probability we are looking for (more than one head in 4 tosses) is $1 - \frac{5}{16} = \frac{11}{16}$.

The probability of getting two number cards in a row is (since there are 40 choices in the deck for the first number card, and 39 for the second):

$$\frac{40 \cdot 39}{52 \cdot 51} = \frac{10 \cdot 13}{13 \cdot 17} = \frac{10}{17}.$$

Alternatively, this probability is given by the number of combinations of 2 out of 40 divided by the number of combinations of 2 out of 52:

$$\frac{_{40}C_2}{_{52}C_2} = \frac{40 \cdot 39}{52 \cdot 51} = \frac{10}{17}.$$

We have

$$\frac{11}{16} > \frac{10}{17},$$

so the probability of getting more than one head in four coin tosses is greater than the probability of drawing two number cards in a row from a deck.

(b) For the sum of two dice to be greater than 8, it should be $9, 10, 11,$ or 12. In these cases we have the following possibilities:

$$\sum = 9 \;: \qquad (3,6)\ (4,5)\ (5,4)\ (6,3)$$

$$\sum = 10 : \qquad (4,6)\ (5,5)\ (6,4)$$

$$\sum = 11 : \qquad (5,6)\ (6,5)$$

$$\sum = 12 : \qquad (6,6).$$

Therefore, $P(A) = \frac{10}{36} = \frac{5}{18}$. The probability of B is $P(B) = \frac{1}{6}$. The outcomes favorable for both A and B are $(4,5)$ and $(4,6)$, therefore $P(A \cap B) = \frac{2}{36} = \frac{1}{18}$. To check if the events are independent, we compute:

$$P(A) \cdot P(B) = \frac{5}{18} \cdot \frac{1}{6} = \frac{5}{108}.$$

Since $P(A \cap B) = \frac{1}{18} \neq \frac{5}{108}$, the events are not independent.

(c) Using the results of (b), we compute:

$$P(A|B) = \frac{P(A \cap B)}{P(B)} = \frac{1/18}{1/6} = \frac{1}{3}.$$

This corresponds to picking the pairs with the sum greater than 8 out of 6 pairs with the first number equal to 4: the last two pairs among $(4,1)\ (4,2)\ (4,3)\ (4,4)\ (4,5)\ (4,6)$.

$$P(B|A) = \frac{P(A \cap B)}{P(A)} = \frac{1/18}{5/18} = \frac{1}{5}.$$

This corresponds to choosing the pairs starting with 4 out of the 10 pairs listed in part (b).

EXAM 4

1. (a) We compute:

$$3.2424242424 = 3 + \frac{24}{100} + \frac{24}{100^2} + \cdots$$

$$= 3 + \frac{24}{100}\left(1 + \frac{1}{100} + \frac{1}{100^2} + \cdots\right) = 3 + \frac{24}{100} \cdot \frac{1}{1 - \frac{1}{100}}$$

$$= 3 + \frac{24}{100} \cdot \frac{100}{99} = 3 + \frac{8}{33} = \frac{107}{33}.$$

(b) If the coffee machine runs forever, the amount of coffee poured out (in ounces) is given by

$$1 + \frac{11}{12} + \frac{11^2}{12^2} + \cdots = \frac{1}{1 - \frac{11}{12}} = \frac{1}{\frac{1}{12}} = 12 \text{ oz.}$$

The first 11 drops amount to

$$1 + \frac{11}{12} + \cdots + \left(\frac{11}{12}\right)^{10} = \frac{1 - \left(\frac{11}{12}\right)^{11}}{1 - \frac{11}{12}} \simeq 7.39 \text{ oz.}$$

This is less than 8 ounces.

2. The compounding coefficient is

$$p = \left(1 + \frac{5}{1,200}\right) \simeq 1.004167.$$

The amount of money to be paid off in 20 years is

$$400,000 \cdot p^{nt} = 400,000 \cdot 1.004167^{12 \cdot 20} \simeq 400,000 \cdot 2.71264$$
$$\simeq \$1,085,056.11.$$

If A amount is paid each month, then the monthly payments sum up to

$$A(1+p+p^2+\cdots+p^{239}) = A\frac{1 - p^{240}}{1 - p} \simeq A\frac{1 - 2.71264}{1 - 1.004167} \simeq A \cdot 411.03$$

Therefore the value of each monthly payment is

$$A = \frac{1,085,056.11}{411.03} \simeq \$2,639.84.$$

3. (a) The order of the members of the group does not matter. Therefore, the answer is given by the product of the numbers of combinations:

$$_4C_2 \cdot _5C_1 \cdot _{10}C_3 = \frac{4!}{2!2!} \cdot 5 \cdot \frac{10!}{7!3!} = \frac{4 \cdot 3}{2} \cdot 5 \cdot \frac{10 \cdot 9 \cdot 8}{3 \cdot 2} = 3,600.$$

(b) Once the participants are chosen, we have a group of 6 people who need to be scheduled to speak. The number of possibilities is given by the number of permutations of 6 out of 6:

$$_6P_6 = 6! = 720.$$

4. (a) For exactly 4 tails in 6 coin tosses, there are $_6C_4$ favorable outcomes; for 5 tails, $_6C_5$ favorable outcomes; and for 6 tails, just 1 favorable outcome. The total number of favorable outcomes is

$$_6C_4 + _6C_5 + _6C_6 = \frac{6!}{4!2!} + \frac{6!}{5!1!} + 1 = 15 + 6 + 1 = 22.$$

The total number of possible outcomes is $2^6 = 64$. Therefore, the probability is
$$P(\geq 4H) = \frac{22}{64} = \frac{11}{32}.$$

(b) The probability of drawing the first card and getting a spade is

$$P(A) = \frac{1}{4}.$$

When we draw the second card from the deck, we do not know what the first card was. To compute the probability of B, we have to consider two possibilities: the first card was a heart (probability $\frac{1}{4}$), or the first card was not a heart (probability $\frac{3}{4}$). In the first case, the probability of drawing a heart is $\frac{12}{51}$. In the second case, the probability of drawing a heart is $\frac{13}{51}$. The probability of B is

$$P(B) = \frac{1}{4} \cdot \frac{12}{51} + \frac{3}{4} \cdot \frac{13}{51} = \frac{12 + 39}{4 \cdot 51} = \frac{1}{4}.$$

Alternatively, you can argue that without knowing what the first card was, the probability of drawing the second and getting a heart is exactly the same as the probability of drawing one card from the deck and getting a heart.

The probability of event $A \cap B$ (drawing first a spade and second a heart) is

$$P(A \cap B) = \frac{13 \cdot 13}{52 \cdot 51} = \frac{13}{204}.$$

This can be computed counting the number of choices for each card, or by using permutations and combinations:

$$P(A \cap B) = \frac{_{13}C_1 \cdot _{13}C_1}{_{52}P_2} = \frac{13 \cdot 13}{52 \cdot 51}.$$

Here we have to use the number of permutations in the denominator, because we need the *first* card to be a spade and the *second* a heart. If we put the number of combinations $_{52}C_2$ in the denominator instead, that would give us the probability of drawing a spade and a heart in any order.

Finally, we compare the probability of $A \cap B$ with the product of the probabilities of A and B:

$$P(A) \cdot P(B) = \frac{1}{4} \cdot \frac{1}{4} = \frac{1}{16} \neq \frac{13}{204} = P(A \cap B).$$

We have that $P(A) = P(B) = \frac{1}{4}$. $P(A \cap B) = \frac{13}{204}$. The events are not independent.

(c) For event A, we have the following favorable outcomes: $(1, 4)$, $(2, 3)$, $(3, 2)$, $(4, 1)$. The total number of outcomes in rolling two dice is 36. The probability of A is

$$P(A) = \frac{4}{36} = \frac{1}{9}.$$

For event B, there are four favorable outcomes for the second die: $3, 4, 5, 6$, and any outcome of the first die is favorable. The probability of B is

$$P(B) = \frac{4}{6} = \frac{2}{3}.$$

For event $(A \cap B)$, there are just two favorable outcomes: $(1, 4)$ and $(2, 3)$.

$$P(A \cap B) = \frac{2}{36} = \frac{1}{18}.$$

Now we can compute the conditional probabilities.

$$P(A|B) = \frac{P(A \cap B)}{P(B)} = \frac{\frac{1}{18}}{\frac{2}{3}} = \frac{1}{12}.$$

This corresponds to picking the pairs with the sum equal to 5 out of all pairs with the second number greater than or equal to 3.

$$P(B|A) = \frac{P(A \cap B)}{P(A)} = \frac{\frac{1}{18}}{\frac{1}{9}} = \frac{1}{2}.$$

This corresponds to picking the pairs with the second number greater than or equal to 3 out of all pairs with the sum equal to 5.

To check independence, we compute

$$P(A) \cdot P(B) = \frac{1}{9} \cdot \frac{2}{3} = \frac{2}{27} \neq \frac{1}{18} = P(A \cap B).$$

Finally, we have $P(A|B) = \frac{1}{12}$, $P(B|A) = \frac{1}{2}$. The events are not independent.[2]

EXAM 5

1. Solve the equation:
$$24 \cdot 2^{5x+1} = 18.$$

   We have
$$2^{5x+1} = \frac{18}{24} = \frac{3}{4}.$$

   Then
$$(5x + 1) = \log_2 \frac{3}{4} \quad \Longrightarrow \quad 5x = \log_2 \frac{3}{4} - 1.$$

   Therefore, $x = \frac{1}{5} \left(\log_2 \frac{3}{4} - 1 \right) \simeq -0.283$.

2. We have the following equation for interest rate r:
$$1 \cdot \left(1 + \frac{r}{1,200} \right)^{12 \cdot 15} = 2.$$

---

[2] "Dead bones, tossed by the living, decide the tosser's fortune." – Leonardo da Vinci.

This equation implies

$$\left(1 + \frac{r}{1,200}\right)^{180} = 2 \quad \Longrightarrow \quad 1 + \frac{r}{1,200} = 2^{\frac{1}{180}} \simeq 1.0038582.$$

Then

$$\frac{r}{1,200} \simeq 0.0038582 \quad \text{and} \quad r \simeq 4.63\%.$$

For (b), we compute

$$X = 1 \cdot \left(1 + \frac{r}{1,200}\right)^{12 \cdot 8} \simeq 1.0038582^{96} \simeq \$1.45.$$

3. We use the exponential growth model. We know that during $t = 3$ weeks, the population increased by a factor of 1.5. Therefore we have the following equation for the relative growth rate k:

$$1.5 = e^{k \cdot 3} \quad \Longrightarrow \quad \ln 1.5 = 3k \quad \Longrightarrow \quad k = \frac{1}{3} \ln 1.5 \simeq 0.13515.$$

To find the time it takes the population to grow from $10,000$ to $60,000$, we use the relative growth rate k and write

$$60,000 = 10,000 \cdot e^{0.13515 \cdot t} \quad \Longrightarrow \quad 6 = e^{0.13515 \cdot t}$$
$$\Longrightarrow \quad \ln 6 = 0.13515t.$$

Then $t = \dfrac{\ln 6}{0.13515} \simeq 13.26$.

Since 0.26 is about a quarter of a week, which corresponds to a little less than 2 days, it will take 13 weeks and 2 days for the population to grow from $10,000$ to $60,000$.

4. (a) The sum of all parts paid to the athletes is

$$\frac{1}{3} + \frac{1}{9} + \frac{1}{27} + \cdots = \frac{1}{3}\left(1 + \frac{1}{3} + \frac{1}{3^2} + \cdots\right) = \frac{1}{3}\frac{1}{1 - \frac{1}{3}} = \frac{1}{3} \cdot \frac{3}{2} = \frac{1}{2}.$$

(b) The organizers get the other half of the prize money: $\$6,561/2 = \$3,280.50$.

(c) There are two ways to find how much money the first five runners get: either by calculating $1/3$, $1/9$, $1/27$, $1/81$, and $1/243$ of the prize fund and summing them up:

$$2,187 + 729 + 243 + 81 + 27 = \$3,267,$$

or by summing a finite geometric series

$$6,561 \cdot \frac{1}{3} \cdot \left(1 + \frac{1}{3} + \cdots + \frac{1}{3^4}\right) = 6,561 \cdot \frac{1 - \frac{1}{3^5}}{1 - \frac{1}{3}} \simeq \$3,267.00$$

5. (a) There are $_{10}C_3$ ways to choose three white marbles and $_{30}C_3$ total ways to choose three marbles out of the bag. The probability of getting 3 white marbles is given by

$$P(3 \text{ white}) = \frac{_{10}C_3}{_{30}C_3} = \frac{10 \cdot 9 \cdot 8}{30 \cdot 29 \cdot 28} = \frac{6}{203}.$$

(b) There are $_{25}C_3$ ways to choose three marbles none of which is red. The required probability is

$$P(\text{none red}) = \frac{_{25}C_3}{_{30}C_3} = \frac{25 \cdot 24 \cdot 23}{30 \cdot 29 \cdot 28} = \frac{115}{203}.$$

(c) There are $_5C_1 \cdot {}_{25}C_2$ ways to choose three marbles so that exactly one of them is red. The required probability is

$$P(1 \text{ red}) = \frac{_5C_1 \cdot {}_{25}C_2}{_{30}C_3} = \frac{5 \cdot 25 \cdot 24}{2} \cdot \frac{6}{30 \cdot 29 \cdot 28} = \frac{75}{203}.$$

6. We need to find $P(\text{cl}|\text{sm})$, where we denote by cl the forecast for a cloudy sky, and sm the event that Mr. Pickwick is carrying his small umbrella. By Bayes' law, we have

$$P(\text{cl}|\text{sm}) = \frac{P(\text{sm}|\text{cl}) \cdot P(\text{cl})}{P(\text{sm})}.$$

The probability $P(\text{sm}|\text{cl}) = \frac{1}{3}$, because when the predicted weather is cloudy, he takes his small umbrella with probability $\frac{1}{3}$. The probability of a cloudy sky being predicted is given to be $\frac{1}{3} = P(\text{cl})$. Let us compute the probability of Mr. Pickwick carrying his small umbrella:

$$P(\text{sm}) = \frac{4}{9} \cdot \frac{1}{4} + \frac{1}{3} \cdot \frac{1}{3} + \frac{2}{9} \cdot \frac{1}{6} = \frac{7}{27}.$$

This is the sum of probabilities of his carrying the small umbrella in case of each of the possible predicted weather conditions times the probability of this particular weather condition. Finally we find the probability that the forecaster predicts a cloudy sky given that Mr. Pickwick is carrying his small umbrella:

$$P(\text{cl}|\text{sm}) = \frac{\frac{1}{3} \cdot \frac{1}{3}}{\frac{7}{27}} = \frac{3}{7}.$$

7. To find the last digit of 447^{447}, first note that only the last digit of 447 matters. We need to compute

$$7^{447} \quad \text{mod } 10.$$

Let us compute congruences modulo 10 for the first few powers of 7, hoping to see a periodic pattern.

$$7^2 \equiv 9 \quad \text{mod } 10, \qquad 7^3 \equiv 3 \quad \text{mod } 10, \qquad 7^4 \equiv 1 \quad \text{mod } 10.$$

Therefore, every power of 7 that is a multiple of 4 is congruent to 1 modulo 10.

$$7^{447} = 7^{444} \cdot 7^3 \equiv 1 \cdot 3 \equiv 3 \quad \text{mod } 10.$$

The last digit is 3.

EXAM 6

1. Solve the equation

$$6^{2x} \cdot 3^{(x+2)} = \frac{\sqrt[7]{36}}{2^{(x+2)}}.$$

Multiplying both sides by $2^{(x+2)}$ and using the properties of exponents, we have

$$6^{2x} \cdot 6^{(x+2)} = 6^{\frac{2}{7}} \quad \Longrightarrow \quad 6^{(3x+2)} = 6^{\frac{2}{7}} \quad \Longrightarrow \quad 3x + 2 = \frac{2}{7}$$

$$\Longrightarrow \quad 3x = \frac{2}{7} - \frac{14}{7} = -\frac{12}{7} \quad \Longrightarrow \quad x = -\frac{4}{7}.$$

2. (a) Let $A(0)$ denote the initial mass of a sample, $A(t)$ the mass at time t, and k the relative rate of decay. First we use the exponential decay model to find the relative rate of decay:

$$A\left(\frac{3}{4}\right) = \frac{3}{4}A(0) = A(0)e^{\frac{3}{4}k} \quad \Longrightarrow \quad \frac{3}{4}k = \ln\left(\frac{3}{4}\right).$$

Therefore,

$$k = \frac{4}{3}\ln\left(\frac{3}{4}\right) \simeq -0.38358/\text{yr}.$$

Then the half-life is

$$T_{1/2} = -\frac{\ln 2}{k} \simeq 1.81 \text{ yr},$$

which corresponds to approximately 1 year and 10 months. For (b), we compute

$$A(1) = A(0)e^{k\cdot 1} = A(0)e^{k} \simeq 0.6814 A(0).$$

A sample will decay to approximately 68.1% of its initial mass in 1 year.

3. (a) Let $p = \left(1 + \frac{r}{n\cdot 100}\right) = \left(1 + \frac{4}{1,200}\right) = 1.003333... = 1.00(3)$ denote the compounding coefficient. Then to find how much money will accumulate after $5 \cdot 12 = 60$ compounding periods, we compute

$$250\left(1 + p + p^2 + \cdots + p^{59}\right) = 250 \cdot \frac{1 - p^{60}}{1 - p} = 250 \cdot \frac{1 - 1.00(3)^{60}}{-0.00(3)}$$
$$\simeq \$16,574.76.$$

(b) Let us denote by n the number of payments it takes to accumulate $\$25,000$. Then we have the equation

$$250 \cdot \frac{1 - p^n}{1 - p} = 25,000 \quad \Longrightarrow \quad (1 - p^n) = 100 \cdot (1 - p).$$

Plugging in $p = 1.00(3)$ and solving for n yields

$$1.00(3)^n = 1 + 0.333333.. \quad \Longrightarrow \quad n\ln(1.00(3)) = \ln(1.(3))$$
$$\Longrightarrow \quad n \simeq 86.45 \simeq 87 \text{ payments}.$$

For (c), we need to express 87 months in years and months. We have

$$87 = 12 \cdot 7 + 3.$$

Therefore, it will take 7 years and 3 months.

4. (a) To find the total distance covered by the horse in infinite time, we have to sum up an infinite geometric series:

$$1 + \frac{10}{11} + \left(\frac{10}{11}\right)^2 + \cdots = \frac{1}{1 - \frac{10}{11}} = \frac{1}{\frac{1}{11}} = 11 \text{ mi.}$$

(b) To find the distance covered in the first hour, we compute

$$1 + \frac{10}{11} + \left(\frac{10}{11}\right)^2 + \cdots + \left(\frac{10}{11}\right)^5 = \frac{1 - \left(\frac{10}{11}\right)^6}{1 - \frac{10}{11}}$$

$$= 11\left(1 - \left(\frac{10}{11}\right)^6\right) \simeq 4.79 \text{ mi.}$$

5. (a) There is $_4C_4 = 1$ way to choose four cherry-filled chocolates out of 4 available, and $_{24}C_4$ total ways to choose four chocolates from the box. The probability of getting 4 cherries is given by

$$P(4 \text{ cherry}) = \frac{_4C_4}{_{24}C_4} = \frac{1}{\frac{24 \cdot 23 \cdot 22 \cdot 21}{4 \cdot 3 \cdot 2 \cdot 1}} = \frac{1}{23 \cdot 22 \cdot 21} = \frac{1}{10,626}.$$

(b) There are $_8C_2$ ways to choose two coffee-filled chocolates, and for each of them, there are $_{16}C_2$ ways to choose two remaining chocolates none of which is coffee-filled. The required probability is

$$P(2 \text{ coffee}) = \frac{_8C_2 \cdot _{16}C_2}{_{24}C_4} = \frac{4 \cdot 7 \cdot 8 \cdot 15}{23 \cdot 22 \cdot 21} = \frac{4 \cdot 4 \cdot 5}{23 \cdot 11} = \frac{80}{253}.$$

(c) Here the event splits into a union of three disjoint events: there can be exactly 2, exactly 3, and exactly 4 coffee-filled chocolates. The number of ways for each of the events is computed similar to (b). Then we have to sum them up. Totally, the required probability is:

$$P(\geq 2 \text{ coffee}) = \frac{_8C_2 \cdot _{16}C_2 + _8C_3 \cdot _{16}C_1 + _8C_4}{_{24}C_4}$$

$$= \frac{4 \cdot 7 \cdot 8 \cdot 15 + 8 \cdot 7 \cdot 16 + 7 \cdot 5 \cdot 2}{23 \cdot 22 \cdot 21}$$

$$= \frac{80}{253} + \frac{64}{23 \cdot 11 \cdot 3} + \frac{5}{23 \cdot 11 \cdot 3} = \frac{240 + 64 + 5}{23 \cdot 11 \cdot 3} = \frac{103}{253}.$$